Anticipation and Medicine

Mihai Nadin
Editor

Anticipation and Medicine

Editor
Mihai Nadin
anté—Institute for Research in Anticipatory
  Systems
University of Texas at Dallas
Richardson, TX
USA

ISBN 978-3-319-83221-0   ISBN 978-3-319-45142-8   (eBook)
DOI 10.1007/978-3-319-45142-8

Printed on acid-free paper

This Springer imprint is published by Springer Nature
The registered company is Springer International Publishing AG
The registered company address is: Gewerbestrasse 11, 6330 Cham, Switzerland

Acknowledgments

From birth to death, medicine accompanies everyone. This in itself testifies to the interest that we all have in medicine as informed by the most significant science and technology. The international conference on Anticipation and Medicine (Delmenhorst, September 28–30, 2015) offered the opportunity for evaluating to which extent anticipation, a subject familiar in the medical field, makes a difference. The *Freshheads* (Die Frischköpfe) of the Metropole Nordwest, the Hanse Wissenschaftskolleg/Institute for Advanced Study, and the University of Texas at Dallas/antÉ—Institute for Research in Anticipatory Systems provided funding for a meeting of practicing physicians, of scientists and technologists, of providers of alternative treatment methods. They succeeded in sustaining a lively conversation on how anticipation informs their activity.

This volume deviates slightly from the program (peer-review stringency, as well as time strictures, explains the deviation), and from what are usually considered conference proceedings. It is representative of the variety of perspectives pursued in medical anticipation research and associated fields. In his message to the participants, Dr. David Daniel (President of the University of Texas at Dallas, before becoming the Deputy Chancellor of the University of Texas System) wrote:

> *Anticipation in Medicine* is frontier science of importance to us all. Here at the University of Texas at Dallas, we are deeply engaged in a broad array of topics that are relevant to medicine, including life science research, biomedical devices, computational methods, ethics, and business models. We can all relate to the importance of this work.

> I can see from the list of conference participants that the world's leaders in this field have come together to share ideas. I am very proud that the Institute for Research in Anticipatory Systems at UT-Dallas was able to play a vital role in bringing together such an esteemed international community.

Dr. Hobson B. Wildenthal, as president *ad interim*, honored the commitment of the University of Texas at Dallas to excellence in research by granting Dr. Nadin an exceptional Special Faculty Development Assignment in support of the Study Group in Anticipation. Dr. Reto Weiler, Rector of the Hanse Wissenschaftskolleg/Institute for Advanced Study, together with the Institute's entire team, deserve recognition for their support of the project. Dr. Dorothee Poggel, in charge of Neuroscience Research at the Hanse Institute, dedicated herself to making this conference a success. Springer—Applied Sciences and Engineering, which assisted in the editorial and production effort beyond the peer review phase, also deserves acknowledgment. Elvira Nadin contributed editorial competence and coordinated the laborious peer-review process. Asma Naz prepared the volume for publication (while simultaneously conducting experiments pertinent to her own Ph.D.).

If this volume will deserve to become a reference title on the subject, all those who made it possible should feel rewarded for their dedication.

Contents

Medicine: The Decisive Test of Anticipation

Mihai Nadin

Abstract Given the life and death extremes at which medicine operates, no other human activity is of higher significance to members of society, and to society itself. Therefore, it is surprising that, instead of aligning itself with the anticipatory condition of life, medical practitioners at all levels approach health from the physics perspective of reaction and reductionism. On the other hand, anticipation—definitory of the living—could prove to be consequential if the perspective it opens would become the backbone of medicine. The study discusses the reported negative effects of healthcare and medical practice based on the mechanical model provided by physics-dominated science. Acknowledging technological progress in medicine, the study also provides actual expressions of anticipation important for the theory and practice of medicine. Complexity is examined as a characteristic of anticipatory systems. Lastly, the study suggests concrete steps towards an anticipation-grounded medical education.

Keywords Anticipation · Complexity · Control · Fractionating · Mechanics-based medicine · Education

"Death by Medicine" is the alarming title of a book published in 2004 [1]. The book is based on sound research from respected medical publications (e.g., *Journal of the American Medical Association, New England Journal of Medicine, National Vital Statistics Report*, among the 152 listed in the Bibliography). The 2016 study appearing in the *British Medical Journal* confirms these findings [2, 3]. They all bring to light some of the consequences of industrial-mechanistic medical practice. There is no denying the spectacular contribution of physics- and chemistry-based technologies in medicine and healthcare. Spectacular successes in mechanistic medicine—organ transplants, prostheses, intra-uterine spina bifida operations—

M. Nadin (✉)
anté—Institute for Research in Anticipatory Systems, University of Texas at Dallas, Richardson, TX, USA
e-mail: nadin@utdallas.edu
URL: http://www.nadin.ws

© Springer International Publishing Switzerland 2017
M. Nadin (ed.), *Anticipation and Medicine*, DOI 10.1007/978-3-319-45142-8_1

1

have never matched the success of the anticipation-grounded polio vaccine. Proactive means and methods—e.g., herbs, teas, exercise, anticipatory diagnoses (based on odor, for instance)—were practiced for millennia in virtue of empirical observations transmitted from one generation to the next. This "self-care" helped reaction-driven medicine to avoid an even worse record of contributing to death as much as, if not more than, cancer and heart disease.

Challenged by conditions beyond the traditional boundary of illness, medicine is discovering that the living is less simple than what the physics paradigm implies. Niels Bohr [4] acknowledged the inadequacy of physics to describe the life. Erwin Schrödinger suggested that a new type of physics is necessary to explain life [5]. Walter Elsasser [6] tried to provide biology with a scientific foundation not in opposition to physics, but rather in its spirit. This study makes the case for integrating physics-based medicine and anticipation-grounded medical practice.

1 Preliminaries

The decisive test of understanding anticipation as definitory of the living [7] is *medicine*. It is also a meaningful test of usefulness: Why study anticipatory processes if they are of no practical consequence? In medicine, where life and death are at stake, the philosophical dispute of whether the condition of the living is different from that of the physical might be of immediate concern. Yet over time, anticipation expression of all kinds—i.e., successful actions that preserve life and help reach desired outcomes—have afforded a rich body of empirical evidence. In recent years, data pertinent to anticipatory processes have been accumulated in a variety of fields of knowledge [8]. This evidence triggered a plethora of explanatory attempts —some anchored in science of unquestionable integrity; the majority, however, rather speculative, usually derived from ill-defined concepts or through less than grounded generalizations. In particular, the notion of "anticipation" itself is usually confused with other forms of dealing with change, such as forecasting, expectation, guessing, etc., and especially prediction. It is rarely understood that they are different in nature from anticipation [9]. Conceptual clarity, more than instrumental obsession (so typical of this particular time) is necessary. When everything is measured—because it can be—in the hope that "big data" technology will reveal "secrets" behind the data, the expectation is that processes are reducible to data—as prediction, forecasting, guessing, etc. are. Totally ignored is the fact that actions informed by anticipation are the expression of *significant* data—usually "little" data, generated ahead of the action. Moreover, the emphasis in anticipatory processes is on *meaning* more than on quantitative descriptions based on the use of numbers.

1.1   A Question of Scientific Legitimacy

It was not an apple falling from a tree (and the equation describing it) that gave legitimacy to the law of gravity. Newton advanced an understanding of physical law that revolutionized science, and continues to have practical relevance. The same can be said in respect to the theory of relativity: Einstein became more famous for $e = mc^2$ than for the revolutionary view of the universe that his mathematics advanced. It might be that quantum mechanics, still at the forefront of science today, qualifies as well as a revolution in thought and discovery. The fact that particular aspects of quantum mechanics (such as entanglement and non-locality) fascinate the public (and even some scientists) should not affect awareness of a totally new view of reality.

Human dedication to understanding the world, within which the living unfolds, eventually crystalizes in revolutionary views—this in itself is worth celebrating. But it invites reflection as well. Nothing comparable, not even the famous DNA helix, is on record in explaining life itself. Newton, Einstein, the quantum mechanics visionaries (Heisenberg, Schrödinger, Bohr, Feynman, among others) are present— and rightly so—in the explanation (as tentative as it still is) of the beginnings of the universe. But the beginning of life is still in the fog of confusion. The attempts to start life from non-life, almost as seductive as alchemy (Newton was one of its proponents) was long ago, persist through new technologies (Venter's claim [10, 11] is only one among many). They entail a rather disconcerting surrender, celebrated as victory: since the living itself is embodied in matter, the more physics we know—and the more physics-based biology we use—the better we will understand life. Nobody ever proved this reduction. The obsession of particle physics (i.e., know the particles and you will know all about the whole they are part of) translated into the hope that molecular biology or genetics will eventually solve all the mysteries of disease and, eventually, life.

Obviously there is much more nuance to all of this. Albeit, science consists mainly of convincing physics theories and their extension into particular phenomena (chemistry, for instance). A large body of generalizations from the physical to biology converges in what seems an inevitable, but false, conclusion: the living is a machine. The material substratum is acknowledged without reservation; omitted is the understanding that the dynamics of the physical and of the living are different. Also absent is the distinct effort to advance a view of the living that defines its own characteristic causality. This might integrate the science of the physical—e.g., physics, chemistry, geology, astronomy, meteorology—without discarding what defines a science of the living proper. The epistemological effort that I argue for might even arrive at the realization that physical causality (explaining change in the nonliving) is ultimately a subset of the extremely rich types of causality that explains the change of life [12].

For accomplishing such a daunting task, the focus should be on dynamics: how and why change takes place. Empirical evidence suggests that change in the

nonliving realm takes place at a timescale different from the timescales characteristic of living processes. Moreover, in the physical, the timescale is relatively
constant, while in the living it varies under the influence of context—sometimes
defined as *Umwelt* [13, 14], i.e., the perceptual world of a specific being:
"Organisms are subjects interpreting their life worlds, not mechanical objects
reacting to external forces" [15]. There is no birth and no death (short of misap-
propriated metaphors applied to stars and black holes) in the physical. And there is,
contrary to poetic license, no intentionality to be either observed or experimentally
documented. In its admirable dedication to exploring levels of reality ever more
deeply, physics (and its close relatives) progressively surrendered a unified view of
reality, while it was searching for a unified theory of its own domain. Those who
study the living with the purpose of understanding its irreducible condition cannot
afford the simplifying effort of ignoring the whole, as does reductionist-based
physics. (On the subject of holism, see Nadin.[1])

A science of the living can only be holistic, because the dynamics of the living is
the expression of its change as a whole over time. As already mentioned, the
realization that physical causality could be a subset of natural causality might entail
the need to understand "Nature" beyond Newton's unifying view that aggregates
the living and the physical and declares the laws of physics—reflecting God's
control over the universe—as universal. Eliminating God from the picture,
Darwin's *Origin of Species* [16] was celebrated as the equivalent of Newton's
foundation of physics (*Philosophiae Naturalis Principia Mathematica*, [17]).
Natural selection describes the implicit dynamics of the living, of a different precision, less precise, but more expressive than that described in Newton's equations.
If the past (as a state) defines the future of physical systems, the future (as a possible
state) expressed in anticipatory processes, is the vector of evolution.

Determinism, the characteristic causality of physical phenomena, is also relevant
to the physics of the living, even though it returns an incomplete explanation of life
as change. Just to present an example along this line of reasoning: physical forces
(e.g., pulls, compressions and stretching, distortions) applied to a cell can further
affect it, probably more than the inherited genetic code does [18, 19]. Taking both
physical forces and the genetic code into consideration affords an understanding of
cell changes that neither can deliver alone. Non-determinism, describing a relation
between cause and effect offering a multitude of possible outcomes, pertains to
change as an expression of something being alive. Indeed, changes due to physical
forces applied on cells (e.g., a cut or a blow)[2] and genetic processes governing all
dynamics are interwoven. There is no way to unequivocally predict whether the cell

---

[1]Nadin, M.: Anticipation and the Brain. In: Nadin, M.: (ed.) Anticipation and Medicine, pp. 135–
162. Springer, Cham (2016).

[2]Genetic manipulation of gene expression for turning mature cells into pluripotent stem cells
brought Shinya Yamanaka a Nobel Prize in 2012. Nevertheless, physical-chemical manipulation
has so far proven to be less successful. In 2014, he had to retract his findings. (See: http://blogs.
wsj.com/japanrealtime/2014/04/28/japanese-nobel-winner-latest-to-apologize-over-stem-cell-
research/).

becomes cancerous or simply divides in a process of self-healing. De la Mettrie's man-machine metaphor [20], to which science since 1748 remains literally enslaved, is but one consequence of a rudimentary view of causality. Only since the advent of quantum mechanics was this view somehow questioned. Stochastic aspects of dynamics were introduced, and indeterminacy accepted as a qualifier for processes less than very precise. Everything that can be fitted to the time series describing the functioning of a machine operates under the expectation of perfect repetition—even though the living is the domain of "repetition without repetition" [21], i.e., non-monotonic change.

1.2 Life Is About Meaning

The question of legitimacy, as it pertains to the anticipatory perspective, transcends the theoretical. It has consequences for the way we conceive of means and methods for maintaining life: the domain of medicine.

To know how the physical (i.e., nonliving entities) changes is to infer from a quantitatively described past state to a future state, under assumptions usually defined as initial conditions (also expressed numerically). To know how the living changes is to integrate inferences from past states with interpretations of the meaning of possible future states. No falling stone will get hurt (not to say die); a living falling (cat, human being) can get hurt (and even die). The framing of change within the respective consequences, not the same in the physical and in the living, is key to understanding their difference. The causality specific to interactions in the physical realm is described in Newtonian laws—action-reaction, in particular. The causality specific to interactions in the living includes, in addition to what Newton's laws describe quantitatively, the *realization of meaning* in connection to the possible future, i.e., anticipation.

Falling on ice or landing on a hot surface are different not in the physics of the process of falling, but in the meaning for the living, that is, the consequences. Evidently, this is pertinent to the hope that understanding change makes possible its description ahead of time, before it actually takes place. For this purpose, physics seeks descriptions, usually in mathematical form, similar to the human construct called *law*: a rule to be obeyed by those adhering to it. The laws of physics encapsulate descriptions of change that are necessary by nature. The fact that the living, in addition to the constraints of physics, is subject to contingent rules of behavior is usually brushed aside.

Reductionism postulates the identity of the physical and of living, to the detriment of a better understanding of the dynamics of the living. The same takes place within the epistemology based on the machine metaphor. Karl Popper [21, p. 224] noticed that in this sense, "...the doctrine that man is a machine has perhaps more defenders than before among physicists, biologists, and philosophers, especially in the form of the thesis that man is a computer." Popper (otherwise an over-rated, opportunistic philosopher) was either unaware of, or unwilling to embrace, the even more radical view of Newell and Simon: "Men and computers are merely two different species of a more abstract genus called information processing systems," [22, p. 234].

Machines, regardless what kind, (from clocks to hydraulic pumps to engines to computers), are constructs meant to function in a predictable manner. Humans make them. If religion postulated that the human being was created in the image of the Creator, *machinomorphism* establishes the religion of the human being made to function like the machines humans conceived. It is a solipsistic view: the making of something (e.g., machine) is the proof of the equivalence between the makers and the made. As already mentioned, for Newton, the mechanism of the universe, whose dynamics were precisely described in his equations, was the proof that everything in this universe behaved, at God's will, like a mechanism. If the clouds (Popper's metaphor) "are highly irregular, disorderly, and more or less unpre-dictable"—examples are molecules in a gas, or gnats—clocks are precise and predictable. Still, for determinists, clouds are clocks. In their view, with enough knowledge, what appears as indeterminate proves to be as determinate as the universe, or as the structure of matter. The language describing their functioning is mathematics built upon the construct we call *numbers*, which ultimately describe quantities. The automation of mathematics (or at least part of it) through compu-tation gave this tendency a new, more specific, though ultimately illusory, viability.

Machines embody the cognitive construct of numbers, i.e., descriptions of quantities. They are representations of the human activity that the machine replaces or augments. The arm and the lever used together to move objects is one simple example. Like any representation, they are, by their condition, incomplete [23]. To ascertain that a representation—the machine—is identical to the represented makes sense only for entities with clear boundaries. A billiard machine is equivalent to a billiard table in which, given the initial and boundary conditions of the billiard balls, the characteristics of the table (size, texture), we can, using the laws of motion, "calculate" the game. Laplace [24] was sure that given the positions and the momenta of all particles in the universe, we could, using Newton's laws, fully describe the past, present, and future of the universe.[3] After all, the universe as a clock, i.e., a machine, is what physics-based determinism ascertains.

But a pump is not equivalent to a heart. For extreme conditions—entailing a number of serious limitations—a pump might be used in order to help resuscitate someone, or extend someone's life—usually in an impaired mode. Moreover, the dynamics of the human being transcends expression in differential equations. Actually, not only isn't Laplace's deterministic view inadequate for describing life, but worse, it leads to aberrations. The fact that biology, and medicine in particular, took the deterministic path is understandable. Explaining away what we don't fully grasp is easier than assuming the responsibility to seek alternatives. Moreover,

---

[3]"une intelligence…rien ne serait incertain pour elle, et l'avenir comme le passée, serait présent à ses yeux." [An intelligence…nothing would be uncertain for it, and the future as the past, would be present to its eyes.]

simpler explanations afford the immediacy of practical methods, sometimes informed more by urgency than by anything else.

The anticipatory perspective is the alternative—a new Cartesian revolution [7]. But it is not as comfortable as the beaten path of physics and its promise for technology. It took over 200 years (more precisely, since Newton, Descartes, and Laplace) for scientists and scholars to realize that the beaten path at best offers partial answers (often wrong) to the question of what change means in the living. One cannot expect abrupt abandonment of the huge investment (time, energy, money, human lives, and the lives of animals used in experiments) in following the wrong path. Against the background of scientific advancement, we can hope for a shorter time for ascertaining a complementary view, and to start applying it to situations for which physics-based medicine is not adequate. The aging of the world population is only one aspect; the degeneration of the species—expressed in, among other ways, systemic disorders and debilitating spectrum conditions—is probably an even more critical problem.

1.3 Man-Machine

Within the medical establishment, a joke is shared with patients complaining about the high cost of medicine. The physician has work done on her BMW. The mechanic, taking note of the expensive car, asks the doctor, "Listen, how come I'm paid so much less than you, when actually we kind of do the same job? You bring the car to me when it's sick. I diagnose the defect and make it run for you." The doctor's already heard the same question. "You're right," she says. "But can you do it while the car is running?"

The punch line (Can you do it while it's running?) only pushes the mechanistic view that has shaped modern medicine to the extreme. Within the same view, the annual check-up can be interpreted as the equivalent of the manufacturer's requirement: scheduled maintenance every 5000 miles and a fluid change (Fig. 1).

It turns out that what makes sense in extending the life of the car's engine, or of the car in general, is at least debatable when it comes to the human being. Modern car maintenance facilities are equipped with automatic diagnostic devices. There are many who believe that the same can be done with the human being (and with pets). The physician becomes a mechanic. The "Precision Medicine Initiative" [25] is based on this belief and promotes medicine as a form of engineering. It emulates the control mechanism model of engines endowed with sensors and extends it to the individual, claiming that it will eventually lead to individualized medicine.

On the other hand, Mehtotra and Prochazka [26] claim:

> Reducing the use of annual physicals could also save money and time. Though on a per-visit basis, the annual physical is not costly, it is the single most common reason that U. S. patients seek care, and cumulatively these visits cost more than $10 billion per year — similar to the annual costs of all lung-cancer care in the United States. Reducing the number of physicals could free up another societal resource — primary care providers' time.

Fig. 1 Most think of it as the human equivalent of a 15,000-mile checkup and fluid change, which can uncover hidden problems and ensure longer engine life

Approximately 10 % of all visits with primary care physicians are for annual physicals, which might be crowding out visits for more urgent health issues. Poor access to primary care has been cited as one reason why patients seek care in emergency departments for low-acuity conditions. Finally, there are large societal costs to asking all 220 million adults in the United States to spend several hours of their lives each year traveling to and waiting for care, when they could use that time productively elsewhere. Given this evidence base, it appears unlikely that annual physicals in their current forms lead to any substantive net clinical benefit.

Missing from the list of arguments is the understanding that medicine as applied physics will continue to be expensive, ineffective, and confusing. The brutality of the "spare parts" understanding of medicine is not only limited to the procedure and the rehabilitation (under heavy use of painkillers that affect overall health), but also to the undermining of whatever health the patient still had before the intervention became necessary. Medicine, in its industrial procedures, "heals" today and produces invalidity of deeper levels tomorrow.

1.4 Holism

Arguing in favor of descriptions appropriate to the functional behavior of biological systems, Rosen [27] stated that such descriptions "bear no simple relation to the structural observables which our physical technique can measure." Without reproducing his arguments here, let us take note of the fact that the scientific methodology of *fractionating*—break what is complex into simpler subsystems— does not even apply universally to the physical. The three-body system problem, notoriously unsolvable, could be fractionated into a variety of two-body and one-body systems. But this does not ultimately produce the knowledge we need to understand the dynamic characteristics of the initial system. Fractionation does not afford the information we seek—and from a holistic perspective, *it does not afford any information*.

All the media-hyped information on genetics that spurs hope in patients (see Garzoni, Centomo, Delledonne[4]) is the result of ignoring a simple principle: health, or its deterioration, is a matter of the whole. Some healing processes can be triggered through identification of what might have caused an imbalance, but only if the fractionation transcends physics and chemistry. Unfortunately, modern industrial medicine is based on a view of the living and of health grounded in physics, and sometimes chemistry. It treats the condition called "disease" with medication and surgery: a cause-and-effect sequence within a reductionist view. Even physical therapy is practiced in this spirit.

From a logical perspective, specialized medicine—which reports spectacular successes never to be underestimated neither in price nor in helping patients— collides with the holistic understanding of what health, or even disease, is. Every year, the medical community celebrates the ten (or however many) greatest accomplishments. There is no way to avoid the feeling of awe. Human lives are saved under extreme conditions and amazing interventions of all kind, some involving new drugs, genetic medicine, prostheses, and highly complicated procedures. The word "miracle" is the first to come to mind. But there is also the dark side, where numbers of a different kind—such as incorrect diagnoses, botched surgeries, questionable medications (to name a few)—add up. Adverse reactions to prescribed drugs (in the millions), needless procedures (close to eight million a year), unnecessary hospitalization (close to 10 million) are documented with the aim of establishing some quality control criteria. Antibiotics—once the miracle treatment for infectious disease—and opioids (hydrocodone, oxycodone, fontanel, codeine, among many others) are rapidly becoming a curse [28] affecting the genetic profile of the entire population (not to mention effects on the environment). It is impossible to predict the long-term consequences of this situation, produced by those who dedicated themselves to serving life, not undermining its viability.

Such examples evince the resistance to understanding healing within an anticipatory perspective. To maintain an individual's viability in the context of change (e.g., aging, styles of work) is quite different from repairing abused bodies within the framework of a mechanic's shop.

---

[4]Garzoni, M., Centomo, C., Avinash, M.V., Delledonne, M.: Next Generation Sequencing for Next Generation Diagnostics and Therapy. In: Nadin, M.: (ed.) Anticipation and Medicine, pp. 78–92. Springer, Cham (2016).

In the absence of a holistic view, the various parameters considered and the threshold values are at best indicative of a measurement method that brings up another anecdote:

- What are you looking for?
- My keys.
- Did you lose them here, around the lamppost?
- No.
- Then why are you looking for them here?
- Because I can see better…there's more light around the lamppost.

The internist takes note of higher blood pressure, the cardiologist prescribes pills (Losartan or Lisinopril), a psychiatrist addresses a stress situation, a practitioner of alternative medicine recommends red beet juice, a Chinese healer initiates a course of acupuncture—each one looking around their own lamppost.

In the absence of a meaningful understanding of change, as it pertains to health, this kind of medicine has a very low predictive performance. Indeed, it is quite surprising that no one seems to notice that while the prediction of physical phenomena is rather successful (and getting better), once the same view is applied to the living, the performance is low—not far from the threshold of sheer guessing. Therefore, despite all the statements to the contrary, medicine driven by physical determinism has a very disappointing proactive success.

For example, tumor–patient–drug interaction remains quite ill defined. Consequently, treatment success in oncology remains low, despite the enormous effort of all involved. While cancer settled in as the main challenge to medicine, only rarely are alternative treatment methods considered. Heart disease belongs to the same area of reductionist-deterministic medicine marred by failure, despite the awareness of its terrible consequences.

1.5 Clarity: The Premise for Dialog

These preliminary notes suggest a conceptual context for framing the discussion of issues in anticipation and medicine. We have to take into account that the notion of anticipation is used currently in medicine with a very precise description attached to it. Anticipation describes a genetic disorder passed from one generation to another, each time at an earlier onset (the so-called trinucleotide repeat disorders, such as Huntington disease, myotonic dystrophy, dyskeratosis congenital, etc.). The operational definition of anticipation advanced in this study explains, after the fact, the choice made by medical practitioners in trying to understand how the trinucleotide repeat occurs and what is involved in the production of the mutant protein.

1.5.1 From Cradle to Grave

Nevertheless, the expression of anticipation is such that it covers the entire life of the individual: from conception to death. In the context of a study that bridges

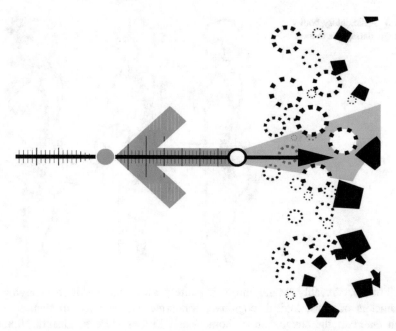

Fig. 2 Anticipatory process: The deterministic arrow of time from a past cause to a present effect and the complementary arrow of time from a possible future to the current state of the system

between anticipation and medicine, my goal is to point to the variety of forms through which anticipation is expressed in action. The various phases of the sexual act, pregnancy, birth, etc. are only one example for the argument that medicine is consubstantial with anticipatory expression. Medicine states that the variety of processes associated with creation—sexual act, impregnation, pregnancy, giving birth, nurturing—can be fully explained in terms of brain activity, the neuro-endocrine systems, hormones, and the like (each taken independently). Before revisiting some of the processes, let us take note of the concrete instantiations of the fundamental thesis I advance: Anticipation, in the sense defined in this paper (see Fig. 2) is definitory of the living. This assertion is connected to yet another thesis: The living is that which reproduces itself: it is its own efficient cause. Rosen's formulation, in reference to Aristotle's typology of causes, is "A material system is an organism if and only if it closed to effective causation" [29, p. 144]. Evolution, as it describes natural selection in the context of change, acknowledges sexuality as it relates to reproduction. The human being, in the new nature it made for itself, added to the reproductive impulse *sexuality as a goal in itself*, most of the time disconnected from reproduction. This defines the human in contradistinction to the rest of the sexually reproducing living. Some of the processes identifiable in the human are not identifiable in the rest of sexually reproducing nature: for example, the cultural aspects of pregnancy and giving birth.

Sexual attraction and the sexual act engage the being's totality. If, according to models inspired by the analogy to the machine, the organism were in a state of

Fig. 3 Anticipatory bodily changes during pregnancy

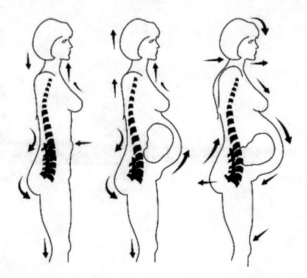

equilibrium, sexuality would throw it entirely away and result in changing all parameters of physiological, cognitive, endocrine, etc. activity. In such case, to even entertain the suggestion of homeostasis [30] is close to absurd. (Bernard suggested this in 1865; Cannon [31] gave it the label in 1926). Think only about how blood pressure changes from one moment to another. The alternative—allostasis [32]—hypothesizing that the brain takes control and acts in anticipation is more adequate in characterizing the dynamics of the living. The anticipation of all that the sexual encounter affords (at times in extreme forms) is not explicitly the reproduction, but the triggering of a process leading to it.

Hormones associated with pregnancy (those otherwise not present) influence the process. The future organism has to be "shaped" in such a manner that it autonomously distinguishes between beneficial and detrimental factors. For example, the placenta expresses an enzyme that prevents exposure to material corticoids. In their details, these processes are of extreme subtlety. During the initial phase of pregnancy, the mother herself "functions" in a manner that protects the forming of the new living entity. It is life from life, but in a context where what is good (or at least not harmful) for the mother might be dangerous to the fetus. Danger itself is implicitly acknowledged as immediate or pertaining to malfunctions in the future. Susceptibility to metabolic imbalance, of the future living, is a long-term projection that leads to anticipatory action to prevent it. Once again: anticipatory processes, being non-deterministic, some succeed (in balance, the majority), some don't (and there is a lot of evidence to document failed anticipation).

Doctors advise pregnant women that their diet influences the entire development of the fetus. The hypothalamus pituitary adrenal activity protects the woman from stress (psychological or physical). This protection is limited in time and is related to neuronal activity. Physicians claim to have evidence that progesterone levels in the pregnant woman's brain are higher: "There is 51 to 40 times more progesterone and estrogen marinating the brains during pregnancy" [33]. They also report on inhibitory activity in the hypothalamus, triggered by allopregnanolone. In non-pregnant women and in men, natural opioids amplify responses to stimuli. In pregnant women, the effect is inhibitory. (Naloxone, on whose basis the drug used to treat opioid overdose is designed, plays an important role in the process described above.) Maternal oxytocin provides a control mechanism in lactation; it promotes parturition (triggering contractions of the uterus) and affects what are described as emotional states. Anticipatory expression in this case is connected to neuro-cortical activity. The manner in which oxytocin is secreted (from the posterior pituitary) is associated with the distension of the birth canal or by sucking. Birth through the birth canal is favorably influenced by the specific biome populating it.

The description of processes here remains more at a macro-level. Let's take note of the following:

1. The body of the pregnant woman changes in anticipation of the pregnancy load (Fig. 3).
2. The physiology changes ahead of the many challenges related to the woman's body and the integrity of the fetus.
3. New processes never before experienced are made possible (the process of contraction or the sucking rhythm described above) in advance of their actual need.
4. Lactogenesis is an anticipatory expression timed to the various phases of pregnancy (early, mid-term, final, i.e., prepartum). Endogenous opioids inhibit some neuronal activity and thus facilitate prolactin activity.
5. Maternal behavior—a term used to describe how a mother's actions (preparations of all kind, nursing, cleaning, etc.) is affected by hormonal priming of the medial pre-optic area. Oxytocin, estrogen, progesterone, prolactin are released in various areas of the brain, in particular the amygdala, which is associated with anticipatory activity. Maternal behavior changes as pregnancy advances. For instance, the activation of oxytocin release processes at giving birth diminishes anxiety but can lead to aggressive behavior (for the management of which pregnant women are sometimes, against the natural course, subject to medication when self-control is not enough).

No doubt, medical experts could provide more details. They discuss a variety of other phenomena related to the creative act we call giving birth (i.e., reproduction). Not discussed is the nature of the very rich and sometimes ambiguous processes that aggregate in pregnancy. A recent study, for example, brings up creatine—an intracellular metabolite partially derived from diet, but also endogenously synthesized. As proved through experiments, the fetus is a "taker" of creatine [34]. Likewise, senescence (which Rosen studied to a certain degree) is the expression of "exhausted" anticipation. But we shall not enter into the details here.

1.5.2 Conceptual Premises

The key to a successful conversation is not the passion of the arguments—and as far as medicine is concerned, it is almost impossible to exclude passion—but the clarity of concepts. Therefore, within the conference to which this volume bears witness, I took it upon myself to define the concepts. Only once we agree upon our understanding of what words mean can we debate whether anticipation is or not significant. Those who practice medicine, and even more those who contribute to a science of medicine meant to overcome the limitations inherent in generalizing physics (and the notion of machine) in the living domain will agree on the need for conceptual clarity.

In the diagram (Fig. 2), the past is solidified in the anamnesis. Nobody can change his or her medical past. The future is one of possibilities, some that will be realized, others that will either disappear or extend further in time. This is where the question of context arises: the environment (in the broadest sense, nature, society, culture, etc.), as well as epigenetic factors. The diagram contains the following definition:

> The current state of an anticipatory system depends not only upon previous states, but also upon possible future states.

The diagram and the definition do not express the holistic view, which entails the fact that the reductionist method will always return a partial understanding of the process. Indeed, health, as well as loss of it, is the expression of the whole called *human being*, the physical embodiment (of the biome shadowing it) and the spiritual expression of the non-physical state (consciousness, preparedness, self-control, etc.). Affecting as little as one cell's condition, or that of the viruses, microbes, and bacteria making up the biome, might, under certain circumstances, trigger a multiplicity of processes, some of extreme consequences, others of episodic nature.

For the sake of the argument, let's take Methicillin-Resistant Staphylococcus Aureus (MRSA). As it is already established, MRSA, caused by the staph bacterium, can affect people who have spent time in "health factories" (e.g., hospitals, nursing homes) or at "health shops" (like dialysis centers). Invasive procedures, such as surgery, intravenous tubing, implantation of artificial knees and hips, and kidney and heart transplantations can trigger infections that prove to be resistant to antibiotic treatment. Physics-based devices of all kinds—all the gadgets and engines we use—also get "sick," that is, they malfunction. But the rate of success in fixing machines is as high as it can get, while the success rate of healing is increasing only slightly (despite the spectacular successes of extremely complicated cases that the media report). The idea that medicine's fundamental perspective might be deficient has not led practitioners to question it, and has not resulted in a vigorous attempt to change it.

Two assumptions ought to be made at this juncture:

1. Medical practitioners will find value in stepping out of their comfort zone (where all they read are medical news and prestigious journals close to their fields of expertise).
2. Those dedicated to research of anticipatory processes will deliver clear assessments of practical consequence to their colleagues in healthcare.

If both are realized, medicine will change. Otherwise, it will take a deeper crisis than the current one before medicine progresses from reaction-driven physics-based practice ("fixing" the patient) to a proactive, anticipation-based dedication to the well-being of the whole person.

Anticipation-based medicine implies a number of possibilities. As opposed to reaction, which is usually swift and short, anticipation unfolds within the timescale of the process involved. The immune system is anticipatory. Vaccination is designed to engage the immune system in order to avoid certain conditions (e.g., the success of the polio vaccine). Its components (antibodies, white blood cells, lymph nodes, T-cells, bone marrow, spleen) evolve according to the specific dynamics of the bacterial and viral expression they address. Nothing is immediate. The possible infection—a future state that the organism would rather avoid—is identified before onset. Immunotherapy is still in its early stages; but it is the closest we have come to an anticipatory perspective in medicine. Among others, Miroslav Radman, expert in DNA repair, alluded to the anticipatory characteristics of the immune system. A knee replacement —of course, sometimes necessary within medicine's emergency model—takes less time to be performed and become the "new knee" than what genetic healing—still more a promise than reality—might one day take. Mechanical interventions introduce the clock of physics where in the reality of the organism many time scales exist.

1.6   A Questionable Ultimate Aim

Is there one identifiable overarching reason for the reductionist-deterministic path taken by medicine? Someone dedicated to understanding anticipatory processes will not be among those searching for the *one* reason, where evidently many factors are involved. If we agree that anticipation is couched in complexity [35], one consequence cannot be avoided: as a result of medicine's surrender to the physics, both medicine and the patient suffer. Indeed, life is the expression of the complex nature of the living. Claude Bernard (1813–1879), iconic figure of modern medicine, echoed Descartes when he wrote, "When faced by complex questions, physiologists and physicians … should divide the total problem into simpler and more and more clearly defined partial problems" [36, p. 72]. In the same spirit, Francis Crick postulated, without any proof, that, "The ultimate aim of the modern measurement in biology is to explain all biology in terms of physics and chemistry" [37]. This thought continues, "Consciousness and mental states can be reduced to chemical reactions that occur in the brain," [38, 39].

Arguing from examples—how often do physicians err in reducing a problem to a smaller one, or how many times the physics and chemistry were right, but the patient died—is at best spurious. The broad image of medicine in these days of spectacular scientific and technological creativity is such that even those inclined to defend its record are not necessarily free of doubt concerning its progress. We landed on the moon, an immense achievement based on physics and chemistry. But we don't know how to handle the flu. (Vaccination is successful at the 50 % level, which is more a qualifier of guessing than of prevention.) We stuffed medical offices with expensive technology and provided the physician with data acquisition and processing capabilities of unprecedented precision. But we still don't know why a patient in a coma might be brought back to consciousness with Zolpiden (a sleep-inducing drug).

To repeat, arguing from examples does not lead to knowledge, rather to questions. These are the outcome of the daily activity of each physician. Very few cases are as clear-cut as the patient, the insurance, the doctor, and society would like them to be. Medicine is about life—such as in assisting a woman in giving birth—but also about what happens when life is subject to change (beneficial or detrimental), when life comes to an end. Physicians cannot avoid seeing themselves in each and every patient. The art and science embodied in the practical world of medicine is the necessary result of the condition of the living. If medicine were like physics or chemistry, society would be entitled to expect perfection. If, to again quote Bernard (justifiably admired for some of his work), "A living organism is nothing but a wonderful machine," we would seek the better mechanic and work on an automated machine-full body diagnostics (which is already in the works, beyond the latest mobile device apps).

1.7   Bringing Together

The dialog between the "mechanics" of human health (the engineers) and those trying to perceive it from a different perspective—healing as art and science—ceased. The *Anticipation and Medicine* conference I organized is one of the very few where at least an effort was made to bring together those who would criminalize, rather seek advice from, each other. To make dialog possible, a question was addressed to all: Why does the same patient prompt many different answers from different practitioners of medicine?

Competence level plays a role here, as do the means used for diagnosis. Culture is important, too; so are the social, economic, psychological, and religious conditions. But even assuming some common denominator—well-trained physicians, good technology, relatively stable societal background—the assessment continues to vary. Depending upon the condition examined—from flu symptoms to a variety of inflammatory conditions, from diabetes or asthma to all kinds of insufficiencies, and up to heart disease and cancer—the variability of evaluations is astonishing. It speaks in favor of the profession that patients are sometimes advised to seek a second opinion. More and more doctors interact, consulting with each other. Still, as opposed to the mechanic (convinced that he does what a medic does), who gets it right because the knowledge domain is limited—and there is no room for ambiguity —physicians, themselves changing as they examine their patients, deal with a subject that does not stay still—not even in a state of coma. The patient is an open system, of a dynamics no physical entity comes even close to reaching.

2   Complexity and Anticipation

Health (or lack thereof) is an expression of the complexity that defines the living. The meaning of the word (its semantics) is as well-defined as that of health. Leibniz [40] seems among the first to examine science from a complexity perspective. In his view, laws should not be arbitrarily complex. If they are, the concept of the law becomes inoperative. Medicine seeks simple definition for medical conditions, so that doctors can use them without difficulty. Poincaré [41], and, closer to our time, Prigogine [42] expressed interest in prediction (relation to future) as it is related to complexity (they are mentioned in connection to "chaos" theory, i.e., dynamic systems theory). From the anticipatory perspective under examination here, Rosen [43] remains the closest reference to the complexity intrinsic in living phenomena. In his view, there is no largest model for complexity. A complex entity is not fractionatable. (For more on this topic, see Staiger et al.[5] and Louie [44].) On the pragmatic level, where medicine is anchored, complexity is associated with efficiency, in particular, efficient treatment—regardless in which form. Patients are not known for patience (no pun intended), although many end up dependent upon life-long treatments.

The reductionist-deterministic model, useful in conceiving chemical means (i.e., pharmaceuticals) of extreme efficacy, is an expression of the attempt to understand which representations of a disease, that is, which symptoms, are easily addressable. The goal of explanations—e.g., how aspirin works—that is, measurements performed to find out which processes are triggered by some medications, under the guidance of reductionism and determinism, is to gain access to knowledge about phenomena otherwise difficult (if not impossible) to explain. "What is a headache?" is such a phenomenon (as common as it is different in its variety). The reductionist-deterministic paradigm indeed led to significant technological and pharmaceutical progress. But this does not eliminate the need to define complexity. A clear criterion (or criteria) for identifying it is more urgent than ever before if we want medicine to overcome the limitations inherent in its mechanistic practice. So far, the focus has remained on scale, i.e., on quantity, while complexity actually defines quality. However, complexity, as consubstantial with the living, is of high-order consequence for medicine. If the living, in particular the human being, is complex, knowing the medical subject in its complexity is of practical importance. (In a different context, I introduced a more general understanding of complexity as it relates to the human being [45].) In what follows, the concepts will be summarily defined and related to medicine.

3   "G-Complexity"—Where Medicine Starts

Let us start with a quote (of more interest to mathematicians than to healthcare professionals):

---

[5]Staiger, T.O., Kritek, P.A., Blakeney, E.L., Zierler, B.K., O'Brien, K., Ehrmantraut, R.H.: Implementing and Evaluating an Anticipatory Systems Model of Complexity for Improving Safety in a Healthcare Organization. In: Nadin, M.: (ed.) Anticipation and Medicine, pp. 27–36. Springer, Cham (2016).

Fig. 4 Neither the whole
body nor parts of it can be
described fully

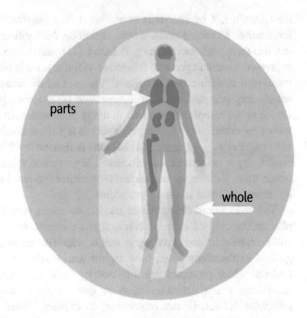

Any effectively generated theory capable of expressing elementary arithmetic cannot be
both consistent and complete. In particular, for any consistent, effectively generated formal
theory that proves certain basic arithmetic truths, there is an arithmetical statement that is
true, but not provable in the theory [46, p. 250]

This relatively simple text is based on Gödel's magnificent work of 1931 [47].
His subject is not the world, not Nature, but the dyophantine number theory, i.e.,
arithmetic—as far as one can imagine from the work of a physician and surgeon.
However, the logical representation justifies the generalization from labels used in
the description of the world (numbers) to existence. Gödel ascertained, and
demonstrated, that some of our descriptions cannot be simultaneously complete and
consistent (Fig. 4).

The doctor acts on partial descriptions. These can be symptoms easily noticeable
even upon superficial examination, or based on elaborate measurement. In the last
ten years only, the number of measurements that technological progress has
facilitated has increased by many orders of magnitude. A high degree of sophis-
tication is reached in discriminating among many parameters, some interrelated (but
of variable interconnectedness), others incidentally correlated. To assume that the
practitioner, who examines thousands of patients, tries to keep abreast of the most
current knowledge in the field, or can keep pace with it, is naïve.
Technology-facilitated data acquisition is way ahead of our full understanding of its
means and methods. In the end, the attempt to fully describe, through data, change
in the living might never succeed. An open system cannot be fully characterized.
Even if it could, that will not change the fact that full description and consistency
are reciprocally exclusive. This is so because phenomena of G-complexity—to

which the dynamics of the human condition belongs—escape both reductionism and determinism as exclusive descriptions of their causality.

There is no effective decomposition rule; the consistency clause ascertains that inferences implicit in determinism (same cause $\rightarrow$ same effect) do not hold for the G-complex. There is no medical practitioner who has not experienced this. Nothing is cause-free; rather, in the living causality is expressed in forms that no longer submit to the time sequence characteristic of determinism. Capturing the dynamics of life's physical substratum, determinism and causality, together with non-determinism and a-causality (i.e., within a condition of complementarity, as Bohr defined it [4]; see also Nadin[6]) afford a more adequate understanding of how life is expressed.

Based on all these arguments (abbreviated in view of the goal of this study), I suggest the following shorthand for the implications of my broader views on the nature of the living:

1. If decidability is the precise criterion for G-complexity, it follows that medicine either settles for the domain of the complicated (a heart transplant, a knee replacement) or triggers natural processes, such as genetic-driven methods of repair and self-repair.

   The knowledge domain of medicine is the undecidable. Above the threshold of complexity, there are no degrees. A system is undecidable or not. If treated under the complexity threshold, the system is equivalent to any physical system—and the criteria for maintenance correspond to this condition. There is no room for equivocation. If treated at the complexity level, healing and self-healing, i.e., the repair function, (which is anticipatory) imply processes characteristic of the living. It is encouraging that medicine is making serious efforts to become more "natural," to align its means and methods with the subject of its concern. Immunotherapy was already mentioned as an example.

2. A G-complex system is characterized by the fact that its information level is always higher than the information received from the environment; that is, a G-complex system generates information. Every patient becomes part of the interaction called treatment. In physics-based interventions, there is only one answer, and therefore this co-participation is minimal. The space of possible futures is where patient and physician can actively search for plausible answers. What a physician "takes in" from medical education (pre-med, residency, research fellowships) is quite different from what, on account of creativity (itself based on interaction), is expressed in the practitioner's activity. If only the effort of individualization were to be considered, this would already confirm the idea. Anticipation-driven medicine is by necessity individualized, because the living is infinitely diverse. All machines are the same; no two persons are. All electrons are the same; no two cells—from as many as the ca. 37 trillion making up the human body—are.

---

[6]Nadin, M.: Anticipation and the Brain. In: Nadin, M.: (ed.) Anticipation and Medicine, pp. 135–162. Springer, Cham (2016).

3. G-complex systems are adaptive systems; physical systems are defined by sameness; that is, they are not affected by context. For medicine this means the understanding that disease itself is related to adaptivity. Moreover, the agents of illness (microbes, viruses, food-born pathogens) are themselves adaptive. Medicine fails when it ignores the complexity of these concurrent processes. The consequences of the deterministic use of antibiotics were brought up; so was the vicious cycle that generated the current opiate addiction (and the mortality associated with it).

4. A G-complex system is not measurable. Understanding this particular condition of living processes will allow the medical community to free itself from the obsession with data and focus on the meaning of change in the patient.

   A G-complex system is represented by its life record. Time series can capture partial knowledge about specific aspects of the dynamics qualified through partial measurements. (For example, after surgery, physicians measure temperature, heart rate, blood oxygen levels, etc. Nonetheless, these variables represent only a limited aspect of the patient's state of health.)

5. G-complex systems have no effective copy procedures; everything in a G-complex system is unique. The knowledge domain of entities and phenomena characterized by G-complexity is the *idiographic*.[7] For any such entity E_i—let's say a patient complaining about back pain, always different from any other, no two back pains are the same—we can define a functional dynamic.

$$y_{Ei} = f_{Ei}\{x_{Ei}(t), \ I_{Ei}(t), \ t\} \tag{1}$$

   Take note that the function is entity specific (f_{Ei}). The simple equation says that the state of the patient experiencing pain depends upon some parameters—let's say a certain movement—and the interaction of the patient in the context. Interactions (I_{Ei}), part of the dynamics, are also specific. The fact that the function is entity specific excludes generalizations. Evidently, the specific dynamic of one identity actually differs in indeterminate ways from the dynamic of any other entity. The aggregate value is therefore meaningless. G-complex entities do not accept *nomothetic*[8] descriptions.

6. G-complex systems are relational.

$$E_i \mathcal{R}_{ij} E_j \tag{2}$$

   Living entities are interrelated. Medical assessments that take relations into consideration have to acknowledge their variety. An easy illustration is the state

---

[7]Of or relating to the study or discovery of particular scientific facts and processes, as distinct from general laws.

[8]Of or relating to the study or discovery of general scientific laws.

of a pregnant woman and that of her husband; so is the parent-child relation. The mirror neurons represent a good example of how relational aspects are expressed. The relational nature of the living translates into practical considerations in the process of medical assessment. It is never the case that a symptom can be examined independent of the relational space in which it manifests itself. Physicians actually report on such factors when they seek correlation: patient's state, family relations, medical history of those constituting the patient's milieu, etc.

7. G-complex systems are endowed with self-evolving anticipatory processes in which past, present and future are entangled.

$$x(t) = f(x(t-\alpha), x(t), x(t+\beta))$$
$$x(t-\alpha) \quad \text{previous state(s)}$$
$$x(t+\beta) \quad \text{future state(s)} \tag{3}$$
$$x(t) \quad \text{present state}$$

Patients embody their history; the future, related to possible interactions (some beneficial, others detrimental) is continuously anticipated. The preparedness of each person is a matter of record, not a dreamed-up hypothesis. Aging, for instance, is the example of the organism's preparing itself for a new state. For healthcare, the anticipatory endowment should translate into awareness of the practical consideration informed by the shared awareness of both the patient and physician. Again—not as an example, but illustration—the changed metabolism associated with aging suggests a different diet, but also an appropriate program for maintaining physical condition. The fact that doctors, eager to "keep the machine going," increasingly overwrite the self-evolving anticipatory process is probably a matter of medical ethics. Age-defined "infertility" has its own significance. A woman's giving birth at age 60 (with the help of fertility drugs) might give her doctor reason to be proud of the performance. However, in the perspective of time, this performance will probably not make anyone happy. Medicine should not compete with machine and drug-enhanced sports for performance that goes against the condition of the living.

A G-complex system is an evolving record of entangled (not quantum entangled, though) past states, current states, and possible future states. For the observer, such as the house physician, the patient's actions are the expression of successful or failed anticipations. G-complex systems are open systems, of unlimited dynamics. Medicine ought to comprehend the non-deterministic nature of both health and disease. "I smoked and drank all my life and lived to be over 100," expresses what we are referring to (obviously in extreme form). Others pay dearly for being only subjected to second-hand smoke.

3.1 Complex Versus Complicated

Complicated systems—such as a replacement knee or a mechanical kidney—are made of simple systems, or can be reduced to a limited number of simple systems. Complicated systems are subject to observation and measurement: the surgeon must inspect the integrity of the implant. To know such a system is to capture its regularity, obvious or hidden. This regularity corresponds to the laws predicting the behavior of such systems: eventually they will have to be replaced. The experimental method as a source of knowledge about the physical—How do such devices perform? What are the consequences for the rest of the body?—is based on the assumption that competing explanatory models (i.e., hypotheses) can be empirically tested. These considerations apply as well to the validation of chemical interventions (e.g., pills, lotions, ointments, injections of fluids).

Experiments are always closed systems, within which variability (of parameters) can be quantitatively described. The dosage for a medicine is such a variable. It is reasonable to rely on them for knowledge acquisition when the dynamics examined and the experimental timeline are congruent. Given the expectation of not doing harm to the human being, many experiments are performed on living substitutes (mice, rats, pigs, monkeys) or on simulations (digital or otherwise). The former case is more realistic, although it implies an equivalent of a sort between the human—who acquired self-awareness—and animals—complex entities, but lacking self-consciousness.

There cannot be complete information about a G-complex system, i.e., about a living entity, since it produces information as it evolves. The living adapts to a variable world, and interacts with it. Life is interaction. Therefore, the experimental model pertinent to the domain of the decidable (we can fully describe, contradiction-free, the physics and chemistry underlying the existence of the living) is not applicable in the G-complexity domain. To repeat: no two individuals are the same, no two medical conditions are identical. The fact that experiments are carried out and presented as trustworthy validations corresponds to the illusion that reductionist-deterministic science generates significant knowledge. This "data-and-experiment cult" is rather a component of the politics of science than an intrinsic part of it. Even generalizations built upon statistical averages and probability distribution defy the nature of the entity subject to knowledge acquisition. A doctor will not better address a patient's health condition based on averaging. (The pitfalls of averaging are discussed in [48].) These are very concrete aspects of practicing medicine without looking through the "eyeglasses" of physics or chemistry. The patient's unique profile should be the source for describing his condition. In particular, I would like to suggest the Anticipatory Profile [49].

The living can be simulated by computation in a nonliving substratum only partially. The entire effort of embedding computation in artificial entities emulating aspects of the living (synthetic neurons, artificial muscles, synthetic DNA, synthetic cells, etc.) deserves respect for the gnoseological, scientific, and technological effort. However, the outcomes of the computation on such substrata can only reflect

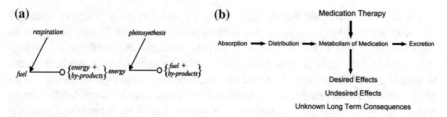

Fig. 5 a Natural processes have multiple outcomes (image courtesy of Louie [44]). **b** Example of multiple outcomes related to medication

the assumptions embedded in the emulated synthetic world. They confirm the physics of the living, not its specific condition. Let us end these considerations with one more observation: Physical and chemical processes have well-defined outcomes. Living processes have multiple outcomes, some antagonistic. It is a known fact that the same medication can be beneficial to some and (highly) detrimental to others: the "paradoxical effect" of medication. One discussion, going back to 1975 [50], deals with benzodiazemines, which trigger aggressiveness instead of acting as tranquilizers. The prevalent practice of medicine-as-deterministic—same drug prescribed → same consequence—has led to serious repercussions in the very young and in the elderly (the two groups usually omitted from testing procedures) (Fig. 5).

4 Turning Ideas into Actions

It would be naïve to believe that practitioners of medicine would from one day to another open up to the thoughts expressed in this study. The volume *Anticipation and Medicine*, the first on record on the subject, should help in providing reference material. It would be even more naïve to expect a rapid cycle of developing methods and means for an anticipation-grounded practice of medical care. The subject is more of economic, political, social, and cultural consequence than it is of science or technology. (The pharmaceutical industry has over one trillion products in the "pipeline.") Having stated that physics-based medicine can unfold as fast as physical interactions, but that anticipation-guided medical care has the same rhythm as life, it is obvious that change, should it take place, will be slow, and will involve not only the medical profession, but also patients. Ultimately, this change is predicated upon a profound necessity: survival of the species. If this assessment sounds exaggerated, consider some of the facts discussed, and especially some of those only alluded to: degeneration of the species, of which aging is only a symptom; sexual conditioning (no longer related only to reproduction); spectrum disorders, the ever-growing catalog of morbidity, rarely only the product of better identification methods.

As an endeavor endemically optimistic, science can only bring to society's attention a worrisome situation, but cannot magically erase it—or postpone its consequences. The place to start for making possible the fundamental change suggested in this study is education. The continuing education of medical practitioners is part of the process. The real chance, however, to change medicine from a mechanistic reactive practice to a proactive, creative activity is connected to schools and universities, to the new generations. Education has to be reconceived from the ground up: solid scientific education in both the physics of the world and in the biology grounded in anticipation is required. This in itself is a high-order endeavor, since schools continue to indoctrinate new generations in the "religion of physics." It is amazing that climate change associated with human activity is reduced to the physics of climate, where sustainability is a matter of choices, and improved technology, instead of being framed in the anticipation perspective: i.e., patterns of behavior, choices people make.

Medicine was always the art and science of healing. The science became more and more a technology; the art was dropped altogether. If some medical practitioners are better than others on account of more than the equipment they have access to, it is considered as incidental. What distinguishes among doctors educated under the same circumstances (same medical school, same internship, etc.) was and should still be identified as talent. Society should offer equal access to medical care, but this social goal does not automatically qualify all providers as equally talented for the profession, or equally dedicated to it. To identify medical talent, dedicated to patient well-being, is a function abandoned, since medical schools are not really lacking applications and the medical establishment defends it turf (read: "return on investment").

Almost 100 years ago, Robert Lovett presented "A Plea for a More Fundamental Method in Medical Treatment" [51]. We can repeat the gesture; we can even use his words, if indeed "Fundamental Method" would be understood as considering the human being from the complementary perspective of physics and biology established within the framework of anticipation. The goals are clear: consider the specific dynamics of the living. This will change the view on what diagnostics is, what it means to relieve suffering, how proactive medicine—maintaining health instead of patching what is broken—engage patient and physician, family, and community. All these amount to a tall order.

The current profile of the physician mimics that of a data processing professional. Patients often mimic their doctors. But do-it-yourself medicine via access to information can sometimes be confusing. Nothing against the patient assuming an active role in healing, as long as it does not mean transferring responsibility (concerning how we live) to machines and chemistry. Family is today more an economic entity; so is community. Their role in addressing an individual's health concerns is diminishing.

Nostalgia for the time when doctors were "magicians," when parents and children were subject to a bond that made the child's suffering a family concern, etc.—illusions of wishful thinking realities—will not do. The answer is not in the past. And in the present, individuals are subjected to the enormous pressure of changes that on the one hand shape us more dependent on society, and on the other less responsible for our choices. It is in this context that the awareness of anticipation will heighten. Or else. Medicine can lead in the process, because after all is said and done, medicine is about life and death.

Acknowledgments The research reported herein took place between 2005 and 2015. Data was acquired in the AnticipationScope and in a variety of experiments using Microsoft Kinect. Among those who assisted during these years are the late Dean Burt Moore (School of Behavioral and Brain Sciences, UT-Dallas), Melinda Andrews, Dr. Navzer Engineer, Dr. Balakrishnan Prabakharan, Robert Fuentes, Dr. Gaurav Pradhan. The research was supported by the University of Texas at Dallas, TZI (Bremen Germany), and Microsoft. The author is indebted to the reviewers for their suggestions.

References

1. Null, G., Dean, C., Feldman, M., Rasio, D., Smith, D.: Death by Medicine. Praktikos Books, Mount Jackson VA. http://www.webdc.com/pdfs/deathbymedicine.pdf (2004). Accessed 22 Mar 2016
2. Makary, M.A., Daniel, M.: Medical error—the third leading cause of death in the US. Br. Med. J. **353**, 2139 (2016). http://www.bmj.com/content/353/bmj.i2139, http://www.bmj.com/company/wp-content/uploads/2016/05/medical-errors.pdf
3. Medical errors now third leading cause of death in United States. https://www.sciencedaily.com/releases/2016/05/160504085309.htm
4. Bohr, N.: Discussions with Einstein on Epistemological Problems in Atomic Physics. Albert Einstein: Philosopher—Scientist Oxford University Press, Cambridge (1949)
5. Schrodinger, E.: What is Life?. Oxford University Press, Cambridge (1944)
6. Elsasser, W.: Reflections on the Theory of Organisms. Holism in Biology. Johns Hopkins University Press, Baltimore (1998) (Originally published in 1987 by ORBIS Publishing, Frelighsburg, Quebec)
7. Nadin, M.: Anticipation—The End is Where We Start From. Lars Müller Verlag, Basel (2003)
8. Nadin, M. (ed.): Anticipation Across Disciplines. Cognitive Systems Monographs. Springer, Cham (2015)
9. Nadin, M.: Anticipation and computation. Is anticipatory computing possible? In: Nadin, M., (ed.) Anticipation Across Disciplines. Cognitive Science Monographs, vol. 26, pp. 163–257. Springer, Cham (2015)
10. Venter, J.C., Gibson, D., et al.: Creation of a bacterial cell controlled by a chemically synthesized genome. Science **329**(5987), 52–56 (2010). http://science.sciencemag.org/content/329/5987/52
11. Callaway, E.: "Minimal" cell raises stakes in race to harness synthetic life. Nature **531**(7596) (2016). http://www.nature.com/news/minimal-cell-raises-stakes-in-race-to-harness-synthetic-life-1.19633
12. Nadin, M.: Anticipation and creation. In: Staicu, V. (ed.) Libertas Mathematica, vol. 35, no. 1, pp. 1–16 (2015)
13. von Uexküll, J.: A Foray into the Worlds of Animals and Humans: With a Theory of Meaning (O'Neil, D.J. (trans.)). University of Minnesota Press, Minneapolis (2010). (Originally appeared as Streifzüge durch die Umwelten von Tieren und Menschen. Verlag Julius Springer, Berlin, 1934)
14. Brentano, C.: Jakob von Uexküll. The Discovery of the Umwelt Between Biosemiotics and Theoretical Biology. Springer, Berlin (2015)

15. Ginn, F.: Jakob von Uexküll.: Beyond bubbles: on umwelt and biophilosophy. Sci. Cult. **23** (1), 129–134 (Taylor and Francis, London) (2014). doi:10.1080/09505431.2013.871245
16. Darwin, C.: On the Origin of Species by Means of Natural Selection, or The Preservation of Favoured races in the Struggle for Life. John Murrary, London (1859)
17. Newton, I.S.: Philosophiae Naturalis Principia Mathematica. S.Pepys, London (1686)
18. Picollo, S.: Developmental biology: mechanics in the embryo. Nature **504**(7479), 223–225 (2013)
19. Picollo, S.: Embracing mechanical forces in cell biology. Differentiation **86**(3), 75–76 (2013)
20. de la Mettrie, J.O.: L'Homme Machine. Elie Luzac, Fils, Leiden (1748)
21. Bernstein, N.A.: On the Construction of Movements. Medgiz, Moscow (1947). (in Russian)
22. Popper, K.: Of Clouds and Clocks. An Approach to the Problem of Rationality and the Freedom of Man, Objective Knowledge: An Evolutionary Approach. Oxford University Press, Cambridge (1972)
23. Newell, A., Simon, H.A.: Human Problem Solving. Prentice Hall, Englewood Cliffs (1972)
24. Laplace, P.S.: A Philosophical Essay on Probabilities (Truscott, F.W., Emory, F.L. (trans.)). Wiley, New York (1902)
25. US Department of Health and Human Services, National Institutes of Health. Precision Medicine Initiative Cohort Program. https://www.nih.gov/precision-medicine-initiative-cohort-program
26. Mehrotra, A., Prochazka, A.: Making a case against the annual physical, N. Engl. J. Med. **373**, 1485–1487 (2015). http://www.nejm.org/toc/nejm/373/16/
27. Rosen, R.: Some systems theoretical problems in biology. In: Laszlo, E. (ed.) The Relevance of General Systems Theory. George Braziller, New York (1972)
28. Counting the Hidden Victims of Medicine. New Scientist, 2953, 25 Jan 2014. https://www.newscientist.com/article/mg22129532-000-counting-the-hidden-victims-of-medicine/
29. Rosen, R.: Life Itself: A Comprehensive Inquiry Into the Nature, Origin, and Fabrication of Life (Complexity in Ecological Systems). Columbia University Press, New York (1991)
30. Schulkin, J.: Rethinking Homeostasis: Allostatic Regulation in Physiology and Pathophysiology. MIT Press, Cambridge (2003)
31. Cannon, W.B.: Physiological regulation of normal states: some tentative postulates concerning biological homeostatics. In: Petit, A. (ed.) A Charles Richet: ses amis, ses collègues, ses élèves. Les Editions Medicales, Paris (1926)
32. Sterling, P., Eyer, J.: Allostasis: a new paradigm to explain arousal pathology. In: Fisher, S., Reason, J.T. (eds.) Handbook of Life Stress, Cognition and Health. Wiley, Chichester, NY (1988)
33. Brizendine, L.: Pregnancy Brain: Myth or Reality. http://www.webmd.com/baby/features/memory_lapse_it_may_be_pregnancy_brain
34. Ellery, S.J., LaRosa, D.A., Kett, M.M., Della Gatta, P.A., Snow, R.J., Walker, D.W., Dickinson, H.: Creatine homeostasis is altered during gestation in the spiny mouse: is this a metabolic adaptation to pregnancy? BMC Pregnancy and Childbirth. http://www.ncbi.nlm.nih.gov/pubmed/25885219 (2015). Accessed 23 Mar 2016
35. Nadin, M.: Anticipation and the artificial. aesthetics, ethics, and synthetic life. AI Soc. **25**(1), 103–118 (Springer, London) (2010)
36. Bernard, C.: An Introduction to the Study of Experimental Medicine (Greene, H.C., trans.). Dover Books on Biology, New York (1957)
37. Crick, F.H.C.: Of Molecules and Men. University of Washington Press, Seattle (1966)
38. Bickle, J.: Philosophy and Neuroscience: A Ruthlessly Reductive Account (form the series Studies in Brain and Mind). Springer, Heidelberg/New York (2003)
39. van Regelmorter, M.H.: Biological complexity emerges from the ashes of genetic reductionism. J. Mol. Recognit. **17**(3), 145–148 (2004). doi:10.1002/jmr.674
40. Calude, C.S.: Randomness and Complexity: From Leibniz to Chaitin. World Scientific, Singapore (2007)
41. Poincaré, H.: Electricité. Sur le dynamique de l'electron, June 5, 1905. Note de H. Poincaré, CR.T 140, 1504–1508 (1905)

42. Prigogine, I.: The End of Certainty. The Free Press, New York (1997)
43. Rosen, R.: Fundamentals of Measurement and Representations of Natural Systems. North-Holland, New York (1978)
44. Louie, A.H.: Robert Rosen's anticipatory systems. Foresight **12**, 18–29 (2010)
45. Nadin, M.: G-Complexity, quantum computation and anticipatory processes. Comput. Commun. Collab. **2**(1) 16–34 (2014). (DOIC: 2292-1036-2014-01-003-18)
46. Kleene, S.C.: Introduction to Metamathematics. Van Nostrand, Princeton, NJ (1950)
47. Gödel, K.: Über formal unenscheidbare Sätze der Principa Mathematica und verwandte Systeme, Monatshefte für Mathematik und Physik, 38, pp. 173–198. The first incompleteness theorem originally appeared as Theorem VI (1931)
48. Rose, T.: The End of Average. HarperCollins, New York (2016)
49. Nadin, M.: The anticipatory profile. An attempt to describe anticipation as process, Int. J. General Syst. **41**(1), 43–75 (Taylor and Francis, London) (2012)
50. Tranquilizers Causing Aggression. Br. Med. J. **1**(5952), 266 (1975). doi:10.1136/bmj.1.5952.266 (Published 01 February 1975)
51. Lovett, R.: A plea for a more fundamental method in medical treatment. Med. Surg. J., 418 (1919)

Part I
Anticipation and Medical Care

A Conceptual Framework for Applying the Anticipatory Theory of Complex Systems to Improve Safety and Quality in Healthcare

Thomas O. Staiger, Patricia A. Kritek, Erin L. Blakeney, Brenda K. Zierler, Kurt O'Brien and Ross H. Ehrmantraut

Abstract Effective anticipation is a fundamental characteristic of highly reliable organizations. In Rosen's anticipatory theory of complex systems, all living systems and virtually all other complex systems require anticipatory models to maintain an organized state. This paper provides an overview of Rosen's anticipatory theory of complex systems and presents a conceptual framework for applying this framework to improve safety and quality in healthcare. Organizational interventions based on this theory could include education of clinicians, patients, and families on how anticipatory complex systems function and improve safety in clinical environments, and systems interventions to promote optimal concordance between a team's model of a clinical situation and the actual clinical situation. Enhanced general understandings of anticipatory complex systems and of their failure modes could help reduce communications failures that are a common cause of serious adverse events.

Keywords Anticipation · Complexity · Safety · Quality · Medicine · Situational awareness · Communication

T.O. Staiger (✉) · P.A. Kritek
Department of Medicine, University of Washington, Box 356330, Seattle, WA 98195, USA
e-mail: staiger@uw.edu

E.L. Blakeney · B.K. Zierler
School of Nursing University of Washington, Seattle, USA

K. O'Brien · R.H. Ehrmantraut
UW Medicine Organizational Development and Training, University of Washington, Seattle, USA

© Springer International Publishing Switzerland 2017
M. Nadin (ed.), *Anticipation and Medicine*, DOI 10.1007/978-3-319-45142-8_2

1 Introduction

Clinical Scenario: *A mother brings her 1-year-old son to an emergency department for evaluation of his respiratory symptoms. Following an evaluation the treating physicians advise her that it is safe for the child to be discharged. The nurses caring for the child express concerns and say they don't think the child should go home. While driving home from the emergency department, the child exhibits increased respiratory distress. The mother returns to the emergency department. The child is admitted, spends several days in the intensive care unit, and makes a satisfactory recovery.*

Substantial efforts are needed to further improve the quality and safety of healthcare [1]. Communication failures are recognized to be an extremely common contributing factor for adverse events [2]. Concepts from safety science are increasingly being employed in efforts to create high reliability systems in healthcare [1]. Some are applying theories from systems and complexity sciences to further improve healthcare systems and processes [3]. Applying complexity principles in healthcare can be challenging in the absence of a generally agreed upon definition of complexity and without a shared understanding of optimal approaches for applying these principles to improve outcomes [4].

A capacity for effective anticipation is recognized as a fundamental characteristic of highly reliable organizations [5]. Theoretical biologist Robert Rosen made the first description of an anticipatory system [6]. Mihai Nadin, another seminal contributor to understanding anticipation [7, 8] has documented a substantial current body of theoretical and applied research in anticipation. In Rosen's anticipatory theory of complex systems, all living systems and virtually all other complex systems require anticipatory models to maintain an organized, far from equilibrium state [9]. This paper provides an overview of Rosen's anticipatory theory of complex systems and presents a conceptual framework and implementation options for applying this framework for improving safety and quality in healthcare.

2 Anticipatory Systems

In *Anticipatory Systems: Philosophic, Mathematic, and Methodological Foundations*, Rosen defined an anticipatory system as containing "a predictive model of itself and/or its environment, which allows it to change state at an instant in accord with the model's prediction pertaining to a later instant," [6]. Nadin's definition: "An anticipatory system is a system whose current state is influenced not only by a past sate, but [especially] by possible future states," [9] connects to the space of possibilities [10]. An anticipatory system cannot produce perfect models of its future internal or external environments since the future is always somewhat uncertain. In an anticipatory system, the system's imperfect models of the future influence how the system changes. Changes occurring in an anticipatory system can range from increased activation of regulatory enzymes to changes in individual or group behavior based on anticipated future conditions.

Anticipatory systems differ fundamentally from recursive (simple/reactive) systems, in which change is entirely due to the influence of forces occurring in the past acting on the present state of the system. Recursive change in physical systems was one of Newton's key discoveries [11]. Recursion in Newtonian mechanics means that what is occurring at a given instant in time (i.e., the positions, velocities, and forces acting on a set of particles) determines what occurs at the next instant in time. As Rosen wrote, "the heart of recursion is the conversion of the present to the future, or the entailment of the future by the present." In describing the Newtonian paradigm and its impact on scientific thought, Rosen said:

> The essential feature of that paradigm is the employment of a mathematical language with a built-in duality which we may express as the distinction between internal states and dynamical laws. In Newtonian mechanics the internal states are represented by points in some appropriate manifold of phases, and the dynamical laws represent the internal or impressed forces. The resulting mathematical image is what is now called a dynamical system [...] Through the work of people like Poincare, Birkhoff, Lotka, and many others over the years, however, this dynamical systems paradigm or its numerous variants, has come to be regarded as the universal vehicle for representation of systems which could not be technically described mathematically: systems of interacting chemicals, organisms, ecosystems, and may others. Even the most radical changes occurring within physics itself, like relativity and quantum mechanics manifest this framework [...] This, then, is our inherited mechanical paradigm, which in its many technical variants or interpretations has been regarded as a universal paradigm for systems and what they do. These variants take many forms: automata theory, control theory, and the like, but they all conform to the same basic framework first exhibited in the Principia [11, p. 78].

Louie, in "Robert Rosen's anticipatory systems" stated:

> Note, in contrast, that a reactive system can only react, in the present, to changes that have already occurred in the causal chain, while an anticipatory system's present behavior involves aspects of past, present, and future. The presence of a predictive model serves precisely to pull the future into the present; a system with a "good" model thus behaves in many ways as if it can anticipate the future. Model-based behavior requires an entirely new paradigm, an "anticipatory paradigm", to accommodate it. This paradigm extends—but does not replace—the "reactive paradigm" which has hitherto dominated the study of natural systems. The "anticipatory paradigm" allows us a glimpse of new and important aspects of system behavior [12].

Figure 1 depicts an anticipatory system. The system enclosed in the dotted box could represent an organism, an organization, or a social system. S, M, and E are components of the system depicted. M is a predictive model of S. Time variables in M run faster than real time. As a result, any observable of M serves as a predictor of a corresponding observable of S at a later instant. In addition, the "system M is equipped with a set E of effectors that operate either on S itself or on the environmental inputs to S" [12]. Louie describes how this system functions as follows:

> We shall now allow M and S to be coupled; i.e., allow them to interact in specific ways. For the simplest model, we may simply allow the output of an observable on M to be an input to the system S. This then creates a situation in which a future state of S is controlling the

Fig. 1 Anticipatory system
(Modified from Louie [12])

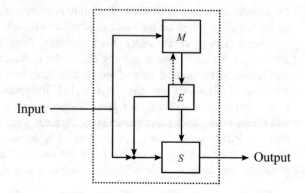

Input

M

E

S

Output

present state transition in S. But this is precisely what we have characterized above as anticipatory behavior. It is clear that the above construction does not violate causality; indeed, we have invoked causality in an essential way in the concept of a predictive model, and hence in the characterization of the system M. Although the composite system (M + S) is completely causal, it nevertheless will behave in an anticipatory fashion [12].

2.1 Feedforward and Feedback

The concept of feedforward is closely linked to anticipatory systems. Feedforward in biochemistry has been defined as: *the anticipatory effect that one intermediate in a metabolic or endocrine control system exerts on another intermediate further along in the pathway; such effect may be positive or negative* [13].

Descriptions of feedforward regulation include this example from glycolysis:

> Control of a metabolic pathway by a metabolite of the pathway that acts in the same direction as the metabolic flux, i.e., downstream or "later" in the pathway, e.g., the activation of pyruvate kinase by fructose 1,6-bisphosphate [14].

If the anticipatory features of glycolysis are represented in Fig. 1, E = pyruvate kinase, M = Fructose 1.6 pyrophosphate, S = phosphoenolpyruvate (a substrate catalyzed by pyruvate kinase). The level of Fructose 1.6 pyrophosphate serves as predictor of subsequent levels of phosphoenolpyruvate. Increased Fructose 1.6 pyrophosphate levels upregulate the effector, pyruvate kinase. Pyruvate kinase then catalyzes the transfer of a phosphate group from phosphoenolpyruvate to adenosine diphosphate (ADP).

Louie has described the role of feedforward in anticipatory systems as follows:

> Anticipatory behavior involves the concept of feedforward, rather than feedback. The distinction between feedforward and feedback is important, and is as follows. The essence of feedback control is that it is error-actuated; in other words, the stimulus to corrective action is the discrepancy between the system's actual present state and the state the system should be in. Stated otherwise, a feedback control system must already be departing from its nominal behavior before control begins to be exercised.

In a feedforward system, on the other hand, system behavior is preset, according to some model relating present inputs to their predicted outcomes. The essence of a feedforward system, then, is that the present change of state is determined by an anticipated future state, derived in accordance with some internal model of the world.

We know from introspection that many, if not most, of our own conscious activities are generated in a feedforward fashion. We typically decide what to do now in terms of what we perceive will be the consequences of our action at some later time. The vehicle by which we anticipate is in fact a model, which enables us to pull the future into the present. We change our present course of action in accordance with our model's prediction. The stimulus for our action is not simply the present percepts; it is the prediction under these conditions. I emphasize again that "prediction" is not prescience, but simply "output of an anticipatory model". Stated otherwise, our present behavior is not just reactive; it is also anticipatory [12].

In the clinical scenario given above, the treating physicians and nurses had different anticipatory models of the likely clinical course of the patient being evaluated. These models strongly influenced their actions. A better-shared anticipatory model by the clinical team would have led to a better outcome.

3 Simple and Complex Systems

While complex systems are described in a variety of ways, it is generally accepted that the components (parts with a function) of a complex system are interrelated and that fragmentation or decomposition of these components causes a loss of information regarding the system [4, 15]. In *Life Itself: A Comprehensive Inquiry into the Nature, Origin, and Fabrication of Life*, Rosen defined complex systems in similar, albeit more precise, terms [11]. Rosen defined a simple system as one in which all of the information can be captured in a single formal model and complex systems as those in which all of the information cannot be represented by any single formal model or any finite sets of formal models. More details about Rosen's extensive work on modeling relationships can be found in Louie's brief summary "Robert Rosen's anticipatory systems" [12], or in Rosen's *Fundamentals of Measurement and Representation of Natural Systems* [16].

A brief summary of Rosen's conceptual approach to simple and complex systems is as follows:

1. Simple systems:

   a. Are non-living
   b. All of the information about the system can be captured in a single model of the system
   c. The system can be fractionated (broken into parts) and reassembled without loss of information about the system
   d. A system may be complicated and composed of many parts (i.e., an aircraft carrier) and still be "simple"
   e. All change is recursive

2. Complex systems:

    a. All living systems are complex systems
    b. No finite set of models can capture all of the information about the system
    c. A complex system can't be fractionated without destroying information about the relationship of the components of the system
    d. A primary characteristic of complex living systems is a capacity to adapt to a changing environment
    e. Complex living systems are anticipatory, meaning they have a capacity for change based on inputs from their anticipatory models

3.1 Stability of Anticipatory Complex Systems

There are important differences governing the stability of simple and anticipatory complex systems. To paraphrase Rosen, in a simple system every global failure results from local failures in the component subsystems; however in a complex system, a global failure is not necessarily associated with a local subsystem failure [17]. Anticipatory complex systems may malfunction when a component in the system fails. What distinguishes an anticipatory complex system from a simple system is that its stability is also dependent upon the accuracy of its models or representations of its present environment and of its anticipatory models of its future environment. While a complex system cannot, by definition, have a perfectly accurate and comprehensive model of its internal and external environments, the more accurate an anticipatory complex system's models are of the states of its present and future environments, the greater the likelihood of the ongoing stability of the system. Conversely, the greater the divergence between the models or representations of an anticipatory complex system and the actual present and future environments, the less adaptive will be the system's responses to the present and future environments, and the greater the likelihood of the system malfunctioning.

4 Application to Improving Safety and Quality in Healthcare

If Rosen's theory of anticipatory complex systems is correct, the following actions would be appropriate approaches for improving the safety and quality of healthcare [18]:

4.1 Education for Clinical Teams

Members of clinical teams, including physicians, advanced practice providers, nurses, pharmacists, social workers, and ancillary care providers, should be educated about the characteristics of anticipatory complex systems in clinical environments. Key concepts would include the following:

a. All patient care encounters involve interactions between complex systems (people).
b. A clinician or clinical team's model/representation of a patient (complex system) can never be absolutely accurate.
c. It is impossible for a team of clinicians, or a team of clinicians, patient, and families to ever have a fully congruent shared mental model of a patient's complex clinical situation.
d. Clinicians, clinical teams, patients, and families are continuously generating anticipatory models representing possible future events and risks [19]. Inputs from the anticipatory models of clinical team members, patients, and families may be useful for identification of and real time mitigation of some clinical situations in which there is an increased risk of a future serious adverse outcome.
e. Clinical experience is likely to contribute to a capacity for generating more accurate anticipatory models. For example, nurses with more than 1 year of experience were significantly more accurate in identifying patients at risk for physiologic deterioration than those with less than 1 year of experience. (72 % vs. 53 %, $p < 0.05$) [20].
f. Clinicians, clinical team members, patients, or family members who believe, even if based only on a "gut sense" or intuition that the diagnosis or plan for a patient is wrong or that a patient is at high risk for an adverse outcome, should be encouraged to speak up.
g. Greater concordance between the models of the current situation and anticipatory models of future states among clinical team members, and among the clinical team, patient, and family increases the likelihood of attaining preferred outcomes.
h. Better concordance between a team's model of a clinical situation and the actual clinical situation will occur if there is an opportunity for meaningful input and for the expression of discrepant views from all team members and from patients and families. A team environment that encourages the expression of divergent input is likely to arrive at a better understanding of the current clinical situation and to have an improved capacity to more reliably anticipate future events.
i. Significantly discrepant present-state or anticipatory mental models between clinical team members or between team and patients/families may indicate an increased risk for an adverse outcome. Recognizing and reevaluating when such divergent opinions exist may provide opportunities to mitigate future risks.
j. Situational Awareness education could be enhanced by including the fact that people (and other complex systems) are continually generating anticipatory predictions of future outcomes and that these predictions can be useful for guiding present decision making. This could help augment and strengthen the approach promoted in situational awareness to "thinking ahead" based on an extrapolation of information known about the current situation [21].

4.2 Education for Patients and Families

Patients and families should be advised that: (1) Care maps represent what we anticipate happening; (2) if they don't agree with the clinical teams' plans that they should alert any care provider; (3) if things don't seem to be going the way the patient/family thought they should (either by a gut sense or something different from the care maps) the patient/family should let any care provider know, because the patient and family know the patient better than any of the clinical team and may notice changes earlier than will members of the clinical team.

4.3 To Do List for Healthcare Institutions

a. Support systems for capturing the clinical team's anticipatory predictions of risk of deterioration or events and to encourage planning based on these predictions. The I-PASS handoff tool, which encourages identification of patients requiring additional monitoring and contingency planning, is an example of this approach [22].
b. Encourage clinical team members to voice concerns about the understanding of the current clinical situation and/or of the risk of future adverse events.
c. Promote a culture in which hierarchy and power gradients do not inhibit clinical team members from sharing anticipatory predictions.
d. Reward clinical team members who express anticipatory predictions.
e. Provide Institutional support for structured clinical rounding, including patients and family members, to create opportunities for information sharing and for raising of questions, concerns, or identification of situations in which divergent models may indicate an increased risk to patient outcomes and safety.
f. Encourage team briefs/huddles as team strategies for creating plan for the day.
g. Support increased availability of clinical decision making tools (e.g., dashboards/reports/summaries to use during rounding that would incorporate real-time results of risk stratification models).

4.4 Measurement Possibilities

An agreement tool, such as one developed to assess a patient's degree of agreement with the clinical team about diagnosis, diagnostic plan, and treatment plan could be administered to patients or patients and families and in one or more of the following ways [23]:

i. Tracking incidence of significant disagreement on an ICU or inpatient unit, providing an alert to clinical team members and assessing the frequency in

which identification of significant disagreement resulted in a clinically important change in management.

ii. To determine if a correlation exists between care coordination sensitive outcomes (LOS, patient satisfaction, show rates at follow-up appointments) and
clinical teams with higher agreement scores.

iii. If a correlation between better agreement and care-coordination sensitive
clinical outcomes exists, feedback to teams about aggregated patient/family
agreement scores could be provided to determine if this resulted in improved
agreement scores and in improved care-coordination sensitive clinical
outcomes.

iv. A modified agreement instrument (do you agree with the team's diagnosis,
diagnostic plan, and treatment plan) could be administered to interdisciplinary
clinical team members to determine if a correlation exists between teams with
higher agreement scores and improved care coordination sensitive outcomes
(LOS, patient satisfaction, show rate a follow-up appointments).

v. An agreement instrument could be employed in an outpatient setting to
determine if better agreement is associated with outcomes such as intention to
adhere to treatment, patient satisfaction, or symptom resolution. (In a pilot
study of 39 patients with new or worsening problems higher agreement scores
were correlated with both better patient satisfaction ($p = 0.029$), and intent to
adhere to treatment ($p = 0.011$) [24].

5 Conclusions

Healthcare organizations are complex adaptive systems in which improving safety
is an ongoing challenge and imperative. If Rosen's and Nadin's anticipatory systems hypotheses are correct, an enhanced understanding of the characteristics and
failure modes of anticipatory systems could supplement existing safety practices
and help an organization further reduce the risk of certain serious adverse events.
Organizational interventions based on an anticipatory theory of complexity could
include: (1) education of clinicians, patients, and families on how anticipatory
complex systems function and contribute to improving safety in clinical environments; (2) systems interventions to promote optimal concordance between a team's
model of a clinical situation and the actual clinical situation; and (3) development of
measurement instruments to provide feedback to help improve performance in these
areas. An enhanced general understanding of anticipatory complex systems and of
their failure modes could help reduce communications failures that are the most
common root cause of serious adverse event.

References

1. Chassin, M.R., Loeb, J.M.: High reliability healthcare: getting there from here. The Milbank Q. **91**, 459–490 (2013)
2. Leonard, M., Graham, S., Bonacum, D.: The human factor: the critical importance of effective teamwork and communication in providing safe care. Qual. Saf. Health Care **13**(Suppl 1), i85–i90 (2004)
3. Sturmberg, J.P., Martin, C.M., Katerndahl, D.A.: Systems and complexity thinking in the general practice literature. Ann. Fam. Med. **12**, 66–74 (2014)
4. Kannampallil, T.G., Schauer, G.F., Cohen, T., Patel, V.L.: Considering complexity in healthcare systems. J. Biomed. Inform. **44**, 943–947 (2011)
5. Weick, K.E., Sutcliffe, K.M.: Managing the Unexpected. Wiley, San Francisco (2007)
6. Rosen, R.: Anticipatory Systems. Philosophic, Mathematic, and Methodological Foundations, 2nd edn. Springer, New York (2012)
7. Nadin, M.: Anticipation (special issue). Int. J. Gen Syst **39**(1), 35–133 (2010)
8. Nadin, M.: Mind—Anticipation and Chaos (Milestones in Research and Discovery). Belser Presse, Stuttgart/Zurich (1991)
9. Nadin, M.: Anticipation—The End Is Where We Start From. Müller Verlag, Basel (2003)
10. Zadeh, L.A.: Fuzzy sets as a basis for theory of possibility. Fuzzy Sets Syst. **1**, 3–28 (1978)
11. Staiger, T.O.: Anticipation in complex systems: potential implications for improving safety and quality in healthcare. In: Proceedings of the First International Conference on Systems and Complexity in Healthcare (In press)
12. Louie, A.H.: Robert Rosen's anticipatory systems. Foresight **12**, 18–29 (2010)
13. Miller-Keane, O'Toole, M.T.: Miller-Keane Encyclopedia and Dictionary of Medicine, Nursing, and Allied Health, 7th edn. Elsevier, Amsterdam (2003)
14. Gen Script Glossary of Biochemistry and Molecular Biology "feed-forward regulation". http://www.google.com/webhp?nord=1#nord=1&q=gen+script+glossary+of+biochemistry
15. Rosen, R.: Life Itself. Columbia University Press, New York (1991)
16. Rouse, W.B.: Health care as a complex adaptive system. The Bridge **38**, 17–25 (2008)
17. Rosen, R.: Fundamentals of Measurement and Representations of Natural Systems. North-Holland, New York (1978)
18. Rosen, R.: Feedforwards and global system failure: a general mechanism for senescence. J. Theor. Biol. **74**, 579–590 (1978)
19. Rock, R.: SCARF: a brain-based model for collaborating with and influencing others. Neuroleadership J., 1–9 (2008)
20. Romero-Brufau, S., Gaines, K., Huddleston, J.: Nurses' ability to identify physiological deterioration of hospitalized patients. In: 11th International Conference on Rapid Response Systems and Medical Emergency Teams. Amsterdam (2015)
21. Parush, A., Campbell, C., Hunter, A. et al.: Situation Awareness and Patient Safety. https://www.ottawahospital.on.ca/wps/wcm/connect/29bd8b804b25b21b8f79df1faf30e8c1/Primer_SituationalAwareness_PatientSafety.pdf?MOD=AJPERES
22. Starmer, A.J., Spector, N.D., Srivastava, R., et al.: I-PASS, a mnemonic to standardize verbal handoffs. Pediatrics **129**, 201–204 (2012)
23. Staiger, T.O., Jarvik, J.G., Deyo, R.A., Martin, B., Braddock, C.B.: Patient-physician agreement as a predictor of outcomes in patients with back pain. JGIM **20**, 935–937 (2005)
24. Zehnder, R., Staiger, T.O.: Association between patient-physician agreement and outcomes in primary care. In: Society of General Internal Medicine. Northwest Regional Meeting. Seattle (2006)

Environment, Genes, and Mental Health

Hans Jörgen Grabe

Abstract The vulnerability stress model, as related to mental disorders, has gained much attention since it captures the multifactorial nature of the disorders. In the last few years, research activities have aimed at identifying biological and genetic components that interact with psychosocial factors, such as stressful life events and city living, which according to this model increase or lower vulnerability to mental disorders. Interplay between environmental factors and biological systems (e.g., stress-axis/hypothalamus-pituitary-adrenal-axis) and clinical outcomes will be analyzed from an anticipatory perspective. Previous studies on brain structure and childhood traumatization have implicated limbic regions like the hippocampus, amygdala, anterior cingulate cortex, and frontal areas. Functional changes in those areas are proposed to mediate the long-term risk of mental disorders. The societal and clinical impact of those findings and models will be discussed.

Keywords Genetics · Childhood adversity · Mental disorders · Individualized medicine · Prevention

1 Anticipating Depressive Disorders

Anticipation of the onset of disease is a very timely but complex topic of high relevance to the course and outcome of psychiatric treatment. Anticipatory medicine is concomitant with preventive medicine: both have the capacity to change the usual approach to mental disorders. In the last decades, numerous risk factors that contribute to the onset of mental disorders have been identified. For severe mental disorders, like schizophrenia and bipolar disorder, several environmental factors—such as antenatal maternal virus infections, obstetric complications comprising hypoxia-related conditions or stress during neurodevelopment—were already

H.J. Grabe (✉)
Department of Psychiatry and Psychotherapy, University Medicine Greifswald,
Ellernholzstraße 1-2, 17475 Greifswald, Germany
e-mail: grabeh@uni-greifswald.de

© Springer International Publishing Switzerland 2017   41
M. Nadin (ed.), *Anticipation and Medicine*, DOI 10.1007/978-3-319-45142-8_3

discovered. Those factors converge in the disturbance of neuro-circuits within the hippocampal [1]. Thus, knowledge of risk factors and their interaction could enable clinicians to anticipate the risk of disorders such as bipolarism and schizophrenia. Needless to state, real life is highly complex, and simple point-by-point associations are rarely observable. The complexity in mood disorders may even be higher since environmental factors seem to interact with the individual's genetic vulnerability. In major depressive disorders, psychosocial stress and traumatization during the perinatal period, infancy, or adulthood are important risk factors. Genetic risk in terms of less adaptive biological systems may put individuals at risk of disease in stressful life situations. On the other hand, protective factors such as resiliency might increase the likelihood of the ability to cope with stressful life situations and thus facilitate better outcomes regarding mental and physical health.

2 Risk Factors

Major depressive disorders (MDD) belong to the prevalent causes of morbidity, disability, and impaired quality of life. Previous studies have also provided robust evidence that depression is a risk factor for type 2 diabetes (T2DM) [2, 3], cardiovascular disease (CVD), and mortality [4–6]. Generally, abuse and mistreatment during childhood (CA) are strong environmental risk factors for the development of depressive disorders [7, 8]. Although a great deal of research has investigated the immediate and delayed pathogenic effects of childhood abuse, only a small body of literature has focused on adaptive outcomes in the aftermath of CA [9, 10]. Yet it seems that in addition to trauma-related factors, such as frequency and intrusiveness, individual biological and psychological factors modify the risk of long-term consequences of childhood abuse [11]. Research has consistently shown that a considerable number of childhood abuse victims show little or no long-term psychological damage [12].

2.1 Resilience

Over the years the vague term "resilience" was established to describe the phenomenon mentioned above. There is no generally agreed upon definition of "resilience". One can assign the attempted definitions to (at least) two mainstream thoughts: (1) the end result of a complex adaptation process; (2) the dispositional ability to access and use resources in the face of traumatic events. Nadin [13] describes resilience:

> "the capacity to cope with unanticipated dangers after they have become manifest, learning to bounce back." Not surprising is the inference that "anticipation seeks to preserve stability: the less fluctuation, the better. Resilience accommodates variability…"

Wagnild et al. [14, 15] related resilience to a "dynamic personality trait" arising in "the aftermath of adversity." They acknowledged the existence of certain inherent resources, which are, however, fluid and alterable rather than determined and inflexible. In a recent empirical study, we investigated 2046 subjects between the ages of 29–89 years (SD = 13.9) from a community-based sample who were free of MDD during the 12 months prior to data collection. They had been examined and diagnosed for *Lifetime diagnosis of MDD* by the Munich-Composite International Diagnostic Interview (M-CIDI) according to *Diagnostic and Statistical Manual of Mental Disorders*–(4th edition) (DSM-IV) criteria [7]. Childhood trauma and resilience were assessed with the *Childhood Trauma Questionnaire* (CTQ) and the *Resilience-Scale* (RS-25). Both childhood maltreatment and resilience were associated with MDD later in life. The detrimental effects of low resilience on MDD were especially prominent in subjects with a history of childhood abuse (odds ratio [OR] = 3.18, 95 % confidence interval [CI] [1.84, 5.50]), but also effective in subjects without CM (OR = 2.62, 95 %CI [1.41, 4.88]). The findings supported the clinical assumption that resilient subjects may be partly protected against the detrimental long-term effects of child abuse and neglect.

2.2   Gene-Environment Interaction

To date, it has been difficult to identify genetic factors that directly put individuals at higher risk of depressive disorders. Although large-scale genetic studies with sufficient statistical power are about to be published, clear-cut genetic risk factors are missing. Nevertheless, research in the field of gene-environment interaction has provided an important model of functional and dysfunctional adaptation to stressful or averse experiences and life conditions.

Nadin [16] distinguishes the various discussions of the concept (including the areas of behavior, cognition, neurology, and in medicine in general). Generally, anticipation is described as behavior defined by possible future states as well as by previous states [17].

In the field of gene-environment interaction research, I would propose a new facet of the concept of "anticipation": In light of the genetic variation at a given genetic risk locus, one genetic variant increases the risk and the other one decreases the risk of disease in stressful conditions. Those stressful conditions can constitute previous states like childhood abuse that still shape behavior and emotional states in adulthood. In addition, biological variation contributes to the anticipation: The environmental stressor determines in interaction with the genetic risk variant the risk of mental disorder or the degree of burden caused by psychopathological symptoms and thereby emotional states and behavior. Traditionally, "anticipation" in the field of genetics denotes to the phenomenon that "the signs and symptoms of some genetic conditions tend to become more severe and appear at an earlier age as the disorder is passed from one generation to the next" [18]. Anticipation is typically seen in disorders that are caused by a type of mutation called a trinucleotide repeat expansion. Those repeats are associated with neurological conditions like Huntington disease, myotonic dystrophy, and fragile X syndrome. This repeat expansion can increase from generation to generation leading to more severe forms of the disorder with each successive generation [18].

This effect modification has been preferentially investigated in genes involved in the physiological regulation of the stress response. A prominent example is the FKBP5 gene, which codes for a co-chaperone regulating the glucocorticoid receptor sensitivity. Previous evidence suggested that subjects carrying the TT genotype of the FKBP5 gene single nucleotide polymorphism (SNP) rs1360780 had an increased susceptibility to adverse effects of experimental stress. We therefore tested the hypothesis of an interaction of childhood abuse with rs1360780 in predicting adult depression in 2076 Caucasian subjects from the general population. We identified significant interaction (p = 0.008) of physical abuse with the TT genotype of rs1360780, increasing the depression scores (BDI-II) to 17.8 (95 % CI 12.3–23.2) compared with 10.6 (8.8–12.4) in exposed CC/CT carriers [T denotes the nucleobase thymine, C denotes cytosine]. Likewise, the adjusted odds ratio (OR) for major depressive disorders (MDD) in exposed TT carriers was 8.9 (95 % CI 2.2–37.2) compared with 1.4 (0.8–2.4) in exposed subjects with CC/CT genotypes [19]. This study revealed interactions between physical abuse and rs1360780 of the FKBP5 gene, confirming its role in an individual's susceptibility to depression.

Given the large effect sizes, rs1360780 could be included into possibility models for depression in individuals exposed to childhood abuse [20]. Figure 1 depicts a simulation of positive and negative predictive values in carriers of FKBP5 TT-genotype exposed to "physical abuse." At a given prevalence rate of MDD, e.g., 30 % the risk prediction (positive predictive value) by genotype and abuse status is about 70 %. Those data are encouraging. However, positive and negative predictive

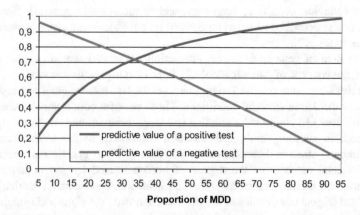

Fig. 1 Simulation of positive and negative predictive values in carriers of FKBP5 TT-genotype exposed to "physical abuse"(TT within abused subjects: Sensitivity = 25 %; Specificity = 95 %; TT within all subjects: Sensitivity = 5.2 %; Specificity = 92 %) (16)

values are still far too low to recommend individual risk prediction via this approach. Moreover, those data need independent replication. A further refinement of this model was possible via the inclusion of the status "high or low resilience." Considering the inclusion criteria "low resilience," the prevalence rate of MDD increased up to 35 % in the selected subjects. Thus the positive predictive value of the screen-positive subjects (TT-carriers and child abuse) increased to 80 %.

2.3 The Serotonin Transporter (5-HTT) Gene

The consequences of childhood trauma and interaction with genetic polymorphisms have been investigated not only in its effects on behavior and health but also on brain structure and function in terms of neuropsychological paradigms elicited by fMRI. A "classical" genetic polymorphism in this respect is the serotonin transporter (5-HTT) gene (SLC6A4, chromosome 17q12). Accordingly, the functional promoter polymorphism of 5-HTT (5-HTTLPR) has often been examined in gene —environment interaction models. In many studies, the short (s) allele of 5-HTTLPR, which is associated with a decreased 5-HTT activity, has been associated with a negative mental outcome after exposure to stress or trauma. A large meta-analysis supported the interaction effect for depressive disorders [21]. Corresponding effects also have been reported for Post-traumatic Stress Disorder (PTSD) [22]. The putative interplay between genes and beneficial environmental conditions like social support on positive outcomes like sense of coherence or resilience has rarely been addressed. "Social support" integrates the perception and actuality that one is cared for, has assistance available from other people, and that one is part of a supportive social network. The questionnaire we have applied assesses social support in dimensions like financial support, care taking in case of illness and emotional support.

We therefore examined the interaction between the serotonin transporter linked polymorphic region (5-HTTLPR) and social support on the sense of coherence (SOC), resilience, and depressive symptoms. Among individuals with the perception of high social support, no significant differences between 5-HTTLPR genotypes regarding all outcome variables were found. However, among those with the perception of low social support, carriers of at least one short allele reported significantly increased levels of SOC and resilience, as well as fewer depressive symptoms than carriers of the l/l genotype. We therefore concluded that in less supportive social environments the impact of distinct genotypes on behavioral outcomes might be more relevant than in supportive environments where social compensation might take place [23]. Morphometrical and functional MRI analyses have demonstrated reduced gray matter volume in short-allele carriers in limbic regions critical for processing negative emotion [24, 25]. Moreover, an amygdala-cingulate feedback circuit critical for emotion regulation was dependent on the polymorphism of the 5-HTTLPR [26]. The magnitude of coupling between amygdala-anterior cingulate cortex inversely predicted almost 30 % of variation in temperamental anxiety [27]. Those data support the view that genetic polymorphisms may alter the brain's structure and function and thereby behavior, too. Environmental exposure to stress may affect those neuro-circuits that are emotionally responsive and may trigger pathological functions in susceptible subjects.

3 Individualized Medicine

The conference *Anticipation and Medicine* prompted several participants (Staiger, Burk, Golubnitschaja, Nadin, among others) to approach this subject. (See respective articles in this volume.) Individualized medicine aims to provide optimal treatment for an individual patient at a given time based on his specific genetic and molecular characteristics [28]. Individualized diagnostic and therapeutic strategies are considered means to improve the efficacy and safety of patient treatment. It might also allow for better individual outcome prediction and risk assessment. Moreover, individualized prevention and early intervention strategies are conceivable. Recently developed high throughput OMICs technologies are thought to enable more targeted diagnostic and treatment approaches [26]. The different conditions and experimental evidence presented so far in this chapter clearly indicate the involvement of psychosocial factors in the prediction of health or disease, even in interaction with biological markers. Therefore, we clearly argue for a broader acknowledgement of traumatization and stress, as well as protective factors such as resiliency, when aiming at anticipating outcomes and modeling the risk structure of diseases. The importance of those factors also advocates early behavioral psychosocial intervention in families with high levels of childhood maltreatment or neglect. Costs to society and the individual's burden of disease could be reduced by early supportive and protective interventions. The consideration of psychosocial and behavioral factors in disease models and treatment algorithms is important not only to psychiatry, but also to various medical disciplines. Recognizing the earliest symptoms and, even better, the pre-symptomatic indicators of disease is anticipatory medicine in practice, leading to more efficient treatment.

References

1. Schmitt, A., Malchow, B., Hasan, A., Falkai, P.: The impact of environmental factors in severe psychiatric disorders. Front. Neurosci. **8**, 19 (2014)
2. Carnethon, M.R., Biggs, M.L., Barzilay, J.I., Smith, N.L., Vaccarino, V., Bertoni, A.G., et al.: Longitudinal association between depressive symptoms and incident type 2 diabetes mellitus in older adults: the cardiovascular health study. Arch. Intern. Med. **167**(8), 802–807 (2007)
3. Knol, M., Twisk, J., Beekman, A., Heine, R., Snoek, F., Pouwer, F.: Depression as a risk factor for the onset of type 2 diabetes mellitus. A meta-analysis. Diabetologia **49**, 837–845 (2006)

4. Vaccarino, V., McClure, C., Johnson, B.D., Sheps, D.S., Bittner, V., Rutledge, T., et al.: Depression, the metabolic syndrome and cardiovascular risk. Psychosom. Med. **70**, 40–48 (2008)
5. Barth, J., Schumacher, M., Herrmann-Lingen, C.: Depression as a risk factor for mortality in patients with coronary heart disease: a meta-analysis. Psychosom. Med. **66**, 802–813 (2004)
6. Carney, R.M., Freedland, K.E., Miller, G.E., Jaffe, A.S.: Depression as a risk factor for cardiac mortality and morbidity: A review of potential mechanisms. J. Psychosom. Res. **53**, 897–902 (2002)
7. Schulz, A., Becker, M., Van der Auwera, S., Barnow, S., Appel, K., Mahler, J., et al.: The impact of childhood trauma on depression: does resilience matter? Population-based results from the Study of Health in Pomerania. J. Psychosom. Res. **77**, 97–103 (2014)
8. Grabe, H.J., Schwahn, C., Appel, K., Mahler, J., Schulz, A., Spitzer, C., et al.: Childhood maltreatment, the corticotropin-releasing hormone receptor gene and adult depression in the general population. Am. J. Med. Genet Part B. **153**, 1483–1493 (2010)
9. Wingo, A.P., Wrenn, G., Pelletier, T., Gutman, A.R., Bradley, B., Ressler, K.J.: Moderating effects of resilience on depression in individuals with a history of childhood abuse or trauma exposure. J. Affect. Dis. **126**, 411–414 (2010)
10. Ungar, M., Liebenberg, L., Dudding, P., Armstrong, M., van de Vijver, F.J.: Patterns of service use, individual and contextual risk factors, and resilience among adolescents using multiple psychosocial services. Child Abuse Negl. **37**, 150–159 (2013)
11. Collishaw, S., Pickles, A., Messer, J., Rutter, M., Shearer, C., Maughan, B.: Resilience to adult psychopathology following childhood maltreatment: evidence from a community sample. Child Abuse Negl. **31**, 211–229 (2007)
12. Werner, E.E., Smith, R.S., French, F.E.: Kauai's children come of age: a longitudinal study from the prenatal period to age ten. In: Werner, E.E., Smith, R.S., French, F.E. (eds.). University of Hawaii Press, Honolulu (1971)
13. Nadin, M.: Anticipation: a spooky computation. In: Dubois, D. (ed.) CASYS, International Journal of Computing Anticipatory Systems, Partial Proceedings of CASYS 99, vol. 6, pp. 3–47. CHAOS, Liege, Belgium (2000)
14. Wagnild, G.M.: The Resilience Scale Users Guide. The Resilience Center, Worden, MT (2009)
15. Wagnild, G.M., Young, H.M.: Development and psychometric evaluation of the resilience scale. J. Nurs. Meas. **1**, 165–178 (1993)
16. Nadin, M.: Annotated bibliography: anticipation. In: Klir, G. (ed.) Special Issue of International Journal of General Systems, vol. 39:1, pp. 34–133. Taylor and Blackwell, London (2010)
17. Nadin, M. (ed.): Anticipation and Medicine. Springer, Cham CH. (2016)
18. US National Library of Medicine. https://ghr.nlm.nih.gov/primer/inheritance/anticipation
19. Appel, K., Schwahn, C., Mahler, J., Schulz, A., Spitzer, C., Fenske, K., et al.: Moderation of adult depression by a polymorphism in the FKBP5 gene and childhood physical abuse in the general population. Neuropsychopharmacol **36**, 1982–1991 (2011)
20. Grabe, H.J., Schwahn, C.: Interaction between psychosocial environments and genes—what is the clinical relevance? Psych. Prax. **38**, 55–57 (2011)
21. Karg, K., Burmeister, M., Shedden, K., Sen, S.: The serotonin transporter promoter variant (5-HTTLPR), stress, and depression meta-analysis revisited: evidence of genetic moderation. Arch. Gen. Psychiatry **68**, 444–454 (2011)
22. Grabe, H.J., Spitzer, C., Schwahn, C., Marcinek, A., Frahnow, A., Barnow, S., et al.: Serotonin transporter gene (SLC6A4) promoter polymorphisms and the susceptibility to posttraumatic stress disorder in the general population. Am. J. Psych. **166**, 926–933 (2009)
23. Reinelt, E., Barnow, S., Stopsack, M., Aldinger, M., Schmidt, C.O., John, U., et al.: Social support and the serotonin transporter genotype (5-HTTLPR) moderate levels of resilience, sense of coherence, and depression. Am. J. Med. Genet Part B. **168**, 383–391 (2015)

24. Fisher, P.M., Grady, C.L., Madsen, M.K., Strother, S.C., Knudsen, G.M.: 5-HTTLPR differentially predicts brain network responses to emotional faces. Hum. Brain. Map. **36**, 2842–2851 (2015)
25. Dannlowski, U., Kugel, H., Redlich, R., Halik, A., Schneider, I., Opel, N., et al.: Serotonin transporter gene methylation is associated with hippocampal gray matter volume. Hum. Brain. Map. **35**, 5356–5367 (2014)
26. Selvaraj, S., Godlewska, B.R., Norbury, R., Bose, S., Turkheimer, F., Stokes, P., et al.: Decreased regional gray matter volume in S' allele carriers of the 5-HTTLPR triallelic polymorphism. Mol. Psychiatry. **16**(471), 2–3 (2011)
27. Pezawas, L., Meyer-Lindenberg, A., Drabant, E.M., Verchinski, B.A., Munoz, K.E., Kolachana, B.S., et al.: 5-HTTLPR polymorphism impacts human cingulate-amygdala interactions: a genetic susceptibility mechanism for depression. Nat. Neurosci. **8**, 828–834 (2005)
28. Grabe, H.J., Assel, H., Bahls, T., Dorr, M., Endlich, K., Endlich, N., et al.: Cohort profile: Greifswald approach to individualized medicine (GANI_MED). J Transl. Med. **12**, 144 (2014)

The Anticipating Heart

Oleg Kubryak

Abstract The perceptions associated with the heartbeat are often interpreted as representative for what is occurring inside the body or as a response to external influences. Older and more recent theories have highlighted autonomous reactions with associated underlying neurological (brain) activity. We report here on research carried on at the Anokhin Research Institute of Normal Physiology. Based on this research, we argue that the heartbeat should be associated with psychological and cognitive processes. This will help overcome the reductionist fixation on localization, which is usually unproductive. Instead of seeking microstructure detail, we should relate to the whole "map" (an idea inspired by Kolmogorov's metaphor of knowledge acquisition). The heart within the dynamic system of mental procedures is a perspective that can lead to progress in understanding how anticipation is expressed.

Keywords Heartbeat · Associative anticipation · Cognitive complexity · Holistic view · Localization

It should be noted here, again, that extremely significant scientific ideas generated by researchers in what used to be the Soviet Union are being gradually rediscovered. Indeed, Anokhin, Bernstein, Beritashvili, Orbeli, Uznadze, Ukhtomsky, and Vygotsky were quite advanced in their views, against a background of dogmatic intolerance in respect to any thought not perfectly aligned with the official dialectic materialism. *Learning from the Past. Early Soviet/Russian contributions to a science of anticipation* [1] and *Anticipation: Russian experimental and empirical contributions informed by an anticipatory perspective* [2], both the result of Mihai Nadin's focus on pioneering work in anticipation were a revelation to scientists in the West, as well as to some Russian researchers in our days. This report on current research at the Anokhin Research Institute of Normal Physiology is a modest testimony to this interest.

O. Kubryak (✉)
Anokhin Research Institute of Normal Physiology, Moscow, Russia
e-mail: o.kubryak@nphys.ru

© Springer International Publishing Switzerland 2017
M. Nadin (ed.), *Anticipation and Medicine*, DOI 10.1007/978-3-319-45142-8_4

1 Background

The theoretical foundation for relating the work of the heart to mental processes is extremely broad. There are historically well-known theories (such as Lange's 1885 theory emotions) and more modern concepts (e.g., Damasio's theory of "somatic markers" [3]). Both of these theories associate autonomous reactions with mental activity. The states of physiological activation ("arousal") determined by the autonomous nervous system represent an integral part of emotional, cognitive, and physical actions. Proceeding from the theory of general activation, Lacey et al. [4] attempted to evaluate physiological activation based on changes in cardio-rhythm and electrical conductivity of the skin [5]. Later, evaluation based on changes in cardio-rhythm was proposed as a universal measure to reflect the interaction of the human organism with the environment [6]. In this view, heartbeat deceleration was explained as the result of perception by the organism of external information and activation related to mental processing of that information. Expectation of a stimulus was emphasized in order to show the relationship between heartbeat deceleration and the influence of the baroreceptors [7, 8]. In more recent studies [9], motive control rather than stimulus expectation was emphasized in some central suppression ("slowing down") of vegetative processes in the course of paying attention. The interrelation between vegetative indicators and psychological processes is being actively researched [10–12].

1.1 Quantitative Perception of the Heartbeat

Attempts to catch the reflection of mental processes in perceptions received from the heart were conducive to the development of quantitative experimental techniques involving the perception of heartbeats. The method proposed by Whitehead et al. [13] in 1977 was possibly one of the first in this respect. In their method, after the ECG-registered R-tooth,[1] an external signal was given with a delay of either 128 or 364 ms. Depending on the signal that a subject chose, a conclusion regarding the better (at 128 ms) or worse (at 364 ms) perception of heartbeats was drawn. The method was called "heartbeat discrimination procedure." Subsequently, classification of subjects was made on the basis of more complex characteristics. It was suggested, for example, that the differentiation of subjects according their emotional reactivity and perception associated with activation [14].

Attention was also directed at the existence of several individual spatial and temporal regions where heartbeats are perceived [15], due to the possibility of stimulation of a different number of receptors in the period between the current and subsequent heartbeats. One more method, called "the method of constant stimuli,"

---

[1]An R-tooth, i.e., R wave, is the first positive deflection in the QRS complex revealed on an ECG readout.

was proposed. It is based on a series of external stimuli given to the subject every 0, 100, 200, 300, 400 and 500 ms after the ECG R-wave [16]. It has also been suggested that sound at different frequencies can be used [17]. Observing stimuli of different modality applied synchronously with heartbeats revealed that the latter are best observed in combined light-and-sound perception, but less so when mechanical signals are given [18]. The approximate band (with an average value of 228 ms) was determined after the ECG R-wave and a sound stimulus when they are perceived as synchronous [19]. Recent versions of the constant stimuli method are described in [20, 21].

A method, in which a subject was asked to mentally count his/her heartbeats over a certain time interval, was proposed by Schandry [22] in 1981. The approach was named "the mental tracking method." The validity of the method was also investigated in comparison with a version of the previously described procedures [23].

1.2   Associative Anticipation

Anokhin's theory of functional systems (considered as classical for the Russian and Soviet school of physiology) identifies a critical component in constructing a behavioral act: the formulation of an aim, in other words, a "prediction" of the future result ensuing from the given act [24]. In relation to this, is his discussions on the universal regularity of brain work, Anokhin noted:

> …in all cases when the brain sends excitations via end neurons to the peripheral working systems, simultaneously with an efferent "command," there is formed some afferent model capable of anticipating the parameters of the future results and of comparing, at the end of the act, that prediction with the parameters of the actual results [24].

(Certain parallels with subsequently developed theories, e.g. "somatic markers" [3] might arise here.) Accordingly, what may be considered as modern interpretations of Pavlov's method of conditioned reflexes might also suggest that there is a connection between the associative anticipation of variously nuanced emotional events and various structures: for example, for various ensembles of dopaminergic neurons in an experiment with primates [25].

2   The Heart-Brain Connection

In cardio-rhythm, the combined effects associated with the anticipated ("ideal") and the actual activities of human beings are reflected. The anatomic-physiological foundation for this is due to the unique "energy" (blood supply) involved in the cardiovascular system in various kinds of activity and the presence of a powerful nerve network that ensures an intense exchange of signals. The system of signal conduction from inner organs can be briefly traced as follows: from the periphery (interoceptors and nociceptors, sympathetic and parasympathetic fibers, metasympathetic nervous system), to the neuraxis (along the sympathetic, and some parasympathetic nerves), to the brain trunk and interior (thalamus, hypothalamus, cerebellar limbic system, amygdala) to the respective cerebral cortex areas [26, 27]. Accordingly, the afferent link of the cardiovascular system is represented only by interoceptors and predominantly nonmyelinized fibers. Heart receptors are represented by dendrite ends of neurons, which are contained in the branches of the pneumogastric and sympathetic cardio-nerves. The cell bodies of interoceptors occur at the bulbar level of the brain trunk and in pneumogastric nerve nodes, in spinal ganglions at the TI to TVII levels, and in the star-shaped sympathetic and cardiac nodes.

Receptors located in various places of the heart participate in launching the reflexes (pseudoreflexes) that regulate the functions of the cardiovascular system [28]. Other complex reflector ties are also operative. According to Pokrovsky [29, 30], there also exists an important efferent mechanism of cardio-rhythm regulation in the efferent structures of the cardiovascular center of the myelencephalon.

> Nerve signals in the form of impulse "volleys" that originate there come to the heart along pneumogastric nerves and interact with structures of the intracardiac pacemaker, thus generating excitation in exact correspondence with the volley frequency [30].

The multi-dimensional organization of cardio-rhythm contributes to the organism's maximal adaptation to current and anticipated events, and also to the optimal correlation with other biological rhythms. Researchers at the Anokhin Research Institute of Normal Physiology suggested that the estimation of the respiratory-cardial synchronization level should be used as an indicator of organism reactivity [31]. They also discovered that changes in cardio-rhythm occur several seconds earlier than the corresponding changes in the electrical activity of the brain when one wakes up from sleep [32].

Based on the study of heart perception in healthy volunteers, researchers in Munich [33] made the following assumptions:

1. Perception of conditions of one's own body is possibly critical for the perception of external stimuli.
2. There exists the capacity to perceive of inwardly directed attention.
3. Signals coming from the body may be associated with the attention system.

It should be noted that attention was considered as a special kind of central-peripheral integrations even earlier [9].

In 2000s, we discovered the enhancement of feedback from the heart when a person mentally counts his heartbeats [34]. This is due to the change in the activity of the object under observation object, i.e., the heart. In our opinion, this confirms the view that the autonomous nervous system plays a part in the development of the near-threshold stimuli into perceived stimuli [35]. New quantitative methods and approaches link changes in cardio-rhythm to changes in types of activity, with the involvement of automatic (subconscious) processes in perception [36]. We also considered the possibility of an ingenious method of objective evaluation of variable stimuli perception based on a special analysis of cardio-rhythm in conjunction with the results of the subject's activity [37].

Some medical researchers believe that heart perception is associated with the feeling of fear—anxiety or "fear" [38]. In our time, depression, panic attacks, and phobias have become a widespread subject of research into cardial perceptions, but which led to no reliable connection with cardiopathology [39, 40]. The opinion that the heart is perceived only when it is beyond normal limits is misleading. Had such an opinion any validity, then the perception of the heart's "hollow thumps," which many people may experience simply by running up a staircase, would be considered abnormal. Cardioception is influenced by various factors: individual traits, e.g., age, physical health condition, gender, peculiarities of physique; personality type (e.g., anxious, melancholy, happy, etc.); the conditions under which observation is made; psychological or emotional state at the time of observation; external signals (noise, light, etc.); artificial biological feedback; number of probes; pharmacological stimulation; body position (sitting, lying down, etc.); training before testing—among other factors. Our data indicate that cardioception of healthy people may also change when shifting the focus on attention. Cardioception studies are often associated with interoception, although the mechanics [41] of its complex multi-dimensional "pulsation" must include somatic components (touching the surrounding tissues by the contracting heart as registered by proprioceptors), at least, in states of intense physical activity (strain) or a specific pose. The various implicit cardial perceptions are not always symptoms of an illness, although "feelings" from the heart are important and represent common clinical material. All this leads us to our interest in studying possible perceptions from the heart in healthy people. The aim is to contribute to the better understanding of how the heart can "feel" and "anticipate."

3 Cardio-Rhythm Evaluation

In a study involving 25 healthy, right-handed volunteers [37], we observed their ability to register their own heartbeats without using any artificial stimuli or feeling their pulse; that is, to register as "naturally" as possible. The volunteers had no training in yoga skills or anything similar. After setting physiological background parameters (I), the volunteer, who was restfully seated, had, as a command was given, to mentally to count his/her own heartbeats (II); the next step was to randomly move the right-hand forefinger with any frequency one wished (III). Finally, one had to show one's cardio-rhythm by moving the forefinger (IV).

Testing to demonstrate heartbeats by finger bending in period IV led to discussion of the characteristics of cardio-rhythm evaluation. For example, volunteers GZ, VN, and L had a close frequency of heart contractions (76, 76, 75, respectively) in period IV. The first tried to mark whole series ("packets") of adjacent heartbeats, the second alternated indication of "packets" with that of single heartbeats, while the third volunteer marked only individual, separated-in-time heartbeats, The distinctions also involved the time parameter of signal giving, For example, volunteer GZ gave the first signal by finger-bending in almost 4 s after the command was given to start. However, two adjacent cardio-intervals were simultaneously marked through his first movements. Subsequently, GZ gave series of signals that corresponded to several adjacent cardio-intervals: up to 6 in a row. Volunteer L gave a first signal to mark heartbeat in less than 1 s after start command. However, both this and all the subsequent signals given by the volunteer only marked individual cardio-intervals while series were absent. The cardio-rhythm tracking scheme of volunteer VN represents a combined version. Different tracking schemes were associated with different results of carrying out instructions, and the ratio of signal number to heartbeat number (as a provisional indicator of "accuracy") for the above volunteers was 0.72 for GZ, 0.53 for VN, and 0.28 for L.

Analysis of the various cardio-rhythm tracking schemes enabled the researchers to identity specific elements that, in our opinion, reflect the character of subjective perception. That is, for one volunteer, subjective perception resulted in a single bending of the finger separated from the previous and subsequent finger-bendings by an omitted cardio-interval; for another volunteer, subjective perception results in a "series;" for the third volunteer, perception is mixed. The number of such specific elements (tracking fragments) varied from 14 to 41 among the 25 volunteers. These individual parameters possibly reflect objective distinctions in the ways and strategies (process, quality) of mental control exercised by the healthy volunteers in respect to their own heartbeats and interrupted perception [37].

4   Heart-to-Brain Signals

Searching present-day technical models for an analogy of how the human brain receives signals from the heart leads to a base model for interaction between open systems: the Open System Interconnection Basic Reference Model (OSI), which assumes a seven-level organization of data transfer [42]. Here, each level serves its own part in the interconnection process. The "levels" organization is an important characteristic of the model. When a query ("live" query) is made, the process of information transfer starts with the upper level of the so-called "stack" (pile, succession) of the protocol. The query is transmitted to the next level, being converted in preparation for the subsequent transmission. The description of such "in-level" organization given here, and in the theory of functional systems developed by Petr Anokhin, results from similar views regarding organization of the information control process. The standard protocol of data transfer in the OSI describes formation of an information "packet" ("data frame" or "window" in technology terms). An information "packet" is a quantity small in comparison to the amount of information of a whole system. Accordingly, the application level can be considered as receiving an instruction; the presentation level, as an orienting reaction, changing the focus of attention and formation of perception codes; the session level can be seen as a modulation of cardio-rhythm (central effects) and an "adjustment" of the central nervous system; the transport level, as a central-peripheral integration, activation of the "nodal" areas of heart perception; the network level, as the formation of an individual spatial-temporal structure, a functional system; the data-link level, as the packing of cardio-rhythm data into an information "packet;" and the physical level, as the transfer of the "packet" into the brain and its registration therein.

4.1   Fragmentation Coefficient

It seems that the fragmentary character of cardio-rhythm as successive groups of heartbeats reflects some fundamental mentality characteristics. In this respect, Evgeny Sokolov, believed that stimuli with similar excitation vectors would group together in the perception space. By contrast, stimuli with different excitation vectors will be discarded and segmented [43]. In the given case, stimuli (heartbeats) constituting one perception fragment are described by a single code. This does not exclude the possibility of determining whether an individual stimulus (heartbeat) pertains to one or another group, described by a known code.

In terms of the information packet analogy, one can conclude that different subject/volunteers with differing cardio-rhythm parameters employ different methods of information coding, which determine the size of "packet" (among other things). In order to develop such an assumption, a specific coefficient ("fragmentation coefficient") would have to be introduced. It would reflect, on an individual time scale, the frequency of change in the stages of cardio-rhythm mental control, and the degree of non-uniformity of the evaluation process itself [37]. Such a coefficient would probably be linked to a coding type. In turn, a coding type would determine the data transfer efficiency.

We attempted to explain the psychological meaning of the "fragmentation coefficient" in the model described as an indicator of the relation between the conscious clearly demanding volitional efforts and the automatic (subconscious) processes in determining one's cardio-rhythm. In other words, a high level of coordination between finger signal movements and cardio-rhythm for some volunteer subjects did not require frequent control, i.e., correction of the "rules" of registration of anticipated perception. When the processing of information took place beyond one's conscious control, a high efficiency of perception was achieved. And the element of conscious registration of heartbeat is not, in this case, paradoxical. In representing the results of heartbeat perception as the product of conscious control exercised by the volunteers, the complexity of the "protocol" of data transfer to the sphere of consciousness, where the "protocol" also includes subconscious mechanisms, should also be pointed out.

If an individual cardio-cycle is considered as the product of multidimensional internal and external influences, then each subsequent heartbeat turns out to be not a certain "average," but a unique event, different from the preceding one. The

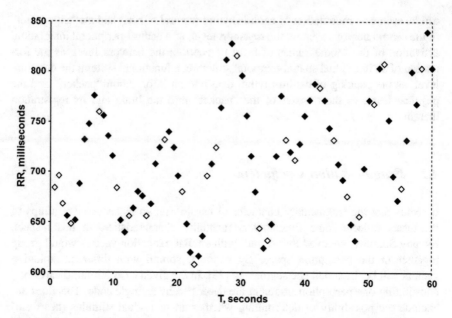

Fig. 1 Marked by finger movement (*dark*) and omitted (*light*) cardio-rhythms of volunteer D during a 60-s period of carrying out the instruction to track heartbeats

simplest method for determining the features of an individual cardio-cycle is to compare it with its neighbors in a "formation." This serves as a basis for many electrocardiographic indicators used in medical practice, as well as in fundamental research. If within the model of heartbeat perception used in healthy volunteers, cardio-cycles with some peculiar characteristics distinguishing them from others are registered, then the identification of such cardio-cycles and their features would facilitate progression towards objectivization of perception. In particular, it would be possible to distinguish, with high probability, a discerned (by the volunteer) heartbeat from an undiscerned one.

Figure 1 shows, against the background of RR-interval[2] fluctuation for volunteer D (37), both marked and missed cardio-cycles when D was carrying out the relevant instruction. One can see that the "omission" often occurs in cardio-cycles that are very close in duration to the immediately preceding ones. Analysis of the differences between the adjacent cardio-intervals in greater detail leads to discernment of even more pronounced differences. This is so for both shorter and longer durations. It is likely that this illustrates the training and adjustment of cardio-rhythm: an increase in tracking accuracy, and the greater differences between the tracked RR-intervals ("convenience of perception") towards the middle of a period, followed by the lowering of the perception threshold, as well as symptoms that seem to indicate tiredness or loss of concentration (an increase in omission number).

---

[2]RR-interval refers to registered r-wave interval.

There are also fragments of uninterrupted tracking: 1, 5, 3, 2 cardio-cycles in succession, etc.

Here a new "stage"—i.e., a change in the perception regime—is always associated with a certain control (or critical) point, which is peculiar to each of the various volunteers during the various observation periods (mental counting of heartbeats and marking them by movements). Apparently, it is exactly in such spatial-temporal regions that error checking, correction of "guiding codes," and "connection" restoration (i.e., conscious control) take place.

If the differences between the cardio-cycles (Fig. 1) are considered as the reflection of some threshold in their perception, then the differences for volunteer subject D constituted, on average, about 25 ms in absolute values; and they constantly varied: up to the maximum 68 ms (also in absolute values). The RR-interval duration characteristic for that volunteer, over the given 60 s, is Me = 714 (668; 757) ms. We assumed that the formation of perception after imparting a stimulus takes about 100–200 ms, while during 200–500 ms, not only formation of images, but also their categorization there takes place [44, 45]. Also assumed is the presence of certain time restrictions on the possibility of spontaneous actions, associated with the formation of subconscious preparedness to act [46]. The presence of "pre-semantically defined windows" for cognitive processes has been studied [47]. Some researchers hold that a shorter interval (80 ms) suffices for signal awareness [48]. Moreover, if one considers perception as a discrete process, then it would be logical to assume that the time necessary for perception stands in relation to the system parameters, the so-called "discretization frequency" (i.e., a temporal matching of a "neurophysiological frame" and an end of a frame) would be registered, and the time for perception would be very short.

As was shown earlier in this text, the absolute length of the R-intervals was poorly related with the accuracy of cardio-rhythm tracking. Volunteer subjects with average heart contraction (beat) frequency demonstrated better determination of cardio-rhythm accuracy determination [37]. This may be considered as the presence of a large "reserve" for adaptive change in the duration of individual cardio-cycles towards either lengthening or shortening. Small differences in cardio-cycle duration —from a few to several tens of milliseconds—may be critical for registration by the volunteers. There is a historical opinion that perception is a continuous process, in which the sensor threshold is small or absent [49, 50]; another opinion supports the series of various degrees clarity. This may also hold true for opinions regarding the differing-in-time coincidence of a stimulus with a current perception "frame," which determines "clarity," and for the negation of the "frame" concept.

5 Cognitive Complexity

The complexity of real cognitive processes frequently transcends the framework of the proposed theories, which gives rise to the greater of smaller discrepancies between actual results and theoretical notions. In this respect, the most correct approach would seem to be viewing perception as a process in which various psycho-physical conditions succeed each other, and that somewhat different rules (whose common features are mentioned in "psycho-physical laws") correspond in each case. Such a view may also account for lucky an unlucky guesses—perception illusions, in the general context of the corresponding central-peripheral integration —specific spatial-temporal structure, and a characteristic mental "pattern."

Let us imagine two hypothetical situations: (1) perception of a stimulus by a healthy person after taking a dose of hallucinogen; (2) perception of the same stimulus by a schizophrenic patient after taking an antipsychotic. Obviously, in both these cases, the peculiarities of perception would be subjected to a very serious correction, in respect to the initial condition (the role of imagination in the perception of the stimuli would be enhanced or diminished), and would be vividly reflected. That is, when considering stimuli perception, the natural (or formed artificially) peculiarities of subjective perceptions they cause one should take into account. Such individuality is, in turn, due to the characteristics of the person's spatial-temporal structure, which is specific to the person's given state. Extraordinary situations excepted, one and the same stimulus—e.g., a light pin prick made without the person's knowledge—may produce, in different healthy persons, different associations and be given different interpretations. Some will perceive the pin prick as painful, some will perceive it as a mosquito bite, and others will not notice it at all. However, in a new environment and in the context of a different situation, perceptions may change.

From the above-mentioned example (Fig. 1), we can verify the frequency of occurrence of the various duration distinctions between the adjacent RR-intervals. Proceeding from the variability of such distinctions to maximum 68 ms, we can distribute all the values of the series in 14 intervals: 0–5 ms; 6–10 ms; 11–15 ms; … and 66–70 ms. The frequencies (occurrence) amount, in total, to 59 units. This is close to the number (58, in Table 1) marked by the volunteer. Using this method for all the other volunteers, we found such conditional boundaries in the periods of mental and motor heartbeat tracking. Thus all the differences between cardio-cycles —with one corresponding to the number marked by the volunteer—can be divided into two parts with sufficient accuracy. For example, for volunteer D, such a "boundary" or a "threshold" for a period of heartbeat tracking, marked by finger bending, shows up as about 16 ms. At the same time, the actual data indicate that sometimes a heartbeat marking fell within a cardio-cycle that differed from the preceding one by less than 16 ms. Here, out of 58 markings, 14 such situations occurred, that is, 24 % of the totality of marks.

Provided that a difference of 16 ms between adjacent cardio-cycles is deemed to be a "threshold", one may refer these 24 % "below-threshold" markings to the following categories: false, imaginable, associated with verification of the motion of indicating, synchronization, erroneous movement, and delay relative to the perceived stimulus (heartbeat). In such case, all differences less than 16 ms for the given volunteer cannot be regarded as reliably determined. Moreover, it is now known precisely which cardio-cycle the volunteer wanted to mark: the preceding, the current, or the predicted. However, these arguments are valid to the extent we

Table 1 The number of real heartbeats compared with the number of given signals according to the instruction for a specific period of observation of 25 volunteers

Volunteer	Number of ECG-registered heartbeats in a period of observation				Number of signals registered by volunteer in observation period		
	I	II	III	IV	II	III	IV
AB	101	99	105	107	51	21	26
FM	94	102	110	97	46	22	40
BL	91	97	99	105	49	65	47
F	92	91	94	93	24	32	24
KS	90	88	90	90	59	54	58
B	84	88	89	94	57	65	54
R	84	80	85	85	27	21	35
OS	81	95	85	92	65	88	71
FN	81	96	90	92	44	81	42
AN	79	91	94	95	89	105	53
S	78	76	82	79	74	105	58
GZ	78	71	78	76	58	61	55
T	74	78	78	84	30	35	21
D	72	81	85	84	60	160	58
K	71	73	74	73	59	19	25
L	71	78	70	75	48	33	21
Z	70	74	72	73	37	26	25
VN	70	87	73	76	39	36	40
KR	70	72	71	78	72	48	55
VL	68	69	78	75	31	72	26
IL	68	67	67	68	34	127	38
GL	66	64	67	66	49	95	32
KL	63	63	66	62	46	56	38
ER	58	62	61	61	40	86	34
M	57	55	54	62	30	40	30

accepted the idea of a stable perception threshold for such a stimulus as one's own heartbeat. Another explanation is that the differences between cardio-cycles become perception-critical—not always, but only in connection with the duration of the cycles themselves. The more realistic view tends towards a fluctuating perception threshold, which results in modifying or adjusting the perception clarity of a number of successive stimuli.

An alternative is to identify (single out) some "averaged over," conditional threshold value, which could, with sufficiently high probability, be accepted as the control: the volunteer's ability to isolate (identify) a given stimulus (heartbeat) in the succession of heartbeats. In the given example, the "averaged" RR-interval length of ± 16 ms or ± 25 ms (average difference value) can be considered as such an averaged threshold value. With greater recognition accuracy, practically all cardio-cycles will turn out suitable for identification, while the critical points will be their difference from the preceding cycle by a certain value. In the case when "packets" were tracked to the point where a definite number of movements corresponded, the differences in parameters of the whole "packets," rather than the individual differences between adjacent cardio-intervals, become significant. For example, fragments of heartbeat tracking by volunteer D begin at the following differences from the preceding ("empty") cardio-cycle: 12, 4, 16, 32, 8, 12, 8, 32, 12, 48, 40, 20, 16,12, 28, 12, 24, 28, 48, 36 ms. The average value of these differences constitutes about 25 ms, although more than half the values of the sequence exceed 16 ms. Since the motive reflection of a "packet" of successive heartbeats is determined, as discussed above, by very complex mutual influences of different systems, the isolation (singling out) of an average value here appears to be more difficult than, say, in a mental cardio-rhythm tracking.

The results of observing the mental and motive tracking of heartbeats by healthy volunteers offer some illustrations upon whose basis one may attempt to explain known conceptions within. In fact, we touch upon the phenomenon of the *relationship* between the configuration of cardio-cycles and their perception, as well as the relationship between the configuration of cardio-cycles and the perception process in general [37].

6 Time Marking

In order to properly calculate proper frequency of the human heart, rhythmic heart contractions reflecting time run should range from 10 to 15 Hz—provided that the value of the excitation propagation rate from the sinoauricular node and up to the pedicle ramules of the atrioventricular bundle is averaged (about 1.2 ms), and while the oscillator length (depending on age, gender, etc.) ranges from 0.08 to 0.12 m. Taking into account the rhythm of intracardial pacemakers, a response of the cardiovascular system should be expected at frequencies coming from the pulsations of various heart structures. These range as follows: 2–3.3 Hz; 1.7 Hz; 1–1.3 Hz; 0.5–0.8 Hz; 0.3 Hz; and 0.2 Hz [51]. Thus, if a heartbeat is considered as a reflection of individual time, and the possibility of a response at some or other frequency is considered as the ability to "perceive," then the above frequencies indicate the "resolving power" of the heart's sensitivity. That is, a segment of 100 ms and less, over which a fluctuation (10–15 Hz) occurs, roughly corresponds to the modern views concerning the time of perception formation after receiving a stimulus.

Of interest are the findings of Frassinetti et al. [52] and Moskvin and Popovich [53], in which they state that in our consciousness, ideas of time are inseparably connected with those of space. Such an assumption seems obvious. For example, a clock dial "converts" the passage of time into the spatial displacement of the hands. Once upon a time, a question about the distance to a certain place could be answered by "two days of travel." Even today, "three-hour flight" is used to describe the distance. Heart rhythm, too, represents a natural marking of interval: e.g., "in a heartbeat." However, unlike a regular clock, the distances between the markings are different and the "time run" can be non-uniform: it can vary—accelerate or slow down—depending on internal and external conditions.

Similar views can be found in Craig's perception model, which includes visceral afferentation and an endogenous timer as important elements [54]. The point is that, in human reality, the possibility of an action—external or internal—is always restricted in time. This fact also determines the effectiveness of any action. The view that cardio-rhythm is a natural mark, as we have suggested here, enables us to correlate the achievement of the required result (perception of a stimulus, random movement, etc.) with definite time "regions," where the required result can be achieved. Central-peripheral integration, vital for every particular case, (formation of specific spatial-temporal structures) provides the physiological basis for achieving a result. Thus, when considering an effective action, one can not only study its somatic and vegetative characteristics, but also evaluate the action itself (i.e., the probability of achieving a result).

Let us take advantage of individual time marking with cardio-rhythm to determine the most likely regions where the goal indicated by an instruction may be achieved: marking by the volunteers of their own heartbeats. To do this, while knowing the result (the number named by a volunteer), we will determine the critical value of differences between the adjacent cardio-interval according to the procedure described above [38]. Figure 2 depicts this situation.

We obtained the following data: the actual time marking of the observation period (1 min, as measured with an electronic timer); the duration of the volunteer's cardio-cycles over that period (ECG-registered); an approximate graph showing the differences between RR-intervals (Fig. 2), which allows one to "sort out" cardio-cycles according to the successful and/or unsuccessful acts of the volunteers (tracking). Figure 3 depicts the individual time markings for volunteer GL (Fig. 3). Here the period that one or another cardio-cycle lasts is classified as either "successful" or "unsuccessful." That time period is represented in individual units: a succession of heartbeats, whose parameters are linked with the "inner" time characteristics [39]. According to the individual time scheme proposed in Fig. 3, for a given volunteer, temporal regions exist, characterized by the different probabilities of achieving the desired result.

In other words, in certain time intervals, it is practically impossible to achieve the required result. This view is well supported, for example, in the practice of sports when the wrong coordination of action in an individual's time (based on breathing) drastically reduces the possibility of successful performance of an exercise or a move. Breathing control shifts not only cardio-rhythm, but also individual time markings, thus assisting in the formation of spatial-temporal

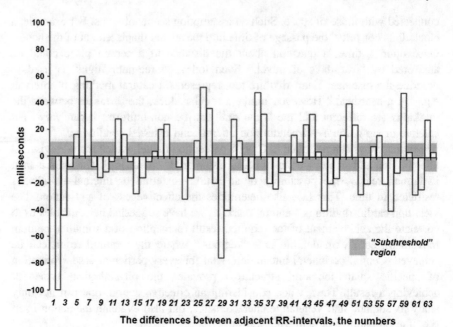

Fig. 2 The supposed critical values of differences between adjacent RR-intervals corresponding to the result of activity (mental counting of heartbeats) of volunteer GL

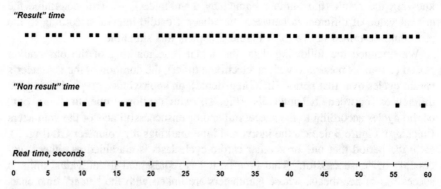

Fig. 3 Individual time marking for subject GL during 60 s of mental counting of own heartbeats (see explanations in the text)

structures that correspond to the possibility that the required result can be achieved. The non-linearity of the internal time "scale" logically accounts for many psychological phenomena: for example, "attentional blinking." Furthermore, the flow of "internal" time is directly related to the dynamics of training and decision-making. Finally, for the success, some specific phenomena are typical (as shown by the analogy with perception "packets"), and are closely related to behavior formation.

The Russian mathematician Andrei Kolmogorov wrote:

> Real objects subject to our study are infinitely complex, but the relations between two really existing objects are exhausted in a simple schematic description. While a geographical map gives us a significant piece of information about a site of the Earth's surface, the microstructure of paper and of the color applied to the paper is in no way related to the microstructure of the depicted Earth surface area [55].

We believe that preparation of such a "geographical map" to show the interrelation of heart rhythm with psychology and cognition should overcome the interdisciplinary barriers and the sometimes excessive focus on searching for some "clear" symptom, parameter, anatomical structure, brain zone, etc. But as the progress of science shows, it is more focused on the microstructure rather than on the whole "map." In keeping with Kolmogorov's metaphor with geography, we also should not forget that the possible discoveries of new "islands" and "continents" on a map describing the heart in the system of mental processes will not always be consistent with current views. However, without scientific search, without trials and errors, these new "territories" will never be discovered.

Acknowledgments We gratefully acknowledge the assistance of Oleg Zilbert in translating this text from the Russian and of Elvira Nadin for preparing the final version. We want to thank Dr. Mihai Nadin for his dedication to making the work of Russian pioneers in anticipation and to those researchers who carry on their work.

References

1. Nadin, M. (ed.): Learning from the past. In: Early Soviet/Russian Contributions to a Science of Anticipation. Cognitive Science Monographs, vol. 25. Springer, Cham (2015)
2. Nadin, M., Kurisma, A. (eds.): Anticipation: Russian experimental and empirical contributions informed by an anticipatory perspective. Int. J. Gen. Syst. **44**, 6 (2015) (Taylor & Blackwell, London)
3. Damasio, A.: Descartes' Error: Emotion, Reason, and the Human Brain. Putnam, New York (1994)
4. Lacey, J.I., Kagan, J., Lacey, B.C., Moss, H.A.: The visceral level: situational determinants and behavioral correlates of autonomic response patterns. In: Knapp, P. (ed.) Expression of Emotions in Man, pp. 161–196. International Universities Press, New York (1963)
5. Lacey, J.I., Lacey, B.C.: Some autonomic-central nervous system inter-relationships. In: Black, P. (ed.) Physiological Correlates of Emotion, pp. 451–457. Academic Press, New York (1970)
6. Lacey, D.A., Lacey, B.C.: The specific role of heartbeat frequency in sensomotor integration. In: Neurophysiological Mechanism of Behavior, p. 434. Nauka Publishers, Moscow (1982)
7. Obrist, P.A.: The cardiovascular-behavioral interaction as it appears today. Psychophysiology **13**(2), 95–107 (1976)
8. Obrist, P.A.: Cardiovascular Psychophysiology: A Perspective. Plenum Press, New York (1981)
9. Danilova, N.N., Astafiev, S.V.: The attention of man as a specific relation of EEG rhythms with wave modulators of cardiorhythm. J. Neuro-psych. Act. **50**(5), 791–804 (2000) (In Russian)

10. Lane, R.D., McRae, K., Reiman, E.M., Chen, K., Ahern, G.L., Thayer, J.F.: Neural correlates of heart rate variability during emotion. Neuroimage **44**(1), 213–222 (2009)
11. Critchley, H.D., Corfield, D.R., Chandler, M.P., Mathias, C.J., Dolan, R.J.: Cerebral correlates of autonomic cardiovascular arousal: a functional neuroimaging investigation in humans. J. Physiol. **523**(1), 259–270 (2000)
12. Koelsch, S., Remppis, A., Sammler, D., Jentschke, S., Mietchen, D., Fritz, T., Bonnemeier, H., Siebel, W.A.: A cardiac signature of emotionality. Eur. J. Neurosci. **26**(11), 3328–3338 (2007)
13. Whitehead, W.E., Dresher, V.M., Heiman, P., Blackwell, B.: Relation of heart rate control to heartbeat perception. Biofeedback Self-Regul. **2**, 371–392 (1977)
14. Katkin, E.S., Blascovich, J., Reed, S.D., Adamec, J., Jones, J., Taublieb, A.: The effect of psychologically induced arousal on accuracy of heartbeat self-perception. Psychophysiology **19**, 568 (1982)
15. Cervero, F., Morrison, J.F.B.: Visceral sensation. Prog. Brain Res. **67**, 186 (1986)
16. Yates, A.J., Jones, K.E., Marie, G.V., Hogben, J.H.: Detection of the heartbeat and events in the cardiac cycle. Psychophysiology **22**(5), 561–567 (1985)
17. Brener, J., Kluvitse, C.: Heartbeat detection: judgments of the simultaneity of external stimuli and heartbeats. Psychophysiology **25**(5), 554–561 (1985)
18. Knapp, K., Ring, C., Brener, J.: Sensitivity to mechanical stimuli and the role of general sensory and perceptual processes in heartbeat detection. Psychophysiology **34**(4), 467–473 (1997)
19. Ring, C., Brener, J.: The temporal locations of heartbeat sensations. Psychophysiology **29**, 535–545 (1992)
20. Wiens, S., Palmer, S.N.: Quadratic trend analysis and heartbeat detection. Biol. Psychol. **58**(2), 159–175 (2001)
21. Mesas, A.A., Chica, J.P.: Facilitation of heartbeat self-perception in a discrimination task with individual adjustment of the S+ delay values. Biol. Psychol. **65**(1), 67–79 (2003)
22. Schandry, R.: Heart beat perception and emotional experience. Psychophysiology **18**(4), 483–488 (1981)
23. Knoll, J.F., Hodapp, V.: A comparison between two methods for assessing heartbeat perception. Psychophysiology, **29**(2), 218–222 (1992)
24. Anokhin, P.K.: Functional system as a basic for physiological architecture of behavioral act. In: Biology and Neurophysiology of Conditioned Reflex, pp. 194–262. Medicine Publishers, Moscow (1968) (In Russian)
25. Matsumoto, M., Hikosaka, O.: Two types of dopamine neuron distinctly convey positive and negative motivational signals. Nature, **459**(7248), 837–841 (2009)
26. Cameron, O.: Interoception: the inside story—a model for psychosomatic processes. Psychosom. Med. **63**, 697–710 (2001)
27. Nichols, J., Martin, R., Valas, B., Fuchs, P.: From neuron to brain. Balaban, P.M. (trans.) p. 672. Editorial URSS, Moscow (2003)
28. Nozdrachew, A.D., Chernysheva, M.P.: Visceral Reflexes, p. 168. University Publishers, Leningrad (1989)
29. Pokrovsky, V.M.: Alternative view on the mechanism of cardiac rhythmogenesis. Heart Lung Circ. **12**(1), 18–24 (2003)
30. Pokrovsky, V.M., Abushkevich, V.G., Gurbich, D.V., Klykova, M.S., Nechepurenko, A.A.: Interaction of brain and intracardiac levels of rhythmogenesis hierarchical system at heart rhythm formation. J. Integr. Neurosci. **7**(4), 457–462 (2008)
31. Dudnik, E.N., Glazachev, O.S.: Formalized criterion of respiratory-cardio synchronization in assessing operative rearrangements of vegetative homeostasis. Physiol. Man **32**(4), 49–56 (2006) (in Russian)
32. Vasiliev, E.N., Uryvaev, Y.V.: Interrelation between fast changes in individual subranges of EEG delta-waves and the cardiorhythm during sleep. Physiol. Man **32**(4), 18–23 (2006) (in Russian)

33. Matthias, E., Schandry, R., Duschek, S., Pollatos, O.: On the relationship between interoceptive awareness and the attentional processing of visual stimuli. Int. J. Psychophysiol. **72**(2), 154–159 (2009)
34. Kubryak, O.V., Bondarev, A.A., Uryvaev, Y.V.: Cardiorhythm characteristics in mental tracking of heartbeats by healthy volunteers. J. Neurophychic. Act. **55**(1), 15–20 (2005) (In Russian)
35. Uryvaev, Y.V.: Perception of a stimulus as a somato-psychic process. Vestnik RAMN **11**, 26–29 (1997) (In Russian)
36. Kubryak, O.V., Uryuvaev, Y. V.: Estimation of the involvement of subconsciousness in heartbeat perception in healthy volunteers. Hum. Physiol. **4**(34), 440–445 (2008). doi:10.1134/S0362119708040075
37. Kubryak, O.V.: Self-perception of Heartbeats and Cognitive Aspects of Cardiorhythm, p. 112. Moscow, LIBROCOM (2010)
38. Broitigham, V., Kristian, P., Rad, M.: Psychosomatic Medicine, p. 129. GEOTAR Medicine, Moscow (1999) (In Russian and German)
39. Ehlers, A., Breuer, P.: How good are patients with panic disorder at perceiving their heartbeats? Biol. Psychol. **5**(42), 165–182 (1996)
40. van der Does, A.J.W., Antony, M.M., Ehlers, A., Barsky, A.J.: Heartbeat perception in panic disorder: a reanalysis. Behav. Res. Ther. **38**(1), 47–62 (2000)
41. Shevchenko, E.V., Lebedinsky, V.Y., Khlopenko, I.A., Dukhanin, A.Y.: Heart biomechanics. Physical nature and mathematical analysis. Siberian Med. J. **1**, 11–13 (1996) (In Russian)
42. ISO/IEC 7498-1:1994. Information technology—Open Systems Interconnection—Basic Reference Model: The Basic Model, http://www.iso.org/iso/catalogue_detail.htm?csnumber=20269
43. Sokolov, E.N., Nezlina, N.I.: Segmentation, grouping and accentuation in stimuli perception. J. Neuro-psych. Act. **59**(1), 15–33 (2009) (In Russian)
44. Ivanitsky, A.M.: Brain Mechanisms of Signal Assessment, p. 264. Medicine Publishers, Moscow (1975) (In Russian)
45. Ivanitsky, A.M.: Brain foundation for subjective emotions: a hypothesis of information synthesis. J. Neuro-psych. Act. **46**(2), 241–252 (1966) (In Russian)
46. Libet, B., Gleason, C.A., Wright, E.W., Pearl, D.K.: Time of conscious intention to act in relation to onset of cerebral activity (readiness-potential). The unconscious initiation of a freely voluntary act (part 3). Brain **106**, 623–642 (1983) http://www.ncbi.nlm.nih.gov/pubmed/6640273
47. Pöppel, E.: Pre-semantically defined temporal windows for cognitive processing. Philos. Trans. R. Soc. Lond. B Biol. Sci. **36**(1525), 1887–96 (2009)
48. Pockett, S.: On subjective back-referral and how long it takes to become conscious of a stimulus: a reinterpretation of Libet's data. Conscious Cogn. **11**(2), 144–161 (2002)
49. Pierce, C.S., Jastrow, J.: On small differences in sensation. Mem. Natl. Acad. Sci. **3**, 75–83 (1884)
50. Urban, F.M.: Psychophysical measurement methods. Psychol. Bull. **10**(5), 180–185 (1913)
51. Khabarova, O.V.: Bio-effective frequencies and their relation to proper frequencies of live organisms. Biomed. Technol. Radioelectron. **5**, 56–66 (2002) (In Russian)
52. Frassinetti, F., Magnani, B., Oliveri, M.: Prismatic lenses shift time perception. Psychol. Sci. **20**(8), 949–954 (2009)
53. Moskvin, V.A., Popovich, V.V.: Phylosophical-psychological aspects of research into the category of time. J. Credo **6** (1998) (In Russian)
54. Craig, A.D.: Emotional moments across time: a possible neural basis for time perception in the anterior insula. Philos. Trans. R. Soc. Lond. B Biol Sci. **364**(1525), 1933–1942 (2009)
55. Kolmogorov, A.N.: Three approaches to definition of the concept of "Amount of Information". Probl. Inf. Transf. **1**(1), 3–11 (1965) (In Russian)

Part II
Evaluating the Risk Factors and Opportunities of New Medical Procedures

Predictive, Preventive and Personalised Medicine as the Medicine of the Future: Anticipatory Scientific Innovation and Advanced Medical Services

Olga Golubnitschaja and Vincenzo Costigliola

Abstract Depending on innovation in medical services and healthcare systems as a whole, two potential scenarios are considered. A pessimistic scenario considers the pandemic of type 2 diabetes mellitus and the dramatic increase of neurological disorders, CVD and cancer diseases, with severe economic consequences to the society. The optimistic scenario considers integrated concepts of early, so-called predictive diagnostics, targeted preventive measures, and treatments tailored to the person as the future of healthcare. Global research and implementation programs in bio-medicine, communication amongst scientific entities, healthcare providers, policy-makers, educators, and patient organisations, together with consolidation of professional groups in the field of personalised medicine, will play a decisive role in driving the situation towards one of the two scenarios. PPPM concepts aim to promote the optimistic scenario in Europe and worldwide. The long-term strategies of the European Association for Predictive, Preventive and Personalised Medicine towards scientific and technological innovation as well as advanced medical services are described.

Keywords Anticipation · Advanced healthcare · Predictive preventive personalised medicine (PPPM) · Scientific and technological innovation

O. Golubnitschaja (✉)
Radiological Clinic, Bonn University, Sigmund-Freud-Str. 25, 53105 Bonn, Germany
e-mail: olga.golubnitschaja@ukb.uni-bonn.de

V. Costigliola
European Association for Predictive, Preventive and Personalised Medicine, Brussels, Belgium
e-mail: vincenzo@emanet.org

© Springer International Publishing Switzerland 2017
M. Nadin (ed.), *Anticipation and Medicine*, DOI 10.1007/978-3-319-45142-8_5

1 Introduction: Time for New Guidelines in Healthcare

As a consequence of the accumulating clinical data and knowledge about the epidemiology and pathological mechanisms of the most frequent causes of morbidity and mortality, medical practitioners are currently reconsidering their view regarding potential causality and progression of cardiovascular, oncologic and neurodegenerative diseases. The majority of these pathologies are chronic in nature: they progress from precursor lesions over one or even several decades of life until a diagnosis is made, which is often too late for effective therapeutic intervention. An excellent example is the epidemic scale of type 2 diabetes mellitus occurring in the European Union. In most industrialised countries and countries with large populations, the permanently growing cohort of people with diabetes creates a serious healthcare problem and a dramatic economic burden. Estimates for diabetes prevalence in the next very few years exceed half a billion patients worldwide [1]. Furthermore, the contemporary onset of the dominant type 2 diabetes has already been observed in the adolescent subpopulation [2]. The costs attached to severe complications secondary to early onset of diabetes mellitus, such as retinopathy, nephropathy, silent ischemia, dementia, and cancer, could lead to collapsing healthcare systems. New guidelines should create the robust juristic and economic platform for advanced medical services utilising the cost-effective models of risk assessment followed by tailored treatments focused on the precursor stages of chronic diseases [3].

2 Healthcare of the Future: Predictive, Preventive and Personalised Medicine

Predictive, preventive and personalised medicine (PPPM) offers great promise for the future practice of medicine. Essential components of this approach include well-organised population screening protocols utilising novel diagnostic biomarkers of disease states, targeted prevention of common human pathologies, optimal treatment planning and personalised medicine—thereby resulting in a substantial improvement in the quality of life. This approach also offers the contingent benefit of delivering care at potentially reduced costs to the population at large, since it addresses social and ethical issues related to access to and affordability of healthcare. A broad distribution and a routine clinical utilisation of advanced technological approaches could enable a significant portion of the population to reach and exceed 100 years of age, while remaining vibrant and in excellent physical and mental health as actively contributing members of the society. One of the central issues discussed is the screening for predisposition of healthy individuals to potential pathologies later in life. The groups at risk identified within the general population can be given a fair chance of a focused diagnostics with multidisciplinary expertise and well-timed preventive measures. The reliability of the blood-tests proposed, potential application to clinical routines, enormous economic and social impacts of the new generation of molecular- and nanotechnology-based diagnostic approaches are all considered from the PPPM perspective [4].

3 Focus on the Patient: Promotion of the Concepts of *Participatory Medicine*

"Nothing about me without me" is the apt slogan of the Society of Participatory Medicine [5], which certainly should be broadly accepted by advanced medical services and society at large. A number of independent evidence-based studies have demonstrated that the efficacy of treatments strongly depends on the level of harmony in doctor-patient collaboration. However, the critical question remains: How can a common language that is understandable to patients and the professionals in the healthcare field and industry be created? The only solution is high-quality educational measures aiming at significant improvements in healthcare knowledge and health related language in the general population. Unfortunately, information retrieved from the Internet is frequently of poor professional quality, providing controversial data that confuse the understanding and slow down the learning process of laymen. People need to be advised of reliable information sources that are well adapted to a corresponding level of understanding (children, adolescents and individual groups of adults) and concrete interests of subpopulations (level of education, groups of professionals, patient cohorts). In the field of education, laboratory medicine may play a leading role in providing up-to-date information, accessible to the layperson, on laboratory tests and their interpretation for individual health and disease conditions. A professional version will provide detailed knowledge about bioactive molecules, enzymatic reactions, molecular and cellular processes underlying the pathomechanisms of individual predispositions and pathologies, as well as medical treatments. These innovations, along with the tight collaboration of organised patient groups, are one of the strongest instruments of more effective promotion of knowledge [6, 7].

4 The Integrated View of PPPM

In his book, *Too Big to Succeed: Profiteering in American Medicine*, Russell J. Andrews, the well-known scientist and neurosurgeon (NASA, USA), criticises currently existing medical services. Moreover, he considers PPPM concepts to be the most advanced for improving healthcare as a whole.

A more comprehensive approach to the use of high-tech medicine for improved healthcare is provided by the European Association for Predictive, Preventive and Personalised Medicine (EPMA). The EPMA is a consortium of physicians and medical researchers with institutional, governmental, and industry participants [...] One should be encouraged by the forward-looking electic, cross-disciplinary, and result-centered (rather then profit-centered) approach advocated by the EPMA [...] By combining the resources of academia, industry, and public and private agencies in the EU countries and beyond (forty-four countries to date), the EPMA appears to be in a much stronger position to have a positive impact on healthcare than a fragmented, individualistic approach that relies upon the interests of for-profit companies.... [8].

Through an integrative view and emphasis on predictive, preventive and personalised medicine, the EPMA is the global leader in PPPM, innovative concepts and prioritised areas. The essential elements making the EPMA the worldwide PPPM leader are explained.

4.1 The Paradigm Shift from "unPPPM" to "PPPM"

How do we adopt such innovative approaches in healthcare systems, while promoting predictive diagnostics, targeted preventive measures, and individualised patient treatment on a global scale? As it is already well known, one size does not fit all. Each patient has the right to receive the best and most appropriate medical care. In contrast to "unPPPM," that is, reactive medicine (discussed in Nadin[1]), the paradigm of the advanced healthcare is to treat the person as an individual case and to provide the appropriate treatment; if medication is called for, to apply the right dose on the right schedule. In order to estimate the overall impact of personalised medicine, the EPMA has created a scientific forum for professionals to discuss this topic. The main objectives of these efforts are to mark stakeholders in the field, to consolidate professional groups, and to elaborate expert recommendations of how to optimise these approaches for the patient [4, 9].

5 Specific Areas of PPPM Application

5.1 Cancer: The Key Question Puzzling PPPM

Detailed autopsy findings reveal that the absolute majority of people are carriers of hardly detectable micro- and asymptomatic tumour lesions which do not necessarily progress into clinically manifested disease. Furthermore, in case of manifested oncologic diseases, less than 1 % of all disseminated and circulated tumour cells

---

[1]Nadin, M.: Medicine: The Decisive Test of Anticipation. In: Nadin, M.: (ed.) Anticipation and Medicine, pp. 1–25. Springer, Cham (2016).

have a potential to form secondary and distanced tumours (metastatic disease)—the phenomenon known as the "metastatic inefficiency" [10]. In this context, the key question puzzling modern predictive, preventive and personalised medicine is how to discriminate between those carriers who are predisposed to a disease manifestation and/or progression and "silent" carriers. Evidence shows that both initial tumours and secondary metastases need a "fertile" microenvironment that effectively supports their growth and progression [11]. What are the mechanisms "fertilising" the microenvironment for a particularly effective cancer advancement? All these questions are effectively addressed by innovative PPPM strategies in cancer [11–14].

5.2 Cardiovascular Disease (CVD) as One of the Major Targets for PPPM

There is a large body of evidence concerning cardiovascular risk factors and preventive strategies at both population and individual levels, but also chronic disease stages that are not adequately addressed because they do not follow the PPPM principles. The promotion of PPPM in CVD management is a global health issue, since the health burden from CVD is currently the most severe in developed countries and is rapidly increasing in most of the developing countries. It is, therefore, of the utmost importance to exchange, on a global scale, scientific insights, knowledge, and skills for risk prediction of cardiovascular disease, and to share and adopt various experiences for preventive measures and for the development of personalised treatment approaches [15, 16].

5.3 Diabetes Pandemic: A 21st Century Disaster and PPPM Solutions

The worldwide increase in the incidence and prevalence of diabetes mellitus (DM) continues to place an alarming burden on healthcare systems. The consequent cost impact poses a major challenge to both developed and developing countries and economies. These prevailing conditions provide the rationale for the concept of PPPM: the prediction of persons at risk should help devise strategies for treatments tailored to the individual and to prevent target organ complications of DM, thereby, reducing morbidity and mortality as well as associated costs. EPMA emphasises the need to address the integrative approach for diabetes care focused on benefits to the patient [17].

5.4 Neurological, Neuropsychiatric and Neurodegenerative Diseases (NNND)

NNND make up the majority of socially and economically devastating disorders and diseases, with multifactorial physical and cognitive disability. They result from individual interplay of epigenetic and environmental risk factors. Insights into molecular pathomechanisms will facilitate the creation of the most effective targeted protective strategies and individualised treatment before pathologies manifest. Multifunctional (multi-drug) therapies should be tailored to individual multi-aetiological aspects of the disorders, in order to advance the healthcare of corresponding patient cohorts. Particular emphasis should be placed on primary prevention by the identification of predisposed individuals early on in life, followed by treatments tailored to the person that altogether need regulations supported by innovative reimbursement programs. This strategy creates a robust platform for the cost-effective medicine of future NNND management [18–21].

5.5 Rare Diseases (RDs): Proof-of-Principles for PPPP Concepts

Although each individual RD is rare, altogether there are 5000–8000 distinct RDs affecting many millions of people worldwide. In Europe alone, there are at least 30 million RD patients. Almost 80 % of RDs have a genetic origin with symptoms appearing in prenatal and early postnatal periods. Currently, there are no appropriate treatment approaches for most of the RDs. The only reasonable approach seems to be a development of methods for early diagnosis of RDs that might lead to the creation of the optimal care management, saving lives and improving life quality within the patient cohort. How the emerging paradigm of PPPM may improve healthcare in RDs? Due to the molecular background of most RD pathologies, it is expected that the multimodal approach (*omics, pharmacogenetics, medical imaging, etc.) with high multidisciplinarity of professionals should be instrumental for "personalisation" in order to diagnose individual RDs, to create effective preventive measures and to develop targeted therapies—the integrative medical approach by PPPM [22].

5.6 Traditional Medicine: Past or the Future?

Traditional medicine (such as Chinese or Indian ones) is several thousands of years old. Does it belong to the past? PPPM provides a platform for innovative strategies in science and healthcare, demonstrating how traditional, complementary, and alternative medicine (TCAM) can enrich modern healthcare. Functionally linked together, the PPPM-TCAM evidence-based approach demonstrates a great potential in person-centered and participatory medicine, predictive diagnostics, targeted prevention, and individualised treatments. It explores tailored care through investigation and treatment of the person as a physical, psychological, and spiritual unity living in dynamic interaction with nature and society. PPPM-TCAM creates a special form of preventive medicine that empowers communities and individuals [23–25].

5.7 Pain Management: Multidisciplinarity and Benefits for All Medical Fields

An integrated vision of PPPM here is a deep diagnostics followed by creating individualised treatment algorithms. This includes topics-relevant animal models, translation research, novel physiological, safe and personalised therapies developed for minimally interventional pain management and physiotherapy. Regenerative therapy, guided by advanced imaging techniques, 3D modeling, robotics, smart prosthetics, etc. are the focuses of innovation in pain management. A variety of syndromes, acute, chronic and systemic disorders are involved, such as acute and chronic pain, musculoskeletal disorders, rheumatologic, orthopaedic, and neurologic conditions, dysfunction of the peripheral nervous system, health conditions considered by rehabilitation and military medicine. Prediction and prevention of a wide spectrum of collateral diseases (NNND, diabetes, cancer) linked to pain management are considered in the context of improved healthcare policy and economic benefits of the societies [26].

5.8 Oral/Dental Health Contributes to the Overall Health and Well-Being of Everybody

A growing body of evidence demonstrates that the manifested dental and oral pathologies are linked to the increased risk of various diseases, including heart and lung disease, vascular pathologies, stroke, diabetes mellitus, neurological disorders, pre-term birth, and even some types of cancer, amongst others. Moreover, certain oral symptoms are considered as early indicators of a spectrum of the mental disorders, such as anorexia, bulimia, anxiety, and depression. On the other hand, dental diseases themselves may be caused by acute and chronic systemic disorders, such as diabetes mellitus. While an association between oral/dental diseases and systemic disorders is well established, the cause-and-effect relationships in these conditions are poorly understood. Investigation of this association is a prerequisite for predictive, preventive and personalised dental medicine [27–29].

5.9   Transplantation and Regenerative Medicine

Currently, very solid research on stem cells creates new perspectives in this area. An integration of the basic sciences at the molecular level and clinical science, together with technological advances, is crucial to progress the area and to satisfy patients' unmet needs in the field. Furthermore, medical ethics, appropriate political regulations and the economy have remarkable impacts on advances in transplantation and regenerative medicine in general. Prediction and personalisation in transplantation are essentials that require an identification of individual pre- and post-transplantation biomarker panels, allowing better donor/recipient matching and assessment of individual risks.

Long waiting lists of patients worldwide reflect major problems and current deficits, which require PPPM solutions. Altogether, improved donor-recipient matching, person-centered immunosuppressive regimens, individual risk assessment for chronic allograft damage, and prediction of graft accommodation may lead to substantially increased allograft survival and decreased patient morbidity, thus advancing this medical area on the global scale.

5.10   A Sensitive Balance Between Health and Disease: The Role of Environment and Clinical Nutrition

The main determinants of health and disease mainly are genetic, environmental, and behavioral; each component merges and interacts with the others. Environment is a still neglected topic in healthcare. Nonetheless, geography, climate, occupation, anthropic modification, urban and rural environments, agriculture and fishing are all subsets that should be considered, along with societal issues and political and economic involvement, for increasing the possibility of successful outcomes in health promotion. An integrative medical approach aims to create professional opinions and to enhance and develop knowledge and skills by taking into account evidence-based scientific achievements in the fields of epidemiology, healthy lifestyle, optimised nutrition, food science/technology/culture, medical ethics, in a framework of cost-effective healthcare and environmental and affordable strategies. Contextually, the goal of PPPM is to produce an evidence-based consensus for sustainable guidelines in predictive medicine together with targeted prevention in healthy individuals, at-risk persons, and stratified patient groups with manifest diseases, and to provide advice to stratified patient groups, institutions, food producers and marketing experts [30, 31].

5.11   PPPM in Body Culture and Sports Medicine (BCSP)

BCSP covers a wide spectrum of topics, ranging from but not limited to, exercise, healthy lifestyle, personalised sleep algorithms, homoeopathy, physical therapy, rehabilitation, amongst others. Anti-doping measures are an essential part of PPPM strategies in Sports Medicine. High-quality research based on measurable effects (including clinical criteria and multi-level biomarker panels) that are associated with modifiable (risk) factors (nutrients, physical activity, lifestyle, etc.) is promoted by PPPM in BCSP, with a particular focus on individually tailored interventions [32–34].

5.12   Translational Medicine Bridges Basic Science and Implementation of PPPM Concepts

With the increasingly complex relationship between basic research and clinical application, there is a pressing need to bridge the translational gap from bench to clinic using integrative methods. The goal is to translate knowledge from studies at the bench side to care at the bedside: from discovery to health application, to evidence-based guideline, to advanced healthcare services, and finally to health impacts for the patient [35, 36].

5.13   Information and Communication Technologies (ICT)

ICT-based holistic presentation of the individual patient and corresponding medical processes imply a redesign of healthcare activities within a given domain of medical discourse, such as cardiovascular, neurological, diabetic, and oncologic disorders. The ICT systems support provided by a medical information and model management system-like architecture, which includes a number of carefully selected diagnostic and therapeutic core functionalities, is the prerequisite for an effective PPPM. With a holistic presentation of a specific patient based on appropriate mathematical modeling methods, such as probabilistic relational models and process models, as well as advanced ICT-enabling tools, the practice of medicine will be substantially transformed towards model-based medical evidence, providing transparency of clinical situations, processes, and decisions for patient and physician. ICT approaches may result in profound and cost-effective modernisation of healthcare. The beneficiaries of these transforming methods and technologies will include patients, healthcare providers, and society at large [37, 38].

5.14   Innovative Technologies (IT)

The aim of PPPM related innovative technologies is to reach advanced healthcare services. The best example of IT is identification, characterisation, and validation of clinically relevant biomarkers. For example, medical imaging, sub-cellular imaging, multi-omics (genomics, transciptomics, proteomics, metabolomics, etc.), and developed hybrid technologies can be used to identify optimal biomarker panels for multi-level diagnostics. If they can detect pre-symptoms in a most timely manner, smart molecular alterations can optimise therapy outcome in thoroughly stratified therapeutic groups. Integrating this information allows selection of personalised targeted treatment regimes, saving unnecessary drug toxicity and decreasing morbidity [12, 39, 40].

5.15   Pharmacogenetics

Currently, the use of genetic information to treat patients is still in its early stages, with some clear successes mostly in the oncology and infectious disease therapy areas. Some successful examples include the targeting of tailored pharmaceuticals developed for the treatment of patients with a particular disease subtype or according to a specific genetic makeup pertaining to the drug's mode of action. In other examples, genetic information is being used to help determine the effective and safe dose of specific pharmaceuticals.

However, implementation of this pharmacogenetic knowledge to the clinic has proven to be challenging, and to require a tight collaboration amongst the various stakeholders throughout the discovery, development, and validation stages so as to ensure the utility of actionable genetic testing in a cost-effective manner. Targeted therapy and reliable prediction of expected outcomes offer patients access to better healthcare management by identifying the therapies effective for the stratified patient group, avoiding prescription of unnecessary treatment, and reducing the likelihood of developing adverse drug reactions [41].

6   The Role of the Laboratory Medicine

Current deficits in medical services, such as delayed intervention, untargeted medication, overdosed patients, and ineffective treatments require a more active and central role of laboratory medicine. Recommendations by the laboratory to assist clinical practice are highly requested. This assistance ranges from advising on the necessity for additional tests to the dynamic analysis of the targets. Novel tests should be considered from the viewpoint of their reasonability, in order to reach an accurate and realistic health-related data interpretation for the individual. The analysis of dynamic changes of the target is essential to evaluate potential health impacts such as an individual predisposition to the disease and/or a predictive diagnosis before a clinical manifestation of symptoms. Laboratory value-added investigation and data interpretation is mandatory in creating an advanced functional relationship between laboratory medicine and clinicians acting hand-in-hand as the responsible decision-makers [7, 42, 43].

6.1 Biomarking and Biobanking

Internationally valid biobanking is currently an ongoing process in PPPM related trends. Considering individual types of biological material (tissue samples, saliva samples, blood samples, DNA, RNA, proteins, metabolites, etc.), the major challenge of this process is how to optimally collect, store, and retrieve samples for sharing and testing. The analytical quality of collections, storage conditions and donation of samples to a *biobank* require strict control both at national and international levels. Disease-focused collections demand that acquired samples be retrospectively valid for development of novel biomarkers and novel drugs/treatments. For disease-specific biobanking, immaculate record keeping regarding patients is vital in order to facilitate optimal clinical decisions. The functional link to reliable clinical data and their interpretation is crucial for the biobank utility [7, 44].

An ideal biomarker does not exist, thus the need for a multi-level biomarker panel is of utmost importance. If novel biomarkers are discovered, are they applicable solely to diagnostics or to the treatment targets and therapy monitoring as well? Are they highly speicific for corresponding pathology? Is the biomarker panel applicable to individuals at risk being useful for targeted prevention? Is a multi-level biomarker panel applicable/available to secure a precise diagnose and therapy targeting? All these questions are crucial to respond by corresponding PPPM related experimental and clinical approaches [45–47].

6.2 Medical Chemistry

Medical chemistry provides a multidisciplinary approach that ranges from the application of innovative active therapeutic medicines to advanced methods for controlled drug delivery. Different areas of interest within the topic include stem cells, rational drug design, new (co)polymers, creation of tailor-made drug delivery systems, incorporation of target molecules to the polymeric structures, encapsulation of approved drugs with the polymers, preparation and characterisation of nanoparticles for in vivo diagnostics and treatment, and evaluation and validation of new systems for in vitro and in vivo studies [48–51].

7   Towards an Effective PPPM Promotion

7.1   Design

Of primary importance is professional design of the PPPM related *Interactome*. The specific challenge for multidisciplinary communication is the design of media to facilitate effective interaction amongst professional groups in PPPM. These groups currently "speak different languages," which reinforces each group's professional perspective while frequently underestimating the added value of the transfer of products between disciplines. The specific output of this design activity is the so-called professional Interactome and the representation of complex networks of information. The Interactome represents the most optimal model of healthcare organisation designed specifically for the implementation of effective interaction amongst professional groups in PPPM [52].

7.2   Education

Education is at the core of PPPM top science and practical implementation. The ultimate goal is to support the creation of a new generation of professionals in medicine who will be able to implement a holistic approach to patient care that recognises the complexity and individuality of any organism, as opposed to treating the patient as a disaggregated "pool of organs" [53, 54]. This requires new training and educational measures, including e-learning tools, in order to ensure the sharing of information important for all PPPM professional groups (medical doctors, industry, students, nurses, etc.), as well as patients and their family members.

In order to promote innovative educational programs, the following worldwide pioneer initiatives have been developed:

The EPMA Journal. This open access, PubMed indexed publication regularly updates information about medical innovations and advanced healthcare providing expert recommendations in predictive diagnostics, targeted preventive measures and personalised treatments of patients [55].

Advances in Predictive, Preventive and Personalised Medicine. [54]. This book series, launched in 2012 [54], provides an overview of multidisciplinary aspects of advanced biomedical approaches and innovative technologies in innovative PPPM fields and healthcare as a whole. Topics focus on cost-effective management of health and disease tailored to professionals, and innovative strategies for standardisation of healthcare services. The book series also includes new guidelines for medical ethics, innovative approaches to early and predictive diagnostics, targeted prevention in healthy individuals, and healthcare economy and marketing. Innovative predictive, preventive, and personalised medicine is emerging as the focal point of efforts in healthcare aimed at curbing the prevalence of common (diabetes mellitus, cardiovascular diseases, chronic respiratory diseases and cancer) and rare diseases. This new book series is intended to serve as a reference source for researchers and the healthcare industry with special emphasis on health promotion in the general population.

7.3 Advanced Business Models for Healthcare

Here the focus is on poor economy of current healthcare systems and delivery. Across Europe, there is a great diversity of systems, and payment and reimbursement schemes. This imposes a highly fragmented market (market access being governed by various public and private organizations), which considerably increases effort (seeking recognition, market authorisation and reimbursement in all different EU-countries and their respective bodies) and costs (each country has different bureaucratic schemes). On the one hand, there is a need for policy dialog in order to achieve some harmony of rules and delivery, but also of access to care and reimbursement to patients across Europe. On the other hand, there is also a great need for more advanced business models (What services are offered/covered? Who is the beneficiary? Who pays?) in order to motivate all stakeholders towards better scientific achievements, more effective implementation, improved medical services, and promotion of interest in the general population to follow the strategies of predictive and preventive medicine for cost-effective healthcare. In view of economic strain and the aging population, this innovation in healthcare systems is critical for keeping the high quality of healthcare in Europe affordable and sustainable [56, 57].

8 From "Passively Performing" to "Actively Advising"

PPPM Centres aim to become the nucleus for advanced healthcare. Successful PPPM implementation requires an unprecedented level of collaboration amongst all stakeholders, long-term multidisciplinary professional partnerships including public-private ones, a robust juristic platform, and intelligent political regulations. It is important that future developments do focus on the integration of all elements of PPPM. Innovative PPPM centres are focused on designing and conducting a new culture in healthcare: high level of multidisciplinarity, innovation, and professional education, well-met patient needs, cost-effective economy of healthcare, etc.

An optimistic versus pessimistic prognosis depends on diagnostics, prevention and treatment approaches that healthcare systems will preferably adopt in the near future. By the 3rd decade of the 21st century, neurodegenerative pathologies (Alzheimer's and Parkinson's diseases, glaucoma, macular degeneration, etc.) could amount to more than 30 % of the global disease burden. Without innovation in healthcare, such developments will pose a serious threat to health economies and may lead to a collapsing healthcare system. By contrast, the effective utilisation of advanced early/predictive diagnostics, together with preventive and personalised medical approaches, could enable a significant portion of the population to reach their senior years in vibrant psychosocial health, with excellent physical and mental well-being, participating actively in society. Global research and implementation programs in bio/medicine, communication amongst scientific societies, healthcare providers, policy-makers, educators, and patient organisations together with a consolidation of professional groups in the branch of personalised medicine, will play a decisive role in driving the situation towards optimal development. PPPM strategies aim at promoting the optimistic scenario in Europe and worldwide [3]. The anticipatory perspective of medicine (See Nadin (see footnote 1) and [58]) and PPPM have several features in common. It was not our intention to compare them, but rather to advance a clear image of our own comprehensive methodology.

References

1. George, B., Cebioglu, M., Yeghiazaryan, K.: Inadequate diabetic care: global figures cry for preventive measures and personalized treatment. EPMA J. **1**, 13–18 (2010)
2. Golubnitschaja, O.: Time for new guidelines in advanced diabetes care: paradigm change from delayed interventional approach to predictive, preventive & personalized medicine. EPMA J. **1**, 3–12 (2010)
3. Golubnitschaja, O., Swanton, C., Danesi, R., Costigliola, V.: Promoting predictive, preventive and personalised medicine: European event of global importance. EPMA J. **2**, 131–136 (2011)
4. Golubnitschaja, O.: Time for new guidelines in advanced healthcare: the mission of the EPMA journal to promote an integrative view in predictive, preventive and personalized medicine. EPMA J. **3**, 5 (2012)
5. Society for Participatory Medicine. http://participatorymedicine.org/
6. Golubnitschaja, O., Costigliola, V.: EPMA.: general report & recommendations in predictive, preventive and personalised medicine 2012: white paper of the European association for predictive, preventive and personalised medicine. EPMA J. **3**, 14 (2012)
7. Golubnitschaja, O., Watson, I.D., Topic, E., Sandberg, S., Ferrari, M., Costigliola, V.: Position paper of the EPMA and EFLM: a global vision of the consolidated promotion of an integrative medical approach to advance health care. EPMA J. **4**, 12 (2013)
8. Andrews, R.J.: Global medical knowledge, local medical implementation: predictive, preventive, and personalised medicine. In: Andrews, R.J. Too Big Succeed—Profiteering in American Medicine, pp. 124–127. iUniverse, Bloomington (2013)
9. Costigliola, V. (ed.): Healthcare Overview—New Perspectives. Springer (2012)
10. Redig, A.J., McAllister, S.S.: Breast cancer as a systemic disease: a view of metastasis. J. Intern. Med. **274**, 113–126 (2013)
11. Cox, T.R., Rumney, R.M.H., Schoof, E.M., Perryman, L., Høye, A.M., Agrawal, A., Bird, D., Latif, N.A., Forrest, H., Evans, H.R., Huggins, I.D., Lang, G., Linding, R., Gartland, A., Erler, J.T.: The hypoxic cancer secretome induces pre-metastatic bone lesions through lysyl oxidase. Nature **522**, 106–110 (2015)
12. Golubnitschaja, O., Sridhar, K.C.: Liver metastatic disease: New concepts and biomarker panels to improve individual outcomes Clin. Exp. Metastasis (2016). doi:10.1007/s10585-016-9816-8

13. Grech, G., Zhan, X., Yoo, B.C., Bubnov, R., Hagan, S., Danesi, R., Vittadini, G., Desiderio, D.M.: EPMA position paper in cancer: current overview and future perspectives. EPMA J. **6**, 9 (2015)

14. Golubnitschaja, O., Debald, M., Yeghiazaryan, K., Kuhn, W., Pešta, M., Costigliola, V., Grech, G.: Breast cancer epidemic in the early 21st century: Evaluation of risk factors, cumulative questionnaires and recommendations for preventive measures. Tumor Biol. (2016). doi:10.1007/s13277-016-5168-x

15. Iso, H.: Promoting predictive, preventive and personalized medicine in treatment of cardiovascular diseases. EPMA J. **2**, 1–4 (2011)

16. Helms, T.M., Duong, G., Zippel-Schultz, B., Tilz, R.R., Kuck, K.-H., Karle, C.A.: Prediction and personalised treatment of atrial fibrillation-stroke prevention: consolidated position paper of CVD professionals. EPMA J. **5**, 15 (2014)

17. Mozaffari, M.S. (ed.): New Strategies to Advance Pre/Diabetes Care: Integrative Approach by PPPM. Springer (2013)

18. Golubnitschaja, O., Yeghiazaryan, K., Cebioglu, M., Morelli, M., Herrera-Marschitz, M.: Birth asphyxia as the major complication in newborns: moving towards improved individual outcomes by prediction, targeted prevention and tailored medical care. EPMA J. **2**, 197–210 (2011)

19. Borisow, N., Döring, A., Pfueller, C.F., Paul, F., Dörr, J., Hellwig, K.: Expert recommendations to personalization of medical approaches in treatment of multiple sclerosis: an overview of family planning and pregnancy. EPMA J. **3**, 9 (2012)

20. Mähler, A., Mandel, S., Lorenz, M., Ruegg, U., Wanker, E.E., Boschmann, M., Paul, F.: Epigallocatechin-3-gallate: a useful, effective and safe clinical approach for targeted prevention and individualised treatment of neurological diseases? EPMA J. **4**, 5 (2013)

21. Mandel, S. (ed.): Neurodegenerative Diseases: Integrative PPPM Approach as the Medicine of the Future. Springer (2013)

22. Özgüç, M. (ed.): Rare Diseases—Integrative PPPM Approach as the Medicine of the Future. Springer (2015)

23. Roberti di Sarsina, P., Iseppato, I.: Why we need integrative medicine. EPMA J. **2**, 5–7 (2011)

24. Roberti di Sarsina, P., Alivia, M., Guadagni, P.: Traditional, complementary and alternative medical systems and their contribution to personalisation, prediction and prevention in medicine-person-centred medicine. EPMA J. **3**, 15 (2012)

25. Wang, W., Russell, A., Yan, Y.: Global health epidemiology reference group (GHERG): traditional Chinese medicine and new concepts of predictive, preventive and personalized medicine in diagnosis and treatment of suboptimal health. EPMA J. **5**, 4 (2014)

26. Bubnov, R.V.: Evidence-based pain management: is the concept of integrative medicine applicable? EPMA J. **3**, 13 (2012)

27. Cafiero, C., Matarasso, S.: Predictive, preventive, personalised and participatory periodontology: "the 5Ps age" has already started. EPMA J. **4**, 16 (2013)

28. Golubnitschaja, O., Costigliola, V.: Dental health: EPMA recommendations for innovative strategies. EPMA J. **5**, A119 (2014)

29. Kunin, A.A., Moiseeva, N.S.: A novel approach for detection of primary tooth caries based on the light influence foundation. EPMA J. **5**, A123 (2014)

30. Trovato, G.M.: Behavior, nutrition and lifestyle in a comprehensive health and disease paradigm: skills and knowledge for a predictive, preventive and personalized medicine. EPMA J. **3**, 8 (2012)

31. Shapira, N.: Women's higher health risks in the obesogenic environment: a gender nutrition approach to metabolic dimorphism with predictive, preventive, and personalised medicine. EPMA J. **4**, 1 (2013)

32. Oja, P., Titze, S.: Physical activity recommendations for public health: development and policy context. EPMA J. **2**, 253–259 (2011)

33. Graf, C.: Preventing and treating obesity in pediatrics through physical activity. EPMA J. **2**, 261–270 (2011)

34. Alivia, M., Guadagni, P., Roberti di Sarsina, P.: Towards salutogenesis in the development of personalised and preventive healthcare. EPMA J. **2**, 381–384 (2011)
35. Younesi, E., Hofmann-Apitius, M.: From integrative disease modeling to predictive, preventive, personalized and participatory (P4) medicine. EPMA J. **4**, 23 (2013)
36. Drucker, E., Krapfenbauer, K.: Pitfalls and limitations in translation from biomarker discovery to clinical utility in predictive and personalised medicine. EPMA J. **4**, 7 (2013)
37. Lemke, H.U., Golubnitschaja, O.: Towards personal health care with model-guided medicine: long-term PPPM-related strategies and realisation opportunities within "Horizon 2020". EPMA J. **5**, 8 (2014)
38. Berliner, L., Lemke, H.U. (eds.): An Information Technology Framework for Predictive, Preventive and Personalised Medicine: A Use-Case with Hepatocellular Carcinoma. Springer (2015)
39. Zhan, X., Desiderio, D.M.: The use of variations in proteomes to predict, prevent, and personalize treatment for clinically nonfunctional pituitary adenomas. EPMA J. **1**, 439–459 (2010)
40. Hu, R., Wang, X., Zhan, X.: Multi-parameter systematic strategies for predictive, preventive and personalised medicine in cancer. EPMA J. **4**, 2 (2013)
41. Grech, G., Grossman, I. (eds.): Preventive and Predictive Genetics: Towards Personalised Medicine. Springer (2015)
42. Waerner, T., Thurnher, D., Krapfenbauer, K.: The role of laboratory medicine in healthcare: quality requirements of immunoassays, standardisation and data management in prospective medicine. EPMA J. **1**, 619–626 (2010)
43. Gahan, P.B. (ed.) Circulating Nucleic Acids in Early Diagnosis, Prognosis and Treatment Monitoring—An Introduction. Springer (2015)
44. Deigner, H.-P.: Challenges in biobanking in personalized medicine—a biobank society's view. EPMA J. **5**, A128 (2014)
45. Mandel, S., Morelli, M., Halperin, I., Korczyn, A.: Biomarkers for prediction and targeted prevention of Alzheimer's and Parkinson's diseases: evaluation of drug clinical efficacy. EPMA J. **1**, 273–292 (2010)
46. Colley, K.J., Wolfert, R.L., Cobble, M.E.: Lipoprotein associated phospholipase A(2): role in atherosclerosis and utility as a biomarker for cardiovascular risk. EPMA J. **2**, 27–38 (2011)
47. Yap, T.A., Swanton, C., de Bono, J.S.: Personalization of prostate cancer prevention and therapy: are clinically qualified biomarkers in the horizon? EPMA J. **3**, 3 (2012)
48. Coelho, J.F., Ferreira, P.C., Alves, P., Cordeiro, R., Fonseca, A.C., Góis, J.R., Gil, M.H.: Drug delivery systems: advanced technologies potentially applicable in personalized treatments. EPMA J. **1**, 164–209 (2010)
49. Coelho, J. (ed.): Drug Delivery Systems: Advanced Technologies Potentially Applicable in Personalised Treatment. Springer, Dordrecht (2013)
50. Spivak, M.Y., Bubnov, R.V., Yemets, I.M., Lazarenko, L.M., Tymoshok, N.O., Ulberg, Z.R.: Development and testing of gold nanoparticles for drug delivery and treatment of heart failure: a theranostic potential for PPP cardiology. EPMA J. **4**, 20 (2013)
51. Ohno, T.: Particle radiotherapy with carbon ion beams. EPMA J. **4**, 9 (2013)
52. Golubnitschaja, O., Lemke, H.U., Kapalla, M., Kent, A.: Design in predictive, preventive and personalised medicine. In: Kuksa, I., Fisher, T. (eds.) Design for Personalisation. Gower Publishing, London (2016) (In press)
53. Polivka, J., Polivka, J., Karlikova, M., Topolcan, O.: Pre-graduate and post-graduate education in personalized medicine in the Czech Republic: statistics, analysis and recommendations. EPMA J. **5**, 22 (2014)
54. Advances in Predictive, Preventive and Personalised Medicine. http://www.springer.com/series/10051
55. EPMA Journal. www.epmajournal.com
56. Ausweger, C., Burgschwaiger, E., Kugler, A., Schmidbauer, R., Steinek, I., Todorov, Y., Thurnher, D., Krapfenbauer, K.: Economic concerns about global healthcare in lung, head and

neck cancer: meeting the economic challenge of predictive, preventive and personalized medicine. EPMA J. **1**, 627–631 (2010)
57. Fischer, T., Langanke, M., Marschall, P., Michl, S. (eds.): Individualized Medicine—Ethical, Economical and Historical Perspectives. Springer, Dordrecht (2015)
58. Nadin, M.: The anticipatory profile. An attempt to describe anticipation as process. Int. J. Gen. Syst. **41**(1), 43–75 (2012) (Nadin, M. (ed.), Taylor & Francis, London)

Next Generation Sequencing for Next Generation Diagnostics and Therapy

Marianna Garonzi, Cesare Centomo and Massimo Delledonne

Abstract DNA sequencing technologies are evolving at a prodigious rate. First-generation approaches have now been largely replaced by second-generation technologies (still known as "next generation sequencing" (NGS) even though they are now current and commonplace), and third-generation technologies (sometimes called "next-next generation sequencing") are starting to arrive. This has led to global boom in whole genome or exome sequencing, boosting the discovery of sequence variants associated with disease that will eventually be translated into new diagnostic, prognostic, and therapeutic targets for individual patients in "precision medicine." Acknowledgement of disease predisposition and specific therapeutic behavior for each individual addresses a more preventive approach. Adoption of such novel means represents an anticipation-relevant outcome as it can affect our healthcare on many different levels, ranging from a simple lifestyle adjustment to a well-defined clinical guideline. In this chapter we summarize current and emerging sequencing technologies for clinical applications, and some of the challenges that lie ahead.

Keywords Precision medicine · Genomics · Sequencing technologies

M. Garonzi (✉) · C. Centomo · M. Delledonne
Department of Biotechnology, University of Verona, Strada Le Grazie 15,
37134 Verona, Italy
e-mail: marianna.garonzi@univr.it

C. Centomo
e-mail: cesare.centomo@univr.it

M. Delledonne
e-mail: massimo.delledonne@univr.it

© Springer International Publishing Switzerland 2017   87
M. Nadin (ed.), *Anticipation and Medicine*, DOI 10.1007/978-3-319-45142-8_6

1   Introduction

Next generation diagnostics and therapy—the object of the conference on Anticipation and Medicine that is the origin of this volume—can be defined as any set of clinical approaches informed by consulting the sequence of a patient's genome. The origins of this approach can therefore be traced to the first medical applications of individual gene sequences. Comprehensive applications required two major developments: the sequencing of the human genome and the arrival of technologies allowing access to individual genomes quickly and inexpensively. We shall focus on the development of technology, connecting it to the anticipatory aspects of genetic research. As we shall see, issues such as individualized treatment, new screening methods, individual genetic disease, among other aspects pertain to the anticipatory perspective (see Nadin[1] and [1]). How technology evolved is not irrelevant to how all the questions related to the role of the "genetic map" are articulated.

The first project to sequence the entire human genome arose from several debates, some doubting whether such a project would be worthwhile in terms of downstream applications, and others arguing that it would expedite cancer research and help to identify medically relevant mutations. The advent of recombinant DNA technology in the 1970s, followed by development of methods to clone larger fragments of DNA in 1980s, provided the tools necessary to isolate and assemble genomic DNA sequences hundreds to several thousands of base pairs in length. Larger DNA inserts, in the megabase range, became accessible following the development of artificial chromosome vectors, allowing the construction of physical maps of the human genome upon which individual clones could be assembled.

The human genome project (HGP) was the first step towards the application of genome technology in medicine, but the initial aim was to assemble a highly accurate reference sequence (one error per 10,000 bases) that spanned the majority of each human chromosome. This sequence was predicted to offer valuable information concerning human biology, thus facilitating applications in other fields, including medicine, drug development and forensics. The HGP officially commenced in 1990 and lasted 13 years, with total funding of US$3.8 billion. A "working draft" of the human genome DNA sequence was completed in June 2000 and published in February 2001 [2].

Alongside the map-based sequencing approach adopted by the publicly funded International Human Genome Sequencing Consortium (IHGSC), Celera Genomics (founded in 1998 by Dr. Craig Venter) declared its intent to sequence the human genome using the comparatively new method of whole-genome shotgun sequencing. This does not require the prior development of a physical map for assembly, but relies instead on the generation of large numbers of overlapping reads that can be assembled de novo. This approach was faster than the IHGSC strategy but much

---

[1]Nadin, M.: Medicine: The Decisive Test of Anticipation. In: Nadin, M.: (ed.) Anticipation and Medicine, pp. 1–25. Springer, Cham (2016).

more computationally demanding, and therefore only became possible towards the end of the publicly funded project when sufficient computing power became available. The Celera Genomics project was able to sequence the human genome in three years at a cost of approximately US$300 million. However, this progress would not have been possible without access to the sequencing data already produced by the IHGSC [3].

Before the analysis of the draft sequence, the human genome was expected to contain $\sim$ 120,000 genes [4], but sequence annotation only revealed $\sim$ 20,500 [5]. Only 1.1 % of the genome was represented by exons, whereas 24 % was represented by introns and 75 % by intergenic DNA. The alignment of the human genome with other sequenced genomes (a relatively new field at the time, known as comparative genomics) revealed vertebrate-specific evolutionary expansions in gene families associated with neuronal functions, tissue-specific developmental regulation, hemostasis, and the immune system [3].

The draft sequences also provided locations for 2.1 million single-nucleotide polymorphisms (SNPs), showing that single-base differences between any randomly selected human genomes occur at an average frequency of 1 every 1250 bp. A SNP map has therefore been integrated with the human genome sequence to highlight how nucleotide diversity varies across the genome, in a manner broadly consistent with the standard population genetics model of human history. This high-density SNP map provided a public resource that defined variation across the genome and identified markers for disease diagnosis and therapy [3].

Both human genome projects relied on several cumulative improvements in the Sanger chain-termination sequencing method to improve accuracy and throughput. This increased the output of a single sequencing machine to about 1.6 million bp per day. But even at that rate it would take 15,000 days of continuous operation for one Sanger sequencer running 96 reactions in parallel to cover the three billion base pairs of the human genome with the minimum eight-fold redundancy required to ensure accuracy. Entirely new sequencing methods were therefore required to gain access to the genomic information of individuals at a cost suitable for standard healthcare practices and rapidly enough to facilitate diagnosis in time for effective therapy, heralding the development of next-generation sequencing (NGS). To reach these goals, a new initiative was founded by the National Human Genome Research Institute (NHGRI) in 2004 aiming to reduce the cost of sequencing a human genome to US$1,000 within 10 years [6].

2 The Advent of NGS Technologies

Next-generation sequencing provided the basis for new diagnostic and therapeutic strategies by accelerating the rate of sequence generation and reducing the cost per base to the extent that individual genomes became accessible for the first time. This step change meant that sequencing technology could be used for the discovery of medically relevant sequences at the level of individual patients, rather that the erstwhile approach of testing short stretches of DNA for previously discovered sequence variants. Several different NGS platforms have been developed but they all share one property that differs from the Sanger method, i.e., the ability to produce millions of short reads in parallel.

The first NGS platform was commercialized in 2005 by 454 Life Sciences [7]. The 454 technology combined emulsion PCR (allowing the amplification of DNA fragments in massively parallel arrays without cloning) with pyrosequencing, a real-time sequencing by synthesis method developed almost 10 years before [8, 9]. Emulsion PCR is based on the in vitro amplification of up to 10 million copies of single DNA fragments attached to the surface of beads encapsulated in water droplets in an oil–water emulsion, thus avoiding time-consuming standard cloning methods. Millions of beads, each decorated with millions of copies of a different genomic fragment, are then placed in picoliter-sized wells where the sequencing reaction takes place, achieving massive parallelization and throughput. The sequencing process is based on the real-time detection of pyrophosphate release following the addition of a nucleotide by DNA polymerase to the growing DNA strand. A sulfurylase converts the pyrophosphate to ATP which acts as a substrate for the luciferase-mediated conversion of luciferin to oxyluciferin thus generating a measurable flash of light.

The 454 sequencing technology overcomes two of the bottlenecks in Sanger sequencing: the need to prepare individual templates, which is avoided by the multiplex emulsion PCR format, and the need to complete a chain-termination reaction and separate the products by capillary electrophoresis, which was addressed by the real-time optical detection of nucleotide insertion in a high density multiwell plate. This early example of NGS increased the throughput by 100-fold compared to Sanger sequencing. In more recent 454 instruments, up to one million reads can be generated per run, each 700–800 bp in length. This remains one of the longest read lengths among all the current NGS technologies, not far short of the ~1,000 bp maximum achieved by Sanger capillary sequencing, and there is a low rate of substitution errors. However, the intrinsic limitations of pyrosequencing mean that 454 sequencing is sensitive to indel errors due to the misinterpretation of homopolymer sequence runs.

Other NGS technologies followed hot on the heels of the 454 method including the Genome Analyzer launched in 2006 by Illumina [10] and *Sequencing by Oligo Ligation Detection* (SOLiD) marketed by Applied Biosystems in 2007 [11]. Illumina offered a novel strategy for preparing the sequencing template, using a "bridge PCR" to amplify the signal directly on the solid surface of a flow cell where the sequencing reaction takes place. The sequencing method is analogous to the Sanger approach because it is based on chain termination. However, it uses reversible terminators and achieves sequencing in real time through cycles of fluorescent nucleotide incorporation, imaging, and cleavage of the terminator group containing the dye. Both technologies have been streamlined and improved to simplify template preparation, remove awkward bead-handling steps and increase the number of reads generated per cycle, thus reducing the per-base cost of sequencing even further. In the original Illumina method, the read length was limited to ~35 bases, although millions of reads were generated per run, but more recent machines can generate billion of reads of up to 250 bases. Compared to 454 sequencing, the Illumina method is less prone to indel errors in homopolymer runs but there is a higher substitution error rate.

Both Illumina and SOLiD use expensive fluorophore-based labeling technologies and optical imaging. In 2010, Ion Torrent sequencing introduced advanced semiconductor technology to detect the hydrogen ions released during nucleotide incorporation, thus further simplifying the overall detection process and providing sequences more rapidly at a lower cost [12]. However, Ion Torrent shares with 454 sequencing the tendency to suffer a high indel error rate in homopolymer runs.

As stated briefly above, the advent of new sequencing technologies that are simpler, faster, and more scalable than the Sanger method has caused the per-base cost of sequencing to fall rapidly. In 2001, when the first draft Human Genome Sequence was published, the cost to sequence 1 Mb was US$5,292.39. During the transition from Sanger-based sequencing to NGS technologies, the cost per Mb fell to ~US$100. During 2008, which is arguably when NGS became "mainstream", the cost per Mb fell from US$100 in January to less than US$4.00 in December. This trend has continued, and as of June 2015, the cost per Mb had declined to a remarkable US$0.015. The cost to sequence a human genome has therefore fallen from US$95,263,072.00 in 2001 to US$1,363.00 in 2015, very close to the US $1,000 target set by the NHGRI in 2004 [13].

3 Limits of NGS Applications

Although NGS has precipitated astonishing advances in the last ten years, all the techniques are limited by the relatively short length of the reads. This is not an issue when sequencing unique or well-characterized regions of the genome, but short reads cannot resolve repetitive regions longer than the read length, such as trinucleotide repeat expansions associated with diseases known as trinucleotide repeat disorders (e.g., Huntington's disease, fragile X syndrome, and neurodegenerative progressive disorders known as ataxias) [14]. For the same reason, only short indels can be detected and the technology struggles with larger structural variants, such as translocations, because of the small spatial resolution achieved by short reads. Structural variants are less common at the population level than SNPs and indels, but recent studies indicate they are associated with a number of human diseases ranging from sporadic syndromes and Mendelian diseases to complex traits, including neurodevelopmental disorders. Chromosomal aneuploidies, such as trisomy 21 (Down syndrome), and monosomy X (Turner syndrome) are well characterized; but de novo copy number variations are now known to be enriched in autism spectrum disorders [15] and structural variations may contribute to other complex traits including cancer, schizophrenia, epilepsy, Parkinson's disease, and immune disorders such as psoriasis [16, 17].

Gene fusions caused by somatic translocations are associated with tumorigenesis, e.g., chronic myeloid leukemia (CML) and acute myeloid leukemia (AML) [18, 19]. Although several strategies based on whole genome sequencing and/or transcriptome sequencing have been used to discover gene fusion events, they remain limited by the high frequency of false positives and low sensitivity of computational approaches based on short sequence reads.

Whole-genome sequencing using NGS platforms also provides little if any haplotype information at the level of an individual genome. Haplotype data facilitates linkage analysis and association studies, and is a key component of population genetics and clinical genetics [20]. Haplotype data can be used to predict the severity and prognosis of certain genetic disorders. For example, intragenic cis-interactions between common polymorphisms and pathogenic mutations in the prion protein (*PRNP*) and cystic fibrosis transmembrane conductance regulator (*CFTR*) genes greatly influence the penetrance and expressivity of hereditary Creutzfeldt-Jakob disease and cystic fibrosis, respectively [21]. Similarly, the gene encoding the protease inhibitor α-2-macroglobulin is located within the Alzheimer's disease (AD) susceptibility locus on chromosome 12p, and a series of studies using SNP markers show that certain haplotypes (especially those containing a 5-bp deletion in intron 18 and a non-synonymous SNP in exon 24) have a high-risk association with AD [22–24]. Although haplotypes can be inferred by population-based methods or by genotyping multiple individuals from the same family, data interpretation can be hindered by low-frequency variants, private variants and de novo variants that are poorly resolved.

4   Third-Generation Sequencing

The second-generation sequencing technologies described above rely on PCR to amplify signals from individual templates. This means that the accuracy of sequencing is dependent on the accuracy of the PCR step and the read length is limited by the need for template amplification. Third-generation technologies overcome this limitation by using ultrasensitive imaging technologies or electrochemical sensors that allow the sequencing of individual molecules without prior amplification. The main advantages of third-generation technologies include minimal sample preparation, the ability to use smaller amounts of biological materials, faster sequence acquisition, increased throughput, longer read lengths and the potential to reduce the cost of sequencing the human genome to US$100.00 within a few years.

The first commercially available third-generation sequencing technology was the Helicos Genetic Analysis Platform [25–28], which achieved single-molecule sequencing by using a high-resolution camera to detect the incorporation of a single fluorophore during DNA synthesis. Although the imaging of fluorescent dyes was reminiscent of NGS, the innovation of the Helicos platform was the use of a single-molecule template, removing the need for an initial PCR amplification step. There was no improvement over NGS in terms of read lengths, throughput, and accuracy, but one unique advantage was the ability to directly sequence RNA, as well as DNA [29]. The Helicos platform is no longer available because the company has ceased trading.

Another example of third-generation sequencing is the Single-Molecule Real-Time (SMRT) technology developed by Pacific Biosciences, which enables the direct observation of a single molecule of DNA polymerase synthesizing a strand of DNA. DNA polymerases are attached to the bottom of ~ 50-nm wells, which act as zero-mode wave guides (ZMWs). The DNA polymerase utilizes γ-phosphate fluorescently labeled nucleotides to synthesize the nascent DNA strand. The narrow width of the ZMW prevents light propagation through the waveguide, but energy penetration over a short distance excites the fluorophores attached to the nucleotides in the area near the DNA polymerase at the bottom of the well. The fluorescence pulse that follows nucleotide incorporation can thus be detected in real time [30].

The SMRT method has been improved by simplifying the sample preparation, scanning, and washing steps to produce results more quickly and with less effort [31]. The absence of template amplification allows the processivity of DNA polymerase to be fully exploited, resulting in reads with an average length of ~ 10 kb and often exceeding 20 kb. This facilitates de novo assembly, the direct detection of haplotypes and the phasing of entire chromosomes. However, one drawback of this technology is that indel errors can exceed 13 % [32].

In 2015, Pacific Biosciences launched a new SMRT-based sequencer claiming higher throughput and lower costs. Although the chemistry has not changed, the SMRT cells have been redesigned to contain one million ZMWs compared to 150,000 in the previous system, increasing throughput 7-fold. Each SMRT cell therefore has a throughput of 5–10 Gb and initial average read lengths of 8–12 kb; both throughput and average read length should increase over time.

Nanopore sequencing is another third-generation technology based on the direct detection of DNA nucleotides passing through a nanoscale pore. The sequence can be recorded directly as a current fluctuation or converted into an optical signal [33]. The Oxford Nanopore Technologies MinION platform is the first available commercial example of nanopore sequencing, following beta-testing in 2014. MinION is a hand-held device which produces much longer reads than other technologies (tens of kilobases) in a short time. Once a sample is charged in the MinION flow cell, initial results are provided in minutes and a run can be completed in a few hours. The main disadvantage of current nanopore sequencing is the high error rate due to the low spatial resolution of the biological pore. New nanopore technologies that aim to overcome such problems replace the protein channels with artificial non-organic pores small enough to report the intervals between consecutive nucleotides. The Oxford Nanopore Technologies MinION platform is simple, portable, much less expensive and capable of producing much longer reads than the other NGS technologies currently available [34, 35].

5   Clinical Analysis of the Human Genome Sequence

There are many technological differences among first-, second-, and third-generation sequencing platforms, but in practical terms there are two main advances that can be brought to bear in clinical diagnosis and therapy—the anticipation-relevant outcomes of such advances. Next-generation sequencing technologies are (i) much faster and (ii) much less expensive than Sanger sequencing, which means a patient could undergo genome sequencing at approximately the same cost and in approximately the same timescale as standard laboratory assays, a consideration that would have been inconceivable 10 years ago. The new sequencing technologies therefore bring more diseases than ever before into the domain of sequence-based diagnosis and therapy.

More than 5,000 human single-gene disorders have been resolved to causative mutations, and others have been associated with structural aberrations or aneuploidies [36]. Although the availability of the human genome sequence has greatly improved our understanding of the genetic basis of disease—including the anticipatory aspects—second- and third-generation sequencing technologies have made the identification of genetic variants feasible on a genomic scale because the sequencing of individual genomes is now possible, making it easier to identify rare SNPs, indels and structural variations with a high degree of confidence. However, detecting variants is only the first part of a complex interpretation process. The number of polymorphisms and rare sequence variants per individual ranges from few hundred thousand in the exome to millions in the entire genome, so comprehensive biochemical characterization and the assessment of a causal link between a gene variant and a disease is not always possible. The comprehensive prioritization of candidate genes prior to experimental testing is therefore necessary, but this requires the screening of genome sequence data to select the most likely clinically relevant variants. Candidates are prioritized using correlative evidence that associates each variant and gene with the given disease based on the integration of molecular, genetic, biochemical, functional and epidemiological data.

Several resources have been developed to facilitate the identification and reporting of variants by collecting human variations and associated outcomes. These resources include ClinVar [37], an archive of relationships between medically important variants and phenotypes along with supporting evidence, and the Human Genes Mutation Database (HGMD) [38], which collates known mutations associated with human inherited diseases. Specific databases have also been developed for variants associated with drug responses, and such resources can be tailored to individual genomic profiles, e.g., the Pharmacogenomics Mutation Database (PGMD) and PharmGKB [39, 40], and the COSMIC database, which collects somatic mutations identified in cancer research [41]. These databases can be screened to determine the relevance of the variants in a given genome sequence; but they have no predictive capability, i.e., they provide no information about novel and unassociated variants.

One way to predict the potential impact of genome sequence variants is to determine their frequencies in population data, as demonstrated by the 1000 Genomes Project discussed in more detail below [42]. At the population level, natural selection removes deleterious alleles, so filtering for low-frequency variants in populations can help to enrich the dataset for potentially dangerous variants. However, a rare allele can also be present for other reasons. It may be a harmless variant that is rare because it has arisen recently or is close to elimination by genetic drift. Indeed, genetic drift may result in a particular variant becoming rare in one population but part of a common polymorphism in others. Therefore, rarity per se is not necessarily evidence for disease association and additional epidemiological and functional evidence should also be sought.

Functional prediction algorithms for non-synonymous variants use different forms of evidence such as sequence conservation (SIFT [43]) or the predicted impact of amino acid substitutions on protein structure and function (PolyPhen2 [44]). Others use a classifier trained with known disease mutations as well as harmless SNPs and indels to predict the likelihood of disease association (MutationTaster [45]). More confidence can be assigned to predictions that are generated using more than one of these tools. Therefore a database of pre-calculated scores obtained from many different predictors for all possible nonsynonymous substitutions in the current human genome sequence has been developed to process queries more rapidly (dbNSFP [46, 47]).

6 Population-Scale Sequencing Projects

The discovery of relevant variations in individual genomes requires data from the analysis of large groups of people because it is necessary to correlate variations with phenotypes in a statistically significant manner. Population-scale sequencing projects thus increase the power of research on diseases and provide the foundations of personalized medicine. This is one of the claims of those who advance the anticipatory perspective (see Nadin [48]).

The first international project aiming to collect and analyze genome data from a large cohort of human subjects was the 1000 Genomes Project mentioned above. This was launched in 2008, and its primary objective was to produce a comprehensive human genetic variation database by identifying all polymorphisms, i.e., genetic variants that have frequencies of at least 1 % in the populations included in the project. The analysis of 2,504 samples from 13 different populations allowed the creation of the first complete catalog of genetic variations and their frequencies, which can be used to filter genomic data from patients afflicted by rare diseases to remove common variants and enrich for rarer variations more likely to be associated with the disease [49]. Many similar projects have been initiated more recently, such as the NHLBI GO Exome Sequencing Project (ESP) which focuses on variants contributing to heart, lung and blood disorders. In this project, the exomes of 6,503 unrelated individuals in diverse well-characterized populations were sequenced and the resulting datasets and frequency tables have been shared with the scientific community [50].

These first massive sequencing projects facilitated the discovery of rare variants by sampling individuals from diverse populations. However, because there is high genetic diversity among populations due to genetic drift and natural selection, population-specific sequencing projects may help with the interpretation of variants in individual genomes. One of the largest studies based on a single population was carried out by the company deCODE Genetics Inc., which sequenced more than 2,600 genomes from the Icelandic population to a median coverage of 20-fold, and used comprehensive national genealogies to accurately impute even rare variants throughout the population [51]. The project discovered novel diseases-associated rare variants such as mutations in *ABCA7* that increase the risk of Alzheimer's disease [52]. A similar example is Genomics England, a UK company owned by the UK Department of Health, which aims to sequence 100,000 whole genomes by 2017 in collaboration with the UK National Health Service and 11 Genomic Medicine Centers across the country. This is the first genomics initiative which is tightly integrated with a national health system to accelerate the translation of research into clinical practice [53].

The main limitation of population studies is that rare variants contributing to quantitative traits can be difficult to identify even in large cohorts. This can be overcome by studying founder populations, in which variants that are rare or absent elsewhere may be more common due to the founder effect. One example of this approach is the analysis of the Sardinian population in Italy, in which 2,120 individual genomes were sequenced with low coverage. The project identified ~ 3.8 million variants that were not detected in previous sequencing-based compilations such as dbSNP 142 and ExAC [54], which are also enriched for predicted functional mutations. The Sardinian project also revealed the presence of 76,286 variants with a frequency exceeding 5 % which are rare (frequency lower than 0.5 %) or absent in other populations.

7 NGS and Precision Medicine

Precision medicine is a new healthcare approach that matches the genomic data and clinical records of individual patients. In this way, treatments are tailored to the patients based on their genetic profile or other molecular and cellular information. The concept of precision medicine is based on the fact that diseases affect individuals in different ways and that different patients show distinct responses to the same treatments. Clinicians have known for many years that individual patients respond differently to the same treatments, but genome analysis now provides data that may allow the development of individualized therapies. Precision medicine encompasses screening for inherited conditions, carrier screening and prenatal testing, through to the identification of targets for cancer treatment, and the diagnosis, and treatment of rare diseases. The possible future that affects a current state [55] is the goal of the PMI initiative [56].

Many of the ~5,000 known genetic disorders manifest during the first 28 days of life, but the full clinical symptoms may not be evident in newborns. Screening at this stage can therefore identify babies with genetic disorders that have silent, heterogeneous, or ambiguous phenotypes at birth, but which benefit from early intervention to avoid an irreversible impact on health. One example is sickle cell disease, which causes blood clotting and a shortage of red blood cells. Early identification can prevent the onset of complications caused by increased susceptibility to infections through proper and timely treatment. This is where the anticipatory perspective plays an important role. Newborn screening has been integrated into postnatal healthcare for many years, but currently it only targets about 50 of the most severe genetic disorders that require urgent clinical decisions [57].

The use of NGS-based newborn screening (NBS) companion to the current biochemical testing regime would dramatically increase the quantity and diversity of information parents and clinicians could derive from screening. A targeted NGS assay based on panels of hundreds of relevant sequence variants could improve diagnostic testing in newborns in a cost-effective manner by selectively sequencing the corresponding genomic regions, mainly exons, following enrichment in a physical DNA capture step [58].

Traditional molecular biology assays such as PCR can be used to identify a limited range of known cancer-related mutations and rearrangements but NGS could reveal comprehensive, individualized mutational landscapes, including both known and novel variations. Integrated high-throughput sequencing of tumor biopsy genomes could also facilitate biomarker-driven clinical trials in oncology [59].

Although individual genetic diseases are rare, they are collectively common, affecting millions of people worldwide. Many rare genetic diseases have escaped traditional gene discovery approaches due to heterogeneity, a limited number of patients or families for analysis, and the loss of reproductive fitness as a result of such diseases. NGS-based gene discovery partially overcomes such limitations and has enabled the discovery of hundreds of novel, rare disease mutations [60].

Finally, NGS is revolutionizing pharmacogenomics as a disease management concept. Pharmacogenomics correlates human genome sequence data with drug responses and aims to improve therapeutic efficacy and reduce side effects by developing qualitatively and quantitatively tailored treatment regimens. Pharmacogenomics has the potential to transform medical practice by replacing broad methods of screening and treatment with a more personalized approach that takes into account both clinical factors and genome data. For example, in cancer treatment, small-molecule inhibitors and antibodies that bind to "druggable" targets are revolutionizing medicine. Personalized anti-cancer therapy requires the identification of cancer-specific driver mutations in each patient. For example, among patients diagnosed with non-small-cell lung cancer, only those harboring the ALK gene fusion (present in less than 5 % of the affected population) respond to treatment with targeted inhibitors such as crizotinib [61]. The identification of patients suitable for this treatment before therapeutic selection would avoid administering ineffective drugs to the >95 % of nonresponsive patients on a trial and error basis, not only hastening the deployment of more suitable treatment regimens, but also reducing the costs associated with wasting the drug.

Another example of the application of pharmacogenomics in healthcare is the prediction of individual responses to warfarin, a commonly prescribed oral anti-coagulant which is used to prevent thromboembolic diseases in patients with deep vein thrombosis, atrial fibrillation, or recurrent stroke [62]. Although warfarin is effective, the optimal dose differs widely among individuals and is often determined on a trial and error basis. However, several studies have shown that the *VKORC1* locus is the single most significant predictor of warfarin tolerance, accounting for ~25 % of the variance in a stabilized warfarin dose [63, 64]. As sequencing technologies become less expensive and more widely available, pharmacogenomics will transition from a niche research area to a main player in drug development and clinical decision making. Whole genome sequencing can reveal rare and even unique markers that would not be detected by conventional genetic screening methods.

8 Future Perspectives

Most current medical treatments are generalized, but one size does not fit all: therapies that are highly successful in some patients may have no effect or even a deleterious effect in others. The outlook for precision medicine has been dramatically improved by the recent development of high-throughput methods to characterize patients individually, including NGS, proteomics and metabolomics, as well as the availability of comprehensive databases containing information derived from large screening projects.

Even so, there remains a lack of broad research programs translating the outcomes from these large-scale projects into clinical practice. In the future, there should be more effort to integrate whole-genome sequencing initiatives with clinical applications, as shown by the Genomics England initiative in the UK and the recently announced Precision Medicine Initiative in the USA [56, 65].

In this future scenario, the tighter integration of research programs, precision medicine initiatives, and health services will allow physicians to consult patient genome data as well as conventional medical records. Whole genome sequencing will be part of routine medical screening, and health insurance companies will cover the (declining) costs of genomic analysis because early investment in preventive medicine would save the greater costs of therapy later down the line (see [66, pp. 75, 101, 111]). New diagnostic and prognostic markers will become available, so physicians will know which diseases present the most risk to their patients and which drugs, at which doses, are likely to be most effective. The details of anticipation-driven medicine were not entered into here, rather, aspects of such an approach were suggested.

References

1. Nadin, M.: The anticipatory profile. An attempt to describe anticipation as process. In: Nadin, M. (ed.) Anticipation (special issue of the International Journal of General Systems), vol. 41, no. 1, pp. 43–75. Taylor and Francis, London (2012). http://www.tandfonline.com/doi/abs/10.1080/03081079.2011.622093, doi:10.1080/03081079.2011.622093

2. Lander, E.S., Heaford, A., Sheridan, A., Linton, L.M., Birren, B., Subramanian, A., Coulson, A., Nusbaum, C., Zody, M.C., Dunham, A., Baldwin, J., et al.: Initial sequencing and analysis of the human genome. Nature **409**, 860–921 (2001)

3. Venter, J.C., Adams, M.D., Myers, E.W., Li, P.W., Mural, R.J., Sutton, G.G., Smith, H.O., Yandell, M., Evans, C.A., Holt, R.A., Gocayne, J.D., et al.: The sequence of the human genome. Science **291**, 1304–1351 (2001)

4. Liang, F., Holt, I., Pertea, G., Karamycheva, S., Salzberg, S.L., Quackenbush, J.: Gene index analysis of the human genome estimates approximately 120,000 genes. Nat. Genet. **25**, 239–240 (2000)

5. Clamp, M., Fry, B., Kamal, M., Xie, X., Cuff, J., Lin, M.F., Kellis, M., Lindblad-Toh, K., Lander, E.S.: Distinguishing protein-coding and noncoding genes in the human genome. Proc. Natl. Acad. Sci. U.S.A. **104**, 19428–19433 (2007)

6. Schloss, J.A.: How to get genomes at one ten-thousandth the cost. Nat. Biotechnol. **26**, 1113–1115 (2008)

7. Margulies, M., Egholm, M., Altman, W.E., Attiya, S., Bader, J.S., Bemben, L.A., Berka, J., Braverman, M.S., Chen, Y.-J., Chen, Z., Dewell, S.B., et al.: Genome sequencing in microfabricated high-density picolitre reactors. Nature **437**, 376–381 (2005)

8. Ronaghi, M., Karamohamed, S., Pettersson, B., Uhlén, M., Nyrén, P.: Real-time DNA sequencing using detection of pyrophosphate release. Anal. Biochem. **242**, 84–89 (1996)

9. Ronaghi, M., Uhlén, M., Nyrén, P.: A sequencing method based on real-time pyrophosphate. Science **281**, 363, 365 (1998)

10. Bentley, D.R., Balasubramanian, S., Swerdlow, H.P., Smith, G.P., Milton, J., Brown, C.G., Hall, K.P., Evers, D.J., Barnes, C.L., Bignell, H.R., Boutell, J.M., et al.: Accurate whole human genome sequencing using reversible terminator chemistry. Nature **456**, 53–59 (2008)

11. Valouev, A., Ichikawa, J., Tonthat, T., Stuart, J., Ranade, S., Peckham, H., Zeng, K., Malek, J.A., Costa, G., McKernan, K., Sidow, A., et al.: A high-resolution, nucleosome position map of C. elegans reveals a lack of universal sequence-dictated positioning. Genome Res. **18**, 1051–1063 (2008)

12. Rothberg, J.M., Hinz, W., Rearick, T.M., Schultz, J., Mileski, W., Davey, M., Leamon, J.H., Johnson, K., Milgrew, M.J., Edwards, M., Hoon, J., et al.: An integrated semiconductor device enabling non-optical genome sequencing. Nature **475**, 348–352 (2011)

13. Wetterstrand, K.A.: DNA Sequencing Costs: Data from the NHGRI Genome Sequencing Program (GSP). www.genome.gov/sequencingcosts

14. Budworth, H., McMurray, C.T.: A brief history of triplet repeat diseases. Methods Mol. Biol. **1010**, 3–17 (2013)

15. Sebat, J., Lakshmi, B., Malhotra, D., Troge, J., Lese-martin, C., Walsh, T., Yamrom, B., Yoon, S., Krasnitz, A., Kendall, J., Leotta, A., et al.: Strong association of de novo copy number mutations with autism. Science **316**, 445–449 (2007)

16. Stankiewicz, P., Lupski, J.R.: Structural variation in the human genome and its role in disease. Annu. Rev. Med. **61**, 437–455 (2010)

17. Girirajan, S., Campbell, C.D., Eichler, E.E.: Human copy number variation and complex genetic disease. Annu. Rev. Genet. **45**, 203–226 (2011)

18. Rowley, J.D.: A new consistent chromosomal abnormality in chronic myelogenous leukaemia identified by quinacrine fluorescence and giemsa staining. Nature **243**, 290–293 (1973)

19. Rowley, J.D.: Identification of a translocation with quinacrine fluorescence in a patient with acute leukemia. Ann. génétique **16**, 109–112 (1973)

20. Browning, S.R., Browning, B.L.: Haplotype phasing: existing methods and new developments. Nat. Rev. Genet. **12**, 703–714 (2011)
21. Lee, J.-E., Choi, J.H., Lee, J.H., Lee, M.G.: Gene SNPs and mutations in clinical genetic testing: haplotype-based testing and analysis. Mutat. Res. **573**, 195–204 (2005)
22. Pericak-Vance, M.A.: Complete genomic screen in late-onset familial Alzheimer disease. Evidence for a new locus on chromosome 12. JAMA **278**, 1237 (1997)
23. Blacker, D., Wilcox, M.A., Laird, N.M., Rodes, L., Horvath, S.M., Go, R.C., Perry, R., Watson, B., Bassett, S.S., McInnis, M.G., Albert, M.S., et al.: Alpha-2 macroglobulin is genetically associated with Alzheimer disease. Nat. Genet. **19**, 357–360 (1998)
24. Saunders, A.J., Bertram, L., Mullin, K., Sampson, A.J., Latifzai, K., Basu, S., Jones, J., Kinney, D., MacKenzie-Ingano, L., Yu, S., Albert, M.S., et al.: Genetic association of Alzheimer's disease with multiple polymorphisms in alpha-2-macroglobulin. Hum. Mol. Genet. **12**, 2765–2776 (2003)
25. Bowers, J., Mitchell, J., Beer, E., Buzby, P.R., Causey, M., Efcavitch, J.W., Jarosz, M., Krzymanska-Olejnik, E., Kung, L., Lipson, D., Lowman, G.M., et al.: Virtual terminator nucleotides for next-generation DNA sequencing. Nat. Methods **6**, 593–595 (2009)
26. Harris, T.D., Buzby, P.R., Babcock, H., Beer, E., Bowers, J., Braslavsky, I., Causey, M., Colonell, J., Dimeo, J., Efcavitch, J.W., Giladi, E., et al.: Single-molecule DNA sequencing of a viral genome. Science **320**, 106–109 (2008)
27. Lipson, D., Raz, T., Kieu, A., Jones, D.R., Giladi, E., Thayer, E., Thompson, J.F., Letovsky, S., Milos, P., Causey, M.: Quantification of the yeast transcriptome by single-molecule sequencing. Nat. Biotechnol. **27**, 652–658 (2009)
28. Tessler, L.A., Reifenberger, J.G., Mitra, R.D.: Protein quantification in complex mixtures by solid phase single-molecule counting. Anal. Chem. **81**, 7141–7148 (2009)
29. Ozsolak, F., Platt, A.R., Jones, D.R., Reifenberger, J.G., Sass, L.E., McInerney, P., Thompson, J.F., Bowers, J., Jarosz, M., Milos, P.M.: Direct RNA sequencing. Nature **461**, 814–818 (2009)
30. Korlach, J., Marks, P.J., Cicero, R.L., Gray, J.J., Murphy, D.L., Roitman, D.B., Pham, T.T., Otto, G.A., Foquet, M., Turner, S.W.: Selective aluminum passivation for targeted immobilization of single DNA polymerase molecules in zero-mode waveguide nanostructures. Proc. Natl. Acad. Sci. U.S.A. **105**, 1176–1181 (2008)
31. Eid, J., Fehr, A., Gray, J., Luong, K., Lyle, J., Otto, G., Peluso, P., Rank, D., Baybayan, P., Bettman, B., Bibillo, A., et al.: Real-time DNA sequencing from single polymerase molecules. Science **323**, 133–138 (2009)
32. Quail, M., Smith, M.E., Coupland, P., Otto, T.D., Harris, S.R., Connor, T.R., Bertoni, A., Swerdlow, H.P., Gu, Y.: A tale of three next generation sequencing platforms: comparison of Ion torrent, pacific biosciences and illumina MiSeq sequencers. BMC Genom. **13**, 1 (2012)
33. Branton, D., Deamer, D.W., Marziali, A., Bayley, H., Benner, S.A., Butler, T., Di Ventra, M., Garaj, S., Hibbs, A., Huang, X., Jovanovich, S.B., et al.: The potential and challenges of nanopore sequencing. Nat. Biotechnol. **26**, 1146–1153 (2008)
34. Mikheyev, A.S., Tin, M.M.Y.: A first look at the Oxford Nanopore MinION sequencer. Mol. Ecol. Resour. **14**, 1097–1102 (2014)
35. Jain, M., Fiddes, I.T., Miga, K.H., Olsen, H.E., Paten, B., Akeson, M.: Improved data analysis for the MinION nanopore sequencer. Nat. Methods **12**, 351–356 (2015)
36. Amberger, J.S., Bocchini, C.A., Schiettecatte, F., Scott, A.F., Hamosh, A.: OMIM.org: Online Mendelian Inheritance in Man (OMIM(R)), an online catalog of human genes and genetic disorders. Nucleic Acids Res. **43**, D789–D798 (2015)
37. Landrum, M.J., Lee, J.M., Riley, G.R., Jang, W., Rubinstein, W.S., Church, D.M., Maglott, D.R.: ClinVar: public archive of relationships among sequence variation and human phenotype. Nucleic Acids Res. **42**, D980–D985 (2014)
38. Cooper, D.: The human gene mutation database. Nucleic Acids Res. **26**, 285–287 (1998)
39. Kaplun, A., Hogan, J.D., Schacherer, F., Peter, A.P., Krishna, S., Braun, B.R., Nambudiry, R., Nitu, M.G., Mallelwar, R., Albayrak, A.: PGMD: a comprehensive manually curated pharmacogenomic database. Pharmacogenomics J. 1–5 (2015)

40. Hewett, M., Oliver, D.E., Rubin, D.L., Easton, K.L., Stuart, J.M., Altman, R.B., Klein, T.E.: PharmGKB: the pharmacogenetics knowledge Base. Nucleic Acids Res. **30**, 163–165 (2002)

41. Bamford, S., Dawson, E., Forbes, S., Clements, J., Pettett, R., Dogan, A., Flanagan, A., Teague, J., Futreal, P.A., Stratton, M.R., Wooster, R.: The COSMIC (catalogue of somatic mutations in cancer) database and website. Br. J. Cancer **2**, 355–358 (2004)

42. Abecasis, G.R., Auton, A., Brooks, L.D., DePristo, M.A., Durbin, R.M., Handsaker, R.E., Kang, H.M., Marth, G.T., McVean, G.A.: An integrated map of genetic variation from 1,092 human genomes. Nature **491**, 56–65 (2012)

43. Ng, P.C.: SIFT: predicting amino acid changes that affect protein function. Nucleic Acids Res. **31**, 3812–3814 (2003)

44. Adzhubei, I.A., Schmidt, S., Peshkin, L., Ramensky, V.E., Gerasimova, A., Bork, P., Kondrashov, A.S., Sunyaev, S.R.: A method and server for predicting damaging missense mutations. Nat. Methods **7**, 248–249 (2010)

45. Schwarz, J.M., Rödelsperger, C., Schuelke, M., Seelow, D.: MutationTaster evaluates disease-causing potential of sequence alterations. Nat. Methods **7**, 575–576 (2010)

46. Liu, X., Jian, X., Boerwinkle, E.: dbNSFP: A lightweight database of human nonsynonymous SNPs and their functional predictions. Hum. Mutat. **32**, 894–899 (2011)

47. Liu, X., Jian, X., Boerwinkle, E.: dbNSFP v2.0: a database of human non-synonymous SNVs and their functional predictions and annotations. Hum. Mutat. **34**, 2393–2402 (2013)

48. Nadin. M.: Anticipation and the brain. In: Nadin, M. (ed.): Anticipation and Medicine, pp. 135–162. Springer, Cham (2016)

49. Sudmant, P.H., Rausch, T., Gardner, E.J., Handsaker, R.E., Abyzov, A., Huddleston, J., Zhang, Y., Ye, K., Jun, G., Fritz, M.H.-Y., Konkel, M.K., et al.: An integrated map of structural variation in 2,504 human genomes. Nature **526**, 75–81 (2015)

50. Fu, W., O'Connor, T.D., Jun, G., Kang, H.M., Abecasis, G., Leal, S.M., Gabriel, S., Altshuler, D., Shendure, J., Nickerson, D.A., Bamshad, M.J., et al.: Analysis of 6,515 exomes reveals the recent origin of most human protein-coding variants. Nature **493**, 216–220 (2012)

51. Gudbjartsson, D.F., Helgason, H., Gudjonsson, S.A., Zink, F., Oddson, A., Gylfason, A., Besenbacher, S., Magnusson, G., Halldorsson, B.V., Hjartarson, E., Sigurdsson, G.T., et al.: Large-scale whole-genome sequencing of the Icelandic population. Nat. Genet. **47**, 435–444 (2015)

52. Steinberg, S., Stefansson, H., Jonsson, T., Johannsdottir, H., Ingason, A., Helgason, H., Sulem, P., Magnusson, O.T., Gudjonsson, S.A., Unnsteinsdottir, U., Kong, A., et al.: Loss-of-function variants in ABCA7 confer risk of Alzheimer's disease. Nat. Genet. **47**, 445–447 (2015)

53. Siva, N.: UK gears up to decode 100 000 genomes from NHS patients. Lancet **385**, 103–104 (2015)

54. Sidore, C., Busonero, F., Maschio, A., Porcu, E., Naitza, S., Zoledziewska, M., Mulas, A., Pistis, G., Steri, M., Danjou, F., Kwong, A., et al.: Genome sequencing elucidates Sardinian genetic architecture and augments association analyses for lipid and blood inflammatory markers. Nat. Genet. **47**, 1272–1281 (2015)

55. Nadin, M.: Anticipation and dynamics: Rosen's anticipation in the perspective of time. In: Klir, G. (ed.) Special issue of International Journal of General Systems, vol. 39, no. 1, pp. 3–33. Taylor and Blackwell, London (2010)

56. Precision Medicine Cohort Initiative. https://www.nih.gov/precision-medicine-initiative-cohort-program

57. Bhattacharjee, A., Sokolsky, T., Wyman, S.K., Reese, M.G., Puffenberger, E., Strauss, K., Morton, H., Parad, R.B., Naylor, E.W.: Development of DNA confirmatory and high-risk diagnostic testing for newborns using targeted next-generation DNA sequencing. Genet. Med. **17**, 337–347 (2014)

58. Saunders, C.J., Miller, N.A., Soden, S.E., Dinwiddie, D.L., Noll, A., Alnadi, N.A., Andraws, N., Patterson, M.L., Krivohlavek, L.A., Fellis, J., Humphray, S., et al.: Rapid whole-genome sequencing for genetic disease diagnosis in neonatal intensive care units. Sci. Transl. Med. **4**, 154ra135 (2012)

59. Roychowdhury, S., Iyer, M.K., Robinson, D.R., Lonigro, R.J., Wu, Y.-M., Cao, X., Kalyana-Sundaram, S., Sam, L., Balbin, O.A., Quist, M.J., Barrette, T., et al.: Personalized oncology through integrative high-throughput sequencing: a pilot study. Sci. Transl. Med. **3**, 111ra121 (2011)
60. Boycott, K.M., Vanstone, M.R., Bulman, D.E., MacKenzie, A.E.: Rare-disease genetics in the era of next-generation sequencing: discovery to translation. Nat. Rev. Genet. **14**, 681–691 (2013)
61. Méndez, M., Custodio, A., Provencio, M.: New molecular targeted therapies for advanced non-small-cell lung cancer. J. Thorac. Dis. **3**, 30–56 (2011)
62. Kamali, F.: Genetic influences on the response to warfarin. Curr. Opin. Hematol. **13**, 357–361 (2006)
63. Tatarunas, V., Lesauskaite, V., Veikutiene, A., Grybauskas, P., Jakuska, P., Jankauskiene, L., Bartuseviciute, R., Benetis, R.: The effect of CYP2C9, VKORC1 and CYP4F2 polymorphism and of clinical factors on warfarin dosage during initiation and long-term treatment after heart valve surgery. J. Thromb. Thrombolysis **37**, 177–185 (2014)
64. Zhang, J., Tian, L., Zhang, Y., Shen, J.: The influence of VKORC1 gene polymorphism on warfarin maintenance dosage in pediatric patients: a systematic review and meta-analysis. Thromb. Res. (2015)
65. Collins, F.S., Varmus, H.: A New Initiative on Precision Medicine. N. Engl. J. Med. **372**, 793–795 (2015)
66. Nadin, M.: Anticipation—The End Is Where We Start From. Müller Verlag, Basel (2003)

Part III
Examining the Brain

Temporal Memory Traces as Anticipatory Mechanisms

Peter Cariani

Abstract Brains can be considered as goal-seeking correlation systems that use past experience to predict future events so as to guide appropriate behavior. Brains can also be considered as neural signal processing systems that utilize temporal codes, neural timing architectures operating on them, and time-domain, tape-recorder-like memory mechanisms that store and recall temporal spike patterns. If temporal memory traces can also be read out in faster-than-real-time, then these can serve as an advisory mechanism to guide prospective behavior by simulating the neural signals generated from time courses of past events, actions, and the respective hedonic consequences that previously occurred under similar circumstances. Short-term memory stores based on active regeneration of neuronal signals in networks of delay paths could subserve short-term temporal expectancies based on recent history. Polymer-based molecular mechanisms that map time-to-polymer chain position and vice versa could provide vehicles for storing and reading out permanent, long-term memory traces.

Keywords Neural timing nets · Neural codes · Pitch · Rhythm · Auditory scene analysis · Temporal codes · Expectancy · Music perception · Engram

1 Introduction

As the aphorism goes, "the purpose of remembering the past is to predict the future." Anticipation involves both predicting future situations and events and preparing for them. Anticipation not only projects *what* will occur but also *when* and *where* it will occur, as well as what to do about it. Anticipatory mechanisms enable organisms to use past experience to act in a manner appropriate for future conditions. This chapter proposes novel anticipatory neural memory mechanisms that are based on neural time codes and temporal pattern memory traces.

P. Cariani (✉)
Hearing Research Center, Boston University, Boston, USA
e-mail: cariani@bu.edu

© Springer International Publishing Switzerland 2017   105
M. Nadin (ed.), *Anticipation and Medicine*, DOI 10.1007/978-3-319-45142-8_7

2   Anticipatory Systems

"Anticipation" is a general notion. The theoretical biologist Robert Rosen coined the specific term "anticipatory system" [1–3] and defined it as a system in which prospective future states determine present behaviors [4]. The papers in this volume are the product of the last of three conferences inspired by Rosen's and Nadin's ideas about anticipatory systems [5, 6].

2.1   Purposive Systems as Functional Organizations

Anticipatory systems are those systems that have embedded goal-states: they are organized in such a way that their action realizes desired future states. The idea has much in common with naturalistically grounded teleologies ("teleonomies") of purposive systems. In the east, Russian physiologists and psychologists (Anokhin, Sudakov, Bernstein, et al.) developed general theories of functional systems [7–9]. In the west, notions of purposive, feedback control systems formed the basis of the early cybernetics movement [10–14].

Rosen and his mentor, Nicolas Rashevsky, were strong proponents of a non-reductionist, relational, theoretical biology that focused on questions of *organization* as explanations of functions, rather than appeal to mechanistic reductionism [15, 16]. Rosen offered the parable of an amoeba in a pot of water, before and after boiling. The live amoeba and the dead one share the same molecular constituents, but the organization of the system has been altered by boiling in such a way that the amoeba was no longer able to regenerate its parts and its organization. Rosen mounted deep criticisms of the machine and computer metaphors for describing living systems, not because the parts in some way violate the laws of physics, but because, in describing living systems solely in terms of trajectories of parts, one misses the organizational relations that make the system a persistent, coherent entity. Knowledge of parts is useful certainly for designing drugs, but it does alone not tell us how to go about building stable, regenerative organisms [17]. For that, as Rashevsky and Rosen foresaw, one needs a theory of mutually stabilizing relations.[1]

---

[1]Such biological system theories have deep implications for medicine. Much of our current understanding of disease in terms of "molecular medicine" is grounded in linear chains of interactions between molecular parts. Many therapies simplistically attempt to control one variable (e.g., blood sugar concentrations) using one or two interventions (insulin) without considering the circular-causal nature of networks of metabolic loops that can stymie such interventions. Only if we are able to model the whole set of systemic interactions and relations can we anticipate what the system will do in the short and long term. Once we have an adequate systems theory of biological organization, we will gain the deeper understanding needed for how to design therapeutic interventions that have self-sustaining effects such that the need for further interventions becomes self-limiting.

2.2   Anticipatory Systems in Animals with Nervous Systems

Although the most obvious examples of anticipation involve animal behaviors mediated by learning and memory, many non-neural examples of biological anticipatory capabilities abound. For example, in many plants and animals, developmental stages are orchestrated to occur at favorable seasons of the year (e.g., seed germination in early spring). The mechanisms for favorable timing of developmental stages arise in environments with strong seasonal variations. The environmental variations create positive selective pressures for anticipation, such that those lineages, whose individual time development enhances survival and reproduction, will tend to persist longer than those that don't. Because of the cyclical, predictable nature of seasonal changes, timing strategies that worked better in previous cycles will continue to work better in present and future cycles. Thus anticipatory timing mechanisms appropriate for coping with the future can evolve, provided that similar situations recur.

Animals are motile organisms that cannot produce their own food. In contrast to fungi, which absorb nutrients, animals ingest and digest their food. As a consequence, most animals must move to find food, such that the immediate environments within which they must orient and transport themselves are ever changing. Animal lineages evolved nervous systems that coordinate the actions of effector organs contingent on the sensed states of immediate surrounds and on current internal goal-related states. Embedded goal-related states include the needs of the organism for survival (e.g., satisfaction of system-goals of homeostasis, self-repair, growth), and reproduction. Those lineages of organisms that evolved more effective embedded goal mechanisms for survival and reproduction tended to persist. In choosing actions contingent on percepts and active goals, organisms in effect anticipate which actions will be most appropriate in satisfying those goals.

On evolutionary timescales, variation, construction, and selection processes yield organisms that are better adapted in their particular ecological contexts for more reliable (survival and) reproduction. During the lifespans of organisms with nervous systems, neural learning processes shape percept-action mappings contingent on past experience and reward. So even in the most primitive kind of adaptive percept-action systems, there is anticipation in the sense that the results of previous experiences and successful performances continually modify system structure and behavior to guide future action.

In both evolution and learning cases, memory mechanisms encode the past and make it available for anticipation of what actions can be most appropriate in the current state. In the evolutionary case, the memory lies (mostly) in the genetic sequences that, shaped by selective pressures and construction constraints, persisted in the lineage. In the learning case, the memory lies in short-term memory traces that guide behavior based on the immediate past and present, and in more permanent long-term memory traces that can guide behavior that is based on the deeper past.

2.3   Organization of Perception and Action

Animals with nervous systems can be characterized in terms of purposive, percept-action systems. They have sensory receptors that permit them to make distinctions on their surrounds; effector organs (mainly muscles) that permit them to influence their environs (action); and nervous systems that permit coordination of action contingent on behavior. Aside from sleep or other dormant states, there is a constant, ongoing cycle of percepts, coordinations, actions, and subsequent environmental changes (Fig. 1, bold arrows).

In parallel with percept-action cycles are internal cyclical neuronal dynamics that steer behavior from moment to moment and over the long run (Fig. 2). These include the neuronal dynamics of competing internal goals, internal modal system-states (e.g., waking/sleeping, affective states), cognitive and deliberative processes, attention, action-selection, and the influence of long-term memories.

Embedded in all nervous systems of animals are feedback-driven goal mechanisms that steer behavior in a manner that reliably satisfy basic organismic imperatives of survival and reproduction (e.g., maintain oxygen/water homeostasis, find food sources, avoid predators, find mates). Competitive dynamics of current goals determine which goals are paramount at any given moment such that their

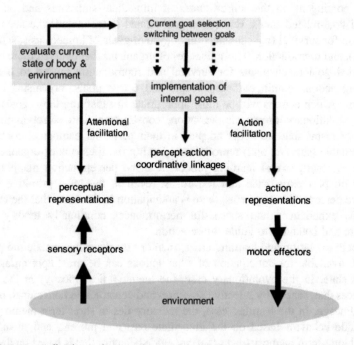

Fig. 1 Percept-coordination-action cycles and goal-directed steering of percept-action coordinations

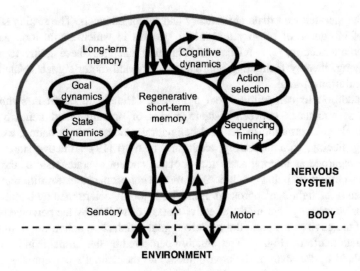

Fig. 2 Functional organization of informational dynamics in brains

drive states partially determine behavioral choices (action selections).[2] This is basically "dominance principle" of the Russian psychologist Ukhtomsky (1845–1942) [9]:

> ... in the nervous system, there is at each moment only one active dominating dynamic structure or constellation of excitation, which is associated with the most actual, urgent ongoing needs and desires. This excitation structure plays the role of a situational nervous network, an agency for organizing the physiological and behavioral response directed to satisfying these needs. At the same time all other goals and desires are suppressed" [18]. What is the difference between a bodily physiological mechanism in an animal and a technical mechanism? Firstly, the former is generated during the course of the reaction itself. Secondly, once chosen, the behavior of a technical mechanism is secured once and for all by its construction, whereas in a reflex apparatus it's possible to successively realize as many different mechanisms as there are available degrees of freedom in the system. Each of the successively realized mechanisms is achieved through the active inhibition of all available processes, except one. (Ukhtomsky, quoted in [18])

The notion of competitive goal dynamics fits extremely well into current conceptions of brains as sets of competing circuits that steer behavior, for better or worse, towards particular sets of goals ("implementation of internal goals" in Fig. 1). The notion of the inhibitory suppression of non-dominant goals fits very well with the emerging conception of the basal ganglia/striatum as a generalized double-inhibition braking system in which brakes are selectively released to

---

[2]Drive states only partially determine actions, because arguably, actions taken also depend upon which action-alternatives (affordances) are perceived by the organism to be immediately available. Those goals that have obvious, apparent means of attainment may be more attractive for action. Perception of options for effective action can feed back to change which drive states become dominant.

facilitate task-relevant thalamic sensory and motor channels. The resting state is a state of inhibition of sensory and motor systems in which neural loop gains are slightly attenuating. The release of inhibition changes these gains to slightly amplifying, thereby facilitating neuronal sensory and motor signals related to the current dominant goal.

In addition to competitive goal dynamics, virtually all animals switch their global system-states between discrete modes of operation and behavior, (e.g., waking-sleep-hibernation cycles and behavioral modes such as eating, excreting, hunting, fleeing, communicating, exploring, mating) [19]. Affective states, which can be regarded as internal assessments of the current overall "state of the organism" also modulate behavior choices by increasing propensities for different modes of action (e.g., fight/flee/approach). Behavior is jointly determined by internal states (goal-drive state, system-mode, affective states), as well as by the perceived current state of the environment and the perceived goal-satisfaction action-possibilities that the situation affords (Fig. 1). The situation perceived by the animal is in turn jointly determined by the state of the environment and the animal's perceptual systems.

2.4 General Types of Functional Organizations

In my view, such animals can be categorized in terms of the functionalities that their organizations afford. Animals are *living systems* because they actively regenerate their organization (material components and relations). This is the core idea underlying conceptions of self-production systems [20], autopoietic systems [21, 22], metabolism-repair systems [23, 24], self-reproducing robots [25], autocatalytic nets [26], self-modifying systems [27], and semantically-closed self-interpreting construction systems [28]. They are *semiotic systems* because their internal operation and resulting behavior can be described in terms of sign distinctions conveyed via neural codes [29, 30]; and they are *autonomous purposive systems* because they are mainly driven by internal goals.

They have their own *agency* to the extent that they have embedded goals (internal motivation), requisite ability (the right mechanisms needed for action), and sufficient freedom of action (autonomy) to reliably achieve particular goals. They are *anticipatory systems* if they have learning and memory mechanisms that allow them to project the past into the present so as to evaluate future consequences of current courses of action. If one defines these different attributes in terms of these specific kinds of material organizations, then a system, such as an autonomous robot, need not be living to be semiotic, autonomous, purposive, or anticipatory, or to exhibit agency. Underlying these different types of functional organization is the Aristotelian notion of *hylomorphism*.

2.5 Hylomorphism as an Ontology for Functional Organization

Hylomorphism is an ontology of functional organizations embedded in matter. Aristotle adopted the hylomorphic framework in formulating his theories of life and mind [31–33].

Life, purpose, meaning, and even conscious awareness are properties of material systems that are organized in particular ways. Explanations based on organization (the system is organized so as to realize a particular goal, i.e., it has a "final cause") are complementary to reductionist, causal explanations based solely on physical properties of parts. Purposive, goal-directed systems are material systems that are organized so as to realize particular goal end-states that in effect become their final causes.

An example of a simple purposive system is a thermostatically regulated heating system that is organized so as to maintain the temperature of a room within a particular range. The "final cause" of the system is the end-state target temperature range that is determined by the thermostat. (Because of its organization and material realization, the system seeks the corresponding temperature states, and, provided that the system is working properly, the thermostat setting "predicts" the final temperature state of the system.)

Hylomorphism is a functionalist ontology to the extent that functional organization can be abstracted from particular material substrates. One can design a thermostatically regulated heating system in terms of functions of and relations between components without specifying exactly how thermostat control mechanisms and heating/cooling elements are to be realized materially (and different material implementations can realize comparable behavioral functions). However, unlike platonic ontologies based entirely on ideal forms, a hylomorphic ontology is materially grounded. In order to realize functions within the material world, organizations must be realized in *some* material form. It is not enough to replicate form; the organization must be fleshed out, embodied, such that it interacts with and changes other parts of the material world.

3 Anticipation and Memory

Memory is a process that entails the maintenance of a distinction through time, and thus it is a semiotic process that is invariant with respect to time. Anticipatory prediction involves estimating the course of future events based on the (remembered) past and present (Fig. 3).

Nervous systems evolved to coordinate behavior. Coordination without memory is possible where mappings between percepts and actions do not change with experience. However, once these coordinative mappings can be modified on the basis of experience, then the effects of past experiences can carry over into present

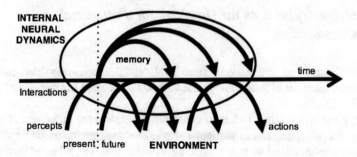

Fig. 3 Memory enables present distinctions to carry over to influence future actions

and future. This kind of simple adaptive modification of behavior does not require explicit storage and retrieval processes.

In biological organisms and nervous systems, anticipation involves not only *what* situations and events are expected, but also *when* and *where* they are expected to occur. In animals with nervous systems, anticipation involves (usually implicit) understanding of the contingent structure of the world (modeling) for deliberative purposes: to decide whether positive action needs to be taken, to determine what actions are available (perceived affordances), and if so, what action is most likely to satisfy those system-goals that are currently of highest priority (goal satisfaction).

3.1 Short- and Long-Term Memory

Standard theories of memory posit a labile, short-term memory coupled with a permanent long-term memory (Fig. 4). There is large literature, old and new, in psychology on the characteristics and nature of memory [34–36]. Many treatments further subdivide different types of memory by modality, the nature of the items stored, temporal processing windows, and temporal persistence, while others seek universal frameworks.

Short-term memory, broadly construed, provides a temporary store of neural signals related to current and recent perceptions, thoughts, affects, motivations, as well items maintained via working memory or recalled from long-term memory. Since the 1930s, the neural mechanisms that subserve short-term memory have been conceived explicitly in terms of neuronal reverberatory processes, i.e., neural activity patterns that are actively maintained and self-sustaining. The sustained firing of neurons that permits these activity patterns to persist is facilitated by activation of N-methyl-D-aspartate (NMDA) receptors that create the biophysical conditions for long-term potentiation (LTP) and spike-timing-dependent plasticity (STDP).

How the specific contents of these temporary memories are coded in neuronal activity patterns is the neural coding problem as it applies to memory. If the

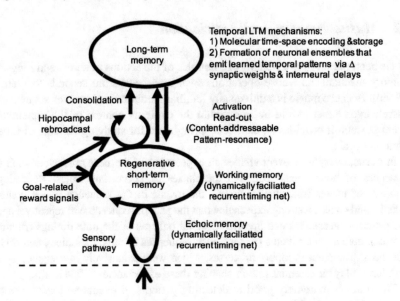

Fig. 4 Short-term regenerative memory and long-term memory. Neuronal signals related to current goal-states are maintained in regenerative short-term working memory and eventually consolidated into long-term memory traces. In this proposed scheme, the contents of short-term memory stores consist of complex temporal spike patterns that are actively regenerated in correlation-facilitated delay paths. Long-term temporal memory traces are activated by corresponding temporal patterns present in working memory, enabling pattern-resonance and content-addressability

information is encoded such that specific subsets of neurons have sustained, elevated firing rates, then the reverberating patterns likely involve persistent activation of these specific neuron subsets. If information is encoded in temporal spike patterns, as is proposed here, then the reverberating patterns likely involve persistent regeneration of spike patterns within neuronal circuits.

The hippocampal formation is the bridge between short-term and long-term memory stores. It appears to be responsible both for maintaining neural activity patterns in short-term memory and for replaying neuronal patterns associated with significant events such that they can be consolidated into long-term memory, mainly during sleep. This work originally came out of animal maze-running experiments, where "place cells," which encode distinct maze locations, were observed to fire in the temporal sequences of the maze running. Recently, "time cells," which encode the timings of events and places, have been found in the hippocampus and elsewhere [37]. During states of sleep or during periods of awake reflection, stored event-sequences can be replayed at faster-than-real-time, enabling them to function as predictive reward mechanisms [38].

Long-term memory is permanent: it can survive sleep, seizure, general anesthesia, and long periods of coma. Once formed, some types of long-term memories can last the entire lifetime of an individual.

3.2 Music, Memory, and Anticipation

Music perception offers a rich set of examples of the actions of short- and long-term memory mechanisms whose operations span timescales from seconds to lifetimes [39, 40]. Because music is a universally familiar medium that involves sequences of discrete events that unfold in time within the evolving context of the remembered recent present, it provides an excellent springboard for studying the role of time in mind and brain.

In music, every note-event creates an expectancy of the next note-event such that sequences of notes create pattern expectancies of succeeding notes. Repeating a sequence of events immediately groups the events in the sequence into a coherent "chunk" and creates a strong expectancy that the pattern sequence will repeat yet again. The pattern expectancies can involve temporal patterns of the note timings (rhythmic patterns), note accent patterns (meter), pitch sequences (motifs, melodies), and timbral sequences. Temporal grouping processes, which were extensively investigated by the Gestaltists, play an essential role in shaping these expectancies [39, 41, 42].

The brain is extremely good at detecting patterns of events and attributes that recur, and it can do this over many timescales.[3] In music, periodic sound patterns whose waveforms rapidly repeat (25–4000 Hz) produce musical pitches (that can carry a melody), whereas slower periodic patterns of sonic events (10 Hz or less), be they musical notes, clangs, or speech fragments, create rhythmic expectancies. Every repeating sound creates a strong expectancy of its continuation, and violations of this expectancy are highly noticeable. The repeating pattern is grouped into a unitary whole (a "chunk"). If the pattern is a fast-repeating acoustic waveform with a repetition rate greater than about 25 Hz, we perceive a pitch at the repetition rate. The tonal quality (timbre) of the sustained sound is determined by the form of the repeated waveform (or in Fourier terms, its spectral distribution and shape).

In music, repeating temporal patterns of onsets and offsets of discrete events with rates of less than roughly 10 per second are perceived as rhythms. Different temporal patterns are heard as different rhythms. Here the repeating events are grouped together into a chunk ("groove"), and after a few repetitions, expectations that each of the events will recur at a given time in the sequence and with the same attributes rapidly develop. This implies that the representation of the rhythmic pattern contains not only information about the sequence of events and their respective attributes (e.g., pitch, timbre, loudness, duration, location), but also a timeline of events. The memory trace of a rhythmic pattern essentially replicates the timeline of the events (the timings of event onsets, mainly, but also offsets that co-encode event-durations).

---

[3]Oliver Selfridge once related to me his experience with early computers with mechanical relays. The programmers would set up the computer and would stay in the room periods of time while the computations ran. After a while they would begin to notice long and elaborate repeating irregular sound patterns that were being produced by the relays as the program executed long, iterated complex instruction loops.

What is the nature of this memory trace? Although it has been conventionally assumed that temporal relations are converted to spatial patterns of neuronal activation, there have been a number of suggestions in the past that the memory traces themselves might be temporal patterns. In his book on the psychology of time, Fraisse [43] references the "cyclochronism" theory of the Russian psychologist Popov. Longuet-Higgins had proposed various nonlocal, holographic memory schemes based on frequency-domain Fourier decompositions (holophone model, [44]). Although these have the merit of using the oscillatory dynamics of neural populations, they had very limited storage capacity. Much more promising was a later time-domain mechanism based on temporal correlations between spikes [45]. Roy John proposed that evoked, induced, and triggered temporal patterns of response could subserve general mechanisms for memory [34, 46, 47].

We propose that temporal patterns of spikes might provide a direct basis for temporal memory trace mechanisms. In essence, temporally patterned stimuli impress their temporal structure on the time structure of neural spikes, such that the temporal patterns of spikes can serve as an iconic time-domain representation of the stimulus.

A great deal of evidence in the auditory system points towards temporal codes for pitch and rhythm, albeit at different levels of the system. The most obvious neural correlates of musical pitch lie in spike timing patterns— distributions of interspike intervals—in the auditory nerve, brainstem, and midbrain [48, 49]. Those for rhythm can be found at all levels of the system, from auditory nerve to auditory cortex and beyond. The onsets and offsets of every note produce well-timed spikes in large numbers of cortical neurons [50], such that temporal patterns of note-events replicate the temporal structures of their rhythms [51, 52]. These evoked neuronal temporal responses provide the necessary inputs for various oscillatory mechanisms of rhythmic and metrical expectancy that have been proposed [53]. (Possible direct time codes for rhythm and their relationship to oscillatory dynamics are discussed a bit more below in Sect. 5.2.2.)

3.3   Mismatch Negativity and Short-Term Expectancies

Short-term musical memory is very sensitive to the timing of events; and there are neural responses that are widely observed in event potentials that appear to be related to short-term temporal expectancies and their violations. Event potentials are averaged electrical or magnetic signals triggered (aligned in time) by the onset of a particular event, such as a note-event or even the change in a note-event pattern.

A so-called "mismatch negativity response" (MMN) is observed when an event is presented repetitively ("standard") and then some change is made in a subsequent event ("deviant"). Such MMN-like responses are widely used to study the dynamics of musical expectancies and their violations [50, 54, 55].

If the event is a musical note, any perceptible change in the physical attributes of the note, such as a change in sound level (loudness), periodicity (pitch), spectrum (timbre), attack (timbre), duration, or location (apparent location), will produce a neural response that differs from the response to the standard event. Here the context of the standard event pattern has created an expectancy that is then violated by the deviant event. The mismatch negativity is computed by subtracting the time-pattern of the averaged response to deviant events (triggered on their onsets) from that of the standards. Depending on the nature of the change in events, the mismatch negativity peaks at a characteristic latency after the beginning of the deviation from the standard, i.e., at the onset of the deviant stimulus. This latency is typically on the order of 100-200 ms for changes in basic auditory perceptual attributes.

The MMN is evoked more or less automatically. It does not require subjects to attend to the stimuli, and indeed can be observed in sleeping subjects, infants, and many animals. MMN responses to a standard metrical musical pattern followed by the same (syncopated) pattern in which one of the beats is omitted are observed in newborn infants. MMN-like responses are observed (sometimes reported under a different name) in many different species, sensory modalities, and brain regions.

MMN-like responses can also be seen for "higher level" patterns and attributes. Complex rhythmic and/or melodic patterns of note-events can be presented as standards, and a pattern that deviates in some respect (e.g., a change in the periodicity (pitch) of one of the notes) will create an MMN closely following the time point of the deviation. MMN-like responses with longer latencies are observed for syntactic and semantic violations and yet other kinds of more abstract attributes.

MMN responses are sensitive not only to changes in the perceptual and cognitive attributes of discrete events, but also to their timings. If a regular metrical sequence is set up as the expected standard pattern, then deviations in the timing of subsequent events (leads or lags relative to the expected timing of the beat) will evoke MMN responses, again with a latency that depends on the timing of the expectancy-violating mismatch. In music, these expectancy violations form the basis for expressive timing, intentional manipulation of the timings of notes to convey and evoke emotions.

MMN phenomena suggest the existence of canonical neural temporal comparison mechanisms. It is as if a timeline of events is being built up, maintained in short-term memory, and compared with incoming temporal patterns. In this chapter, I propose a complex neural time code in which both the attributes of events and their relative timings are encoded and simple neuronal delay-and-compare mechanisms (recurrent timing nets) that would produce similar kinds of behaviors. What is needed is a mechanism that both builds up an expectancy of patterns of events when they recur, and computes the deviation of the incoming stream of new events from what was expected at that particular time point.

There is an ongoing debate about the nature and meaning of the MMN [56]. Some current theories of the MMN hold that a memory trace is formed when the standard is repeated and that there is a comparison of incoming neural activity patterns against this memory trace that was constructed from very recent experience. Others hold that the memory trace itself may be embedded in the responses of ensembles of cortical neurons. Still others reject the notion of organized memory traces in favor of explaining the MMN in terms of neuronal adaptation processes.

It is notable that early studies of electrical conditioning in single neurons, conducted before MMN was discovered, found that cortical neurons assimilated rhythmic patterns (10 Hz flashes of light) that were presented in conjunction with correlated electrical pulses similar in many respects to a reward signal [34, 57, 58]. The stimuli were presented over and over, paired with electrical pulses; and over tens of trials the temporal response pattern of the neurons came to resemble that of the stimulus (10 Hz firing pattern). When the stimulus was then abruptly changed to 1 Hz, the slower flashes evoked the 10 Hz pattern for many repetitions, but eventually the assimilated rhythm was extinguished.

3.4 Temporal Theories of Associative Memory

Predictive timing is a key element of anticipatory behavior. It is often important to know when a reward will come. The relative timing of rewards and the events that lead up to them has been an ongoing concern of theories of learning.

It has also been observed that spike timings of dopaminergic neurons reflect discrepancies between anticipated and observed courses of the neural concomitants of events associated with rewards [59, 60]. This discovery has spawned a host of adaptive temporal prediction models.

Many studies in animal and human conditioning suggest that the timings of all correlated events relative to the arrival of a reward are implicitly and intrinsically stored in both short and long term memory, such that any of them can serve as anticipatory temporal predictors [61]. This temporal coding of memory hypothesis thus proposes that "the temporal conditions (e.g. the CS-US interstimulus interval) are not merely catalysts in the formation of associations, but are also a part of the content of learning" [62]. The hypothesis further asserts that animals can build temporal maps from relationships between events that were never physically paired, "that is, temporal information from different training situations which have a common element can be integrated based on super-positioning of the common element in different temporal maps" [63]. This means that systematic maps of temporal relations between events can be built up from separate experiences of subsets of events. Such maps of temporal relations can then subserve anticipatory prediction—each event becomes a predictor for other temporally correlated events [64].

4 A Temporal Theory of Brain Function

Brains implement anticipatory predictions that subsequently guide behavior. In this paper we propose a high-level theory of brain function based on temporal pulse pattern codes that can be actively regenerated, stored, and retrieved. In this theory, prediction and steering of behavior are achieved by encoding and retrieving temporal patterns of spikes associated with internal events connected to percepts, actions, rewards, and punishments. Although it shares many assumptions with mainstream connectionist theory, (e.g., recurrent connections, distributed coding and processing), this proposed theory differs from connectionism in its fundamental neural coding assumptions. Whereas connectionism is based entirely on channel-activation codes (which neurons fire at which average rates), the alternative neural architectures envisioned here rely on temporal codes, i.e., neural codes that encode distinctions in temporal patterns of spikes.

A general theory of brain function requires specification of several basic aspects.

1. *Neural codes.* What are the system's signals? A neural coding scheme based on spike timing patterns must be capable of representing all the distinctions that we make (e.g., encoding all of the attributes of objects, events, and their relations and their compositions).
2. *Neural networks.* What processing architectures are needed to realize the informational signal processing operations that the system performs? A neural processing architecture capable of operating on temporal patterns in order to carry out informational operations—such as detections, discriminations, pattern recognitions, invariances, transformations, and groupings—is required.
3. *Memory mechanisms.* How are informational distinctions encoded and decoded in memory? Here mechanisms of short- and long-term memory that can store and recall temporally coded temporal patterns in a content-addressable manner are needed.

These different aspects of the system need to be compatible with each other. The nature of the neural codes that bear informational distinctions heavily determines the nature of neural signal processing architectures and the memory mechanisms needed to utilize them. Conversely, the available neural mechanisms for processing, storing, and retrieving information heavily constrain what kinds of codes the system can use.

5 A Universal Coding Framework Based on Complex Temporal Spike Patterns

5.1 The Neural Coding Problem

Understanding the nature of the neural code (the "neural coding problem") in central circuits is arguably the most fundamental problem facing neuroscience today. Without an understanding of the precise nature of the "signals of the system," we cannot have a firm grasp of the nature of information processing and storage in brains. Neuroscience today is in a situation comparable to cellular biology and genetics before DNA nucleotide sequences came to be understood as the primary vehicle for inheriting and expressing genetic information. However, neural coding is rarely explicitly mentioned as an unsolved problem in neuroscience. Although interest in neural coding has undergone cyclical changes from decade to decade, it has never yet risen to the forefront of neuroscience.

A common tacit assumption in mainstream neuroscience is that the coding problem has already been solved, that the brain is a large, complex connectionist network. The instantaneous functional states of the system are thought to be patterns of average firing rates across neurons, and its structural informational states are thought to be characterizable in terms of interneural connection weights. The instantaneous functional state (mental state), taken together with the structural state (the "connectome"), is thought to determine how the system behaves. Note that these assumptions only hold to the extent that spike timing and temporal relations between neurons (intra- and inter-neuronal time delays) have no significant role in informational functions. Solving the neural coding problem is critical for interpreting the functional significance of specific neuronal connection patterns.

Neural codes, as discussed here, are the functional signals of the system, those neuronal activity patterns that have functional significance for information processing in the brain. Different kinds of informational distinctions (e.g., perceptual attributes, thoughts, desires, affective states) and their specific alternatives (e.g., for the visual attribute of color, the distinctions of red vs. blue vs. green vs. yellow) are mediated through specific patterns of neural spiking activity. These specific types of patterns and different patterns within a type constitute the neural codes [65–69].[4]

This notion of coding is related to the conception of a sign in semiotics as a distinction that has functional significance for its user. A sign, in Gregory Bateson's phrase, is a "difference that make a difference." A neural code is a pattern of activity that makes a functional difference in the brain, i.e. a difference that alters internal functional states and subsequent behavior. To go further, some spike patterns may be meaningful, having consequences for internal states or overt behaviors, whereas others that are not interpretable within a coding scheme may not constitute a coherent internal message, and so would not be meaningful to the system. Examples of the latter might include uncorrelated "spontaneous" spike patterns, spikes generated during epileptic seizures, and incoherent firing patterns produced under general anesthesia.

5.2   Types of Neural Codes: Channel Codes and Temporal Codes

Neural pulse codes can be divided into two types: channel codes and temporal codes. In channel codes, activation of particular subsets of neurons (channels)

---

[4]The functional definition of neural coding is different from formal, information-theoretic Shannonian estimates of channel capacities that are independent of whether or not the system makes use of the different states.

conveys distinctions (e.g., through across-neuron firing rate profiles), whereas in temporal codes particular temporal patterns of spikes convey distinctions. Temporal codes can be further divided into those codes that depend on temporal patterns of spikes irrespective of their times-of-arrival (spike latency) and those codes that depend on the relative latencies of spikes. Temporal pattern codes based on inter-spike intervals appear to subserve auditory pitch and cutaneous flutter-vibration sensations, whereas relative time-of-arrival codes appear to subserve various stimulus localization mechanisms based on temporal cross-correlation in audition, somatoception, and electroception.

Connectionist theory adopts the assumption that the central codes operant in brains are firing-rate channel codes, whereas the neural timing net theory outlined here posits that these are complex temporal pattern codes. Thus, for most of the history of modern neuroscience, channel coding has been adopted as the conventional, default assumption and with it, connectionism as the default neural network assumption. However, a significant minority opinion has involved proposal of various kinds of temporal codes as alternatives [65]. Early examples include Rutherford's "telephone theory of neural coding," Troland's temporal modulation theory of hearing, Wever's temporal volley theory, the Jeffress' model for binaural localization, and Licklider's duplex model for pitch perception. There has likewise always been an alternative tradition for temporal processing architectures as well [70–74]. We have discussed many of these various neural coding schemes elsewhere [69, 75, 76].

5.2.1 Types of Temporal Codes

In simple temporal codes (Fig. 5a), one temporal parameter conveys one perceptual distinction. For example, in the early auditory system, times between spikes (interspike intervals) carry information about the periodicities of sounds. At these early auditory stations, the neural code for pitch involves the mass statistics of interspike intervals amongst whole populations of neurons [49, 69, 79]. The pitch that is heard corresponds to the most common interspike intervals that are produced by the population. The sensation of flutter-vibration has a similar basis in simple interspike interval patterns. Binaural auditory localization in the horizontal plane utilizes sub-millisecond spike timing differences produced by corresponding neurons in neural pathways that originate in the two ears. Analogous examples exist in nearly every sense modality [76].

Complex temporal pattern codes can be formed from combinations of simple temporal-pattern primitives (Fig. 5a–c). Here different orthogonal *types* of temporal patterns encode different independent primitive features. For example, for a musical note-event, the different dimensions of pitch, loudness, duration, location, and tonal qualities (timbral distinctions) would be conveyed via different types of temporal patterns of spikes that were produced concurrently within some population of auditory neurons (Fig. 5c). Combinations of specific temporal patterns present at any given time form a feature-vector whose dimensions are determined by the pattern types.

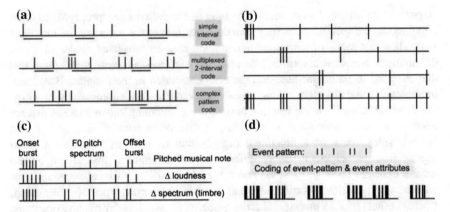

Fig. 5 Temporal codes. Idealized spike trains illustrate different coding schemes. **a** Simple and complex temporal pattern codes: *top* interspike interval code (e.g., for pitch or flutter-vibration); *middle* multiplexing of two different interspike intervals related to different types of information; *bottom* complex pulse pattern code. **b** Multiplexing of three complex temporal patterns associated with different types of cutaneous sensation (after [76]); *bottom* spike train shows interleaving of the different signals. **c** Hypothetical spike latency pattern scheme for encoding different attributes of auditory events (pitch, loudness, timbre, duration). **d** Hypothetical universal scheme for encoding event-patterns that includes event attributes and the relative timings of events

Complex temporal codes can potentially be conveyed in single units or in the pattern-statistics of spike time patterns in ensembles and populations. The various patterns can be interleaved or embedded in other patterns of spikes (Fig. 5b), permitting signal multiplexing (concurrent transmission of multiple types of signals over the same neuronal transmission lines). A complex temporal coding scheme that was proposed in the past for multiple cutaneous sensory qualities [78] provides a concrete example of how such codes might be organized.[5]

A neural code must convey two types of information: the type of distinction the signal conveys (e.g., pitch or color), and the attribute distinction itself (e.g., a particular pitch or color). In channel codes, the identity of the channel (which neuron, as determined by its place in the network, its interconnectivities) conveys informational type, whereas patterns of channel activations convey different attribute values. For channel-coding schemes, neural channel-identities maintained via specific interconnections are absolutely critical for function. A neural firing rate is meaningless to the rest of the system without the identity of the neuron that is firing. Connectionist networks are completely dependent on the maintenance of highly regulated connectivities ("synaptic weights").

In contrast, because complex temporal codes embed the type of the signal in the form of the message, highly specific interneural connectivities and signal transmission paths are no longer essential for function. Temporal codes thus permit

---

[5]As far as we know, Emmers' findings have not been replicated experimentally by others, so the neuronal reality of this powerful, multidimensional coding scheme is unproven.

"signals to be liberated from wires." As long as the patterns can propagate to other populations, the specific paths they take does not change the nature of the messages sent. Thus this mode of neural communication enables broadcast modes of signal distribution. Neuronal assemblies downstream can be selectively tuned such that they respond to particular temporal patterns embedded in their inputs. Broadcast, multiplexing, and selective tuning enable decentralized communications in which tuned neural assemblies can respond only to those incoming signal patterns that are relevant to their functional roles. Here a "neural assembly" is used in the Hebb-Lashley sense of a functional organization of a set of neurons—a given neuron may participate in many different neural assemblies that are organized around different tasks [80]. Some codes may be restricted to local neuronal populations (e.g., restricted cortical regions) that handle specific types of information, whereas others may be propagated more globally, re-broadcast by the hippocampus, and then consolidated into long term memory.

Finally, complex temporal codes can represent patterns of events that occur over different time scales. If a sequence of musical notes is played, such as a motif or a melody, a temporal code that can encode all of the attributes associated with the individual events can also encode the timings of the events (Fig. 5e). Thus rhythmic pattern and musical meter can be encoded on coarser timescales and the same coding framework can handle different levels of musical organization.

5.2.2   Evidence for Temporal Codes

Temporal codes are found in a very wide range of sensory systems [66–69, 75–79]. Temporal codes in sensory systems have been found with sub-microsecond spike timing precisions (electroception), microsecond precisions (bat and cetacean echolocation), sub-millisecond precisions (auditory, somatoception, vision), and still coarser (olfaction, gustation).

Generally speaking, auditory systems tend to have the highest frequencies of the synchronization of spikes with stimuli. In the auditory systems of barn owls, who use spike timing to localize their prey in the dark, spikes fire in time with the fine structure of sounds up to periodicities of 10 kHz. In humans and cats, primary auditory neurons phase lock up to roughly 5 kHz. Next, spike synchronizations to electrical shocks delivered to the skin approach rates up to about 1 kHz. Visual neurons lock to modulations in luminance up to roughly 100 Hz.

Temporal codes are found not only in sensory systems in which spikes follow the fine time structure of the stimulus, such as hearing and touch, but also in sensory systems such as vision, where eye movement transforms spatial luminance patterns into correlated temporal patterns at the retina. Even in the chemical senses of smell and taste, and in color vision, differences in the temporal response properties of sensory receptors produce corresponding characteristic temporal patterns of spiking.

In general, the most obvious temporal codes are found near sensory receptor surfaces. In early stations of the auditory system, spike timing is most abundant and its functional role most obvious. However, as one proceeds up ascending sensory pathways, stimulus-related timing information is mixed with other kinds of information and in some cases smeared out such that stimulus-related fine timing information above roughly 100 Hz becomes successively less apparent as one proceeds to the cortical level [75]. In the auditory system, despite considerable progress, cortical representations for basic auditory attributes such as pitch and loudness are still poorly understood [81].

In lieu of strong coding hypotheses, it is difficult either to confirm or entirely rule out prospective candidate codes at the cortical level. Thus the complex temporal pattern-coding scheme outlined here is a very tentative hypothesis. If complex temporal codes do exist in central circuits that subserve the representation of all simple sensory attributes associated with events (e.g., the timing, duration, loudness, pitch, and timbre of a single musical note), they may involve timing patterns that are difficult to observe because they are distributed across neurons and/or not rigidly synchronized either with the stimulus or each other.

Although temporal patterns associated with fine temporal structure above a few hundred Hz are not abundant at the cortical level, temporal patterns of neuronal response associated with the slower successions of onsets and offsets of discrete sensory and motor events (periodicities <10 Hz) are very prominent. The precision of the coding of these onsets, on the order of 100–200 μs, is maintained all the way up the auditory pathway. Temporal patterns related to patterns of musical events, flashing lights, electric shocks, and tactile pulses are widely observed in evoked electrical and magnetic activity over large parts of the cerebral cortex [34, 46].

Musical rhythm is thus a prime candidate for temporal coding. Patterns of event onsets are seen widely in averaged electrical and magnetic auditory evoked potentials and also more recently even in single trial stimulus presentations. At present, it is possible to determine which of two auditory streams a listener is attending on the basis of the corresponding rhythmic pattern of neuronal response in an EEG or MEG signal. Even beats that are expected, but not presented or heard, as in a syncopated meter, produce observable responses at the times when the acoustically sounded beat was expected to arrive [82–84]. Actively imagining a rhythmic pattern facilitates the pattern in neuronal cortical populations such that it can be observed in EEG recordings.

Rhythms in speech, though less regular and not as well defined as their counterparts in music, likewise produce corresponding temporal patterns that reflect acoustic contrasts [52, 85]. Recently rhythmic patterns of neuronal response associated with different levels of sentential organization (e.g., syllables, words, phrases, sentences) have been observed at the cortical level [86].

All of these phenomena argue for direct temporal coding of rhythm in music and speech at all levels of auditory processing. Coarse rhythmic patterns (<10 Hz) are supra-modal, with neural temporal responses that are very widely distributed across cortical regions [46]. The ubiquity of these correlated neural response patterns may explain how musical rhythm can provide a cross-modal temporal scaffold for movement and memory [87].

In the last decade, there has been a renewed interest in brain rhythms, and many of these phenomena have been interpreted in terms of oscillatory dynamics of neuronal populations [86, 87–89], rather than under the rubric of temporal codes [49, 51, 90]. It is important to distinguish between evoked, exogenous rhythms that reflect driving stimulus periodicities both from endogenous rhythms produced by intrinsic dynamics of neuronal excitation and recovery and from induced rhythms that are triggered or released by external stimulus events. The focus is usually on entrainment and induced rhythms rather than stimulus-locked synchronization or stimulus-related periodicities. Synchronization is often regarded as a passive process, in contrast to "active modification of ongoing brain activity" in entrainment [51]. Note, however, that for every observed oscillation, there is an associated temporal organization of spiking activity. Recent evidence points to substantial functional roles for both evoked and induced rhythms in the grouping and analysis of speech and musical events. Oscillatory dynamics of cortical neurons may govern temporal processing windows for music and speech [89, 91, 92] in a manner that limits the rate at which events and their various attributes can be accurately represented. In terms of neural coding, the durations of these windows may place constraints on how fast complex spike latency codes (such as the code of Fig. 5c), which require different readout times for different attributes, can be produced and processed.

6 Neural Architectures for Temporal Processing

What kinds of neural architectures would be needed to utilize a temporal coding framework such as the one outlined above?

6.1 Basic Plan of the Brain

The basic structural plan of animal brains [93] is well-conserved phylogenetically. Despite its apparent neuroanatomical complexity, brains consist of a relatively small number of component subsystems and neuro-computational architectures. First and foremost, as neuroscientists have understood for more than a century now, the brain is a network of recurrent pathways. These have variously been called loops, neural circuits, re-entrant paths, and nets with circles.

The brain can thus be regarded as a network of neuronal circuits, i.e., large numbers of interconnected neuronal loops that contain excitatory and inhibitory neurons with local connections, and excitatory (and sometimes inhibitory) neurons with longer range projections. Many different canonical neural circuits have this recurrent organization (e.g., thalamocortical loops, cortico-cortico re-entrant pathways, and hippocampal loops). The systematic sets of recurrent pathways in the hippocampus have been often regarded as computational substrates for an auto-associative memory mechanism. Different regions of the cerebral cortex are reciprocally connected to neighboring ones by local connections and to more distant cortical regions by white-matter long-range axonal tracts. However, closed cyclical chains of excitatory neurons without inhibition create positive feedback loops that quickly saturate. By regulating the amount of inhibition, the loop-gains of these circuits can be modulated from their resting, slightly attenuating state, in which incoming neuronal activity patterns die out, to states of attention.

For the most part, neuroscience has been more focused on the neuroanatomy of connections, i.e., "the connectome," than on temporal relations within and between the various neuronal populations. However, for timing theories of brain function, each connection between neural elements has not only a synaptic weight, but also time delays associated with axonal and synaptic transmission. Each of these various loops have characteristic time delays associated with them: short-delays for local circuits and longer ones amongst more distant neuronal ensembles. The more numerous unmyelinated axons, with their slow conduction velocities and long conduction times, yield much longer transmission delays than myelinated axons. In addition, there are delay processes that are inherent in the recovery dynamics of the neuronal elements that can perform many of the same neural signal processing functions as transmission delays.

6.2 Neural Time-Delay Architectures

Purely connectionist networks do not represent time explicitly, except as sequences of changing spike rates. Time-delay networks, on the other hand, include time delays between elements that allow them to interconvert temporal patterns and spatial activation patterns. Early time-delay neuro-computational architectures were proposed for that utilized binaural time disparities for localizing sounds and monaural interspike intervals for perceiving periodicity pitches (Jeffress and Licklider models, [52, 59]).

Time-delay networks use coincidence-detector elements with short time integration windows for handling temporal patterns of spikes, and rate-integrator elements with long time integration windows for converting spike coincidences to average firing rates. This "coincidence counting" allows them to interface with connectionist architectures. In effect, temporal codes are converted to rate-channel codes. Time-delay architectures can be flexible in their ability to handle both temporal and spatial information: by tuning delays, one can change synaptic efficacies, and vice-versa.

6.3 Neural Timing Network Architectures

After many years of searching for alternatives to both connectionist rate-place and time-delay neural network schemes, I proposed yet a third kind of neural network, which I called *neural timing nets*. For the most part, early work on these networks involved demonstrating the various time-domain operations that could be elegantly carried out [94–96].

Temporal pattern codes allow simple form transformations (position shift invariance), and time warping of patterns yields tempo-invariance of rhythms, transposition invariance of pitch sequences, and magnification invariance of spatial forms. Using these kinds of temporal representations, separation of independently moving or changing forms can be effected. One can easily separate objects on the basis of invariant relational patterns of elements within objects (fusion, grouping) vs. the changing relations between elements of different, independently moving objects (separation).

Neural timing networks consist of arrays of delays and coincidence elements that operate on temporally-coded inputs and produce temporally coded outputs. Essentially, everything is kept in the time domain, and neural signals can interact with each other to sort out common temporal subpatterns. Whereas connectionist and most time-delay architectures are "connection-centric" (all informational function depends on particular synaptic connection weightings), neural timing networks are "signal-centric" (action lies in interactions between signals: "signal dynamics").

The signal-centric nature of the networks (and networks processing based on signal dynamics) sidesteps many of the problems of connectionist and time-delay network architectures, in that precise and elaborate point-to-point connections are not needed for such networks to function. It is enough to bring the various neural signals into the same regions at approximately the same time. By operating on the temporal pattern statistics of ensembles of neurons, as long as there are some points of interaction, it no longer matters whether this or that neuron produced this or that output.

Feedforward timing nets (FFTNs) are arrays of coincidence detectors and delay lines that cause temporally patterned signals to interact. Various correlation and convolutional operations can be carried out, enabling multiplexing and demulitplexing of signal primitives. In FFTNs the spike train signals collide, interact, interfere, and/or mutually amplify each other, essentially performing correlation-like filtering signal processing operations in the time domain.

Compared to connectionist networks, the temporally coded representations and signal processing operations are more iconic and analog in character and more parallel in implementation. Template matching can be realized by injecting a temporal pattern archetype into the network, which will serve to amplify any incoming temporal pattern signals that have significant correlations with it. Content-addressable search can likewise be realized by injecting temporal patterns related to the features that one is interested in. Other neural signals circulating within the network will interact with the search signal if and only if they have feature-related temporal subpatterns in common. Essentially complex temporal pattern signals can implement a vectorial representation in which the signals themselves can sort out those dimensions that they have in common. The informational operations involve "pattern resonances" [97]. The processing scheme as it currently stands is provisional and still in a rudimentary state of development.

Nonetheless, it appears to be much more flexible than any connectionist scheme we have seen to date.

7 Temporal Mechanisms for Short-Term Regenerative Memory

Most high level accounts of brain function posit networks of recurrent pathways (re-entrant loops, neural circuits) that support a dynamic, working short-term memory coupled to a more permanent long-term memory mechanism that permits storage and retrieval of relevant patterns (Fig. 2). Neuronal activity patterns that have hedonic salience for the animal (i.e., are part of a string of events that leads to significant reward or punishment) are rebroadcast by the hippocampal formation such that the patterns are maintained in working memory and later consolidated and fixed in long-term memory. This rebroadcast can replicate event sequences at faster-than-real-time rates.

Conventionally, the nature of short-term memory is commonly assumed to involve subsets of specific neurons in recurrent neural circuits that maintain higher rates of activity, whereas long-term memory is thought to involve changes in the effective connectivities between neurons at synapses (in neural network terms, "synaptic weights"). Thus short-term memory is conceived in terms of a complex reverberation pattern of neuronal activations, while long-term memory is thought to entail more permanent synaptic changes.

In addition to feedforward timing nets, there also can be recurrent timing nets (RTNs), in which there are arrays of delay loops that span a wide range of recurrence times. RTNs were initially conceived as models for pitch- and rhythm-based grouping and separation mechanisms [94–96]. For both pitch and rhythm, repeating waveforms and temporal event patterns respectively create strong temporal expectancies and groupings. Our auditory systems easily separate concurrent sounds with different fundamental frequencies (F0 s), such that we are able to hear out different musical instruments and voices when they are mixed together. RTNs, which act in a manner similar to adaptive comb filters, separate out the respective temporal patterns of multiple speakers with different voice pitches and of multiple musical instruments playing different notes.

Recurrent timing nets are perhaps the simplest kind of reverberating, temporal memory that can be imagined (Fig. 6a). Here an incoming pattern is compared with a delayed circulating pattern; and if the recurrence time of the delay loop is equal to the repetition time of a repeating pattern, then the pattern is facilitated (builds up) in that particular loop. If the difference between the circulating and the incoming pattern is also computed, then the difference signal can be fed into the array. Each delay loop creates an expectation of what the next incoming signal fragment will be. (The expectation is a primitive anticipation in that the recent past is used to predict the near future.) In the case of a repeating pattern of auditory events, each

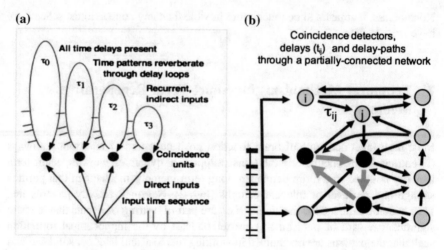

(a)

All time delays present

τ_0

Time patterns reverberate through delay loops

τ_1

Recurrent, indirect inputs

τ_2

τ_3

Coincidence units

Direct inputs

Input time sequence

(b)

Coincidence detectors, delays (t_{ij}) and delay-paths through a partially-connected network

τ_{ij}

Fig. 6 Recurrent timing nets (RTNs). **a** An array of delay loops and correlation-facilitated coincidence detectors. A periodic pulse pattern fed into the network maximally facilitates itself in the delay loop, whose recurrence time is equal to the duration of the repeated pulse pattern, not unlike a time-domain implementation of a comb filter. Such a network stores temporal patterns in the delay loops. Multiple periodic patterns sort themselves out in the different delay loops, which then function as complex pulse pattern-oscillators. **b** Alternative implementation of a neural timing net in a richly interconnected network of coincidence-facilitating elements. A pulse pattern is fed into the leftmost, input layer and delay paths corresponding to the internal delay structure of the stimulus are facilitated (*gray arrows, black coincidence elements*). The reverberating patterns are regenerated within the network (synfire cycles)

delay loop is making a temporal prediction about when the next event will occur. And if the event-attributes (timing, loudness, pitch, timbre, duration, location) are also encoded in time via a complex, multidimensional time code, then it creates expectancies for those also.

The RTN delays can be either monosynaptic (recurrent collaterals) or polysynaptic (delay paths through networks, Fig. 6b). The number of recurrent monosynaptic 2-element paths in a fully connected network of N elements is on the order of N^2. The number of paths increases combinatorially with maximum path-length M, roughly as M^N. Although the brain is thought to be more like a small world rather than a fully interconnected network [98], all neurons are thought to be interconnected by at most three or four interneurons. The number of delay paths is still astronomical, far greater than even the combinatorics of individual synapses (because each synapse connecting two neurons has a delay associated with it).

If the synapses are spike-timing dependent (inhibited/facilitated by recent spike correlation history), then the repeating pattern will flow through those delay-paths that have recurrence times equal to the pattern repetition time. Other paths with other delays that are not in the pattern will be temporarily inhibited. In this manner, to whatever degree there is repetition, the recurring time structure of the rhythm input will build up in the network. If there are only locally repeating patterns, say ABCABCCABBCDEFAB, which has a maximum repeated pattern length of 2 (AB), then the network will revert to the probabilities of shorter sequences. This becomes a neural implementation of a variable-order Markov chain that can adapt to variable N-gram lengths. The system thus predicts specific, longer sequences when those have been presented in the not-too-distant past, but in lieu of repeating sequences reverts to Bayesian statistics.

Timing-dependent synapses support competition between signals and winner-take-all dynamics (facilitation of one set of signals inhibits others), as well as the possibility of regenerative, self-facilitating "synfire cycles." These differ from synfire chains [73, 99, 100] in that the temporal pattern statistics, rather than which neurons are firing, encode particular attribute distinctions. The regenerative cycles in effect would constitute a temporal echoic and working memory buffer that would hold the temporal patterns (maintaining the statistics of the patterns) such that they could be compared with incoming ones.

8 Temporal Mechanisms for Long-Term Memory

Lastly, a general theory of brain function needs to account for a second memory mechanism by which permanent, stable long-term memories are laid down and their contents retrieved. The nature of the storage mechanisms operant in brains, Lashley's "engram" [80], is a fundamental problem that is intimately related to the neural coding problem. The competing theories of memory parallel those of neural coding and neural architectures [34, 101].

As Lorente observed, "permanent circulation of impulses in neural chains" cannot be the basis for long-term memory because these memory traces survive the cessation of neuronal activity that occurs under anesthesia, deep shock, and hibernation [102]. Like short-term memory, long-term memory is content-addressable. Neuronal activity associated with any attribute can be used to activate long-term memory traces that encode that attribute. Two alternative types of temporal long-term memory mechanisms can be envisioned that could store and retrieve events encoded via complex temporal spike patterns.

8.1 Formation of Time-Delay Networks for Recognizing and Producing Temporal Patterns

The first possibility tunes up time-delay networks to produce the complex temporal patterns that are to be stored. First the temporal patterns are repetitively presented to neuronal populations such that synapses with offsetting time delays that produce

spike-timing correlations within the local network are strengthened via
spike-timing-dependent plasticity (STDP). The result would be that the repetitive
complex patterns presented to the network would be assimilated in a manner similar
to what was observed under the electrical conditioning experiments discussed
earlier [58]. Subsequent activation of such a network might reproduce the consolidated delay pattern, thereby recreating the stored complex temporal pattern and
injecting it into the rest of the network such that it can interact with signals that are
currently circulating in regenerative short-term memory.

This explanation has the merit of relying on the same kinds of synaptic changes
as connectionist theory. There is a great deal of accumulated biophysical evidence
of semi-permanent changes in synaptic efficacy with use, and this process could
involve those mechanisms. In effect, the neuronal assemblies become time-delay
networks that are configured from past experience. These can become activated by
incoming temporal patterns to facilitate channel activations (connectionist account)
and/or to emit temporal patterns (timing net account).

8.2   Polymer-Based Time-Space Molecular Memory
       for Storing Temporal Patterns

A second possibility is that of a molecular memory that stores time patterns.
Molecular memory mechanisms have been discussed for some time [103], and
many are inspired by the power of the genetic nucleotide sequence code.
A temporal molecular memory is attractive because it does not depend on particular, highly specific synaptic connections. Theories of RNA-based molecular
memory ("memory RNA") that were inspired by planaria memory transfer experiments were popular in the early 1960s, but due to failures to clearly replicate the
basic transfer phenomena, this entire field was defunded at the NIH for a generation. In recent years, research on possible molecular memory mechanisms has been
revived [64].

Complex temporal patterns lend themselves to instructional "tape recorder"
memory mechanisms [33, 101, 104] that preserve temporal relations between
events. If the attributes of the events are also temporally coded (e.g., sensory
features of a particular place in a maze), then such temporal memory traces can
serve as universal memory mechanisms whose form need not be radically transformed in the storage and retrieval process. I have previously proposed a mechanism similar to the scheme in Fig. 7 [105].

One potential molecular mechanism is that time patterns of intracellular ionic
fluxes could be converted to spatial patterns of markers on polymers [106]. We

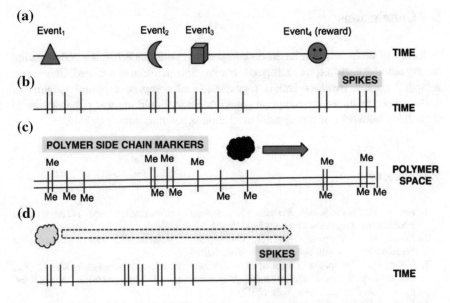

Fig. 7 Hypothetical scheme for long-term molecular storage of temporal patterns. **a** A sequence of temporally correlated events (1–3) that lead to a reward event (4). **b** Internal spike patterns produced by the internal events (e.g., encoding event-feature attributes). **c** Polymer (*double lines*) consisting of a backbone chain plus side-chains that can be chemically modified (e.g. methylated). A polymerase moves along the polymer backbone, adding a side-chain marker whenever the neuron is depolarized (or ionic concentrations change). The polymerase moves at a constant speed down the chain, thereby converting temporal pulse patterns to spatial patterns of markers on the polymer. **d** A second type of polymerase that moves down the chain at a faster rate, triggering ionic fluxes and generation of action potentials when a marked side chain is encountered. Such a mechanism would enable faster-than-real time readout that could subserve anticipatory steering of action

know that there are polymerases that move down the length of polymers, presumably at a constant average speed, and that there are also mechanisms for reversibly labeling the side chains of these polymers (e.g., methylation). Thus it is conceivable that temporal patterns of ionic fluxes (e.g., related to local calcium concentrations) could be laid down along the length of a polymer by a writing enzyme (there are much higher concentrations of DNA methylase in the nervous system than in other tissues). A second reading polymerase could scan the polymer at the same (or even faster) rate than the writing enzyme, thereby reading out the pattern. If all neural information related to relative event timings, attributes, and resulting reward or punishment is stored, as conditioning studies suggest, then a faster-than-real-time readout mechanism becomes a relatively simple means of predicting the future hedonic consequences (reward or punishment) of a present situation or course of action.

9 Conclusions

A theory of brain function based on complex temporal pattern spike codes, neural timing net architectures, and temporal memory mechanisms is outlined. Short-term temporal pattern memory entails regeneration of complex temporal patterns of spikes, whereas long-term temporal pattern memory could involve either tuning of time-delay networks or a polymer-based time-space molecular mechanism.

References

1. Rosen, R.: Anticipatory Systems: Philosophical, Mathematical, and Methodological Foundations. Pergamon Press, Oxford, New York (1985)
2. Rosen, R.: Anticipatory Systems: Philosophical, Mathematical, and Methodological Foundations, 2nd edn. Springer, New York (2012)
3. Nadin, M.: What speaks in favor of an inquiry into anticipatory processes? In: Klir, G. (ed.) Anticipatory Systems: Philosophical, Mathematical, and Methodological Foundations, pp. xv–lvii. Springer, New York (2012)
4. Louie, A.H.: Robert Rosen's anticipatory systems. Foresight **12**, 18–29 (2010)
5. Nadin, M. (ed.): Anticipation—Learning from the Past: The Russian/Soviet Contributions to the Science of Anticipation, vol. 25. Springer, Cham, CH (2015)
6. Nadin, M. (ed.): Anticipation Across Disciplines, vol. 29. Springer, Cham CH (2016)
7. Tsagareli, M.G.: I.S. Beritashvili and psychoneural integration of behavior. In: Nadin, M. (ed.) Anticipation: Learning from the Past: The Russian/Soviet Contributions to the Science of Anticipation, vol. 25, pp. 395–414. Springer, New York (2015)
8. Vityaev, E.E.: Purposefulness as a principle of brain activity. In: Nadin, M. (ed.) Anticipation: Learning from the Past: Russian/Soviet Contributions to the Science of Anticipation, vol. 25, pp. 231–254. Springer, New York (2015)
9. Zueva, E.Y., Zuev, K.B.: The concept of dominance by A.A. Ukhtomsky and Anticipation. In: Nadin, M. (ed.) Anticipation: Learning from the Past: The Russian/Soviet Contributions to the Science of Anticipation, vol. 25, pp. 13–35. Springer, New York (2015)
10. Rosenblueth, A., Wiener, N., Bigelow, J.: Behavior, purpose and teleology. Philos. Sci. **10**, 18–24 (1943)
11. Rosenblueth, A., Wiener, N.: Purposeful and non-purposeful behavior. Philos. Sci. **17**, 318–326 (1950)
12. de Latil, P.: Thinking by Machine. Houghton Mifflin, Boston (1956)
13. Ackoff, R.L., Emery, F.E.: On Purposeful Systems. Aldine-Atherton, Chicago (1972)
14. George, F.H., Johnson, L.: Purposive Behaviour and Teleological Explanations. Gordon and Breach Science Publishers, New York (1984)
15. Rashevsky, N.: Mathematical Biophysics: Physico-Mathematical Foundations of Biology, vols. I & II. Dover, New York (1960)
16. Rosen, R.: Biological systems as organizational paradigms. Int. J. Gen. Syst. **1**, 165–174 (1974)
17. Rosen, R.: Life Itself. Columbia University Press, New York (1991)
18. Kazansky, A.B.: Agental anticipation in the central nervous system. In: Nadin, M. (ed.) Anticipation: Learning from the Past: The Russian/Soviet Contributions to the Science of Anticipation, vol. 25, pp. 108–117. Springer, New York (2015)
19. Kilmer, W., McCulloch, W.S.: The reticular formation command and control system. In: Leibovic, K.N. (ed.) Information Processing in the Nervous System, pp. 297–307. Springer, New York (1969)

20. Mingers, J.: Self-Producing Systems. Plenum Press, New York (1995)
21. Maturana, H., Varela, F.: Autopoiesis: the organization of the living. In: Maturana, H., Varela, F. (eds.) Autopoiesis and Cognition (1980), vol. 42. D. Reidel, Dordrecht, Holland (1973)
22. Maturana, H.R.: Autopoiesis. In: Zeleny, M. (ed.) Autopoiesis: A Theory of the Living. North Holland, New York (1981)
23. Rosen, R.: Some realizations of (M, R) systems and their interpretation. J. Math. Biophys. **33**, 303–319 (1971)
24. Rosen, R.: What does it take to make an organism? In: Rosen, R. (ed.) Essays on Life Itself, pp. 254–269. Columbia University Press, New York (2000)
25. von Neumann, J.: The general and logical theory of automata. In: Jeffress, L.A. (ed.) Cerebral Mechanisms of Behavior (the Hixon Symposium), pp. 1–41. Wiley, New York (1951)
26. Kauffman, S.: The Origins of Order. Oxford University Press, New York (1993)
27. Kampis, G.: Self-Modifying Systems in Biology and Cognitive Science. Pergamon Press, Oxford (1991)
28. Pattee, H.H., Raczaszek-Leonardi, J.: Laws, language and life Howard Pattee's classic papers on the physics of symbols with contemporary commentary by Howard Pattee and Joanna Raczaszek-Leonardi. Biosemiotics vol. 7. Springer, Dordrecht, New York (2012)
29. Cariani, P.: The semiotics of cybernetic percept-action systems. Int. J. Signs Semiotic Syst. **1**, 1–17 (2011)
30. Cariani, P.: Sign functions in natural and artificial systems. In: Trifonas, P.P. (ed.) International Handbook of Semiotics, pp. 917–950. Springer, Dordrecht (2015)
31. Graham, D.W.: Aristotle's Two Systems. Oxford University Press, New York (1987)
32. Modrak, D.K.: Aristotle: The Power of Perception. University of Chicago, Chicago (1987)
33. Favareau, D.: The evolutionary history of biosemiotics. In: Barbieri, M. (ed.) Introduction to Biosemiotics, pp. 1–67. Springer, Dordrecht (2008)
34. John, E.R.: Mechanisms of Memory. Wiley, New York (1967)
35. Milner, B., Squire, L.R., Kandel, E.R.: Cognitive neuroscience and the study of memory. Neuron **20**, 445–468 (1998)
36. Eichenbaum, H.: The Cognitive Neuroscience of Memory: An Introduction. Oxford University Press, New York (2012)
37. Eichenbaum, H.: Memory on time. Trends in Cognitive Sciences **17**, 81–88 (2013)
38. Atherton, L.A., Dupret, D., Mellor, J.R.: Memory trace replay: the shaping of memory consolidation by neuromodulation. Trends Neurosci. **38**, 560–570 (2015)
39. Snyder, B.: Music and Memory. MIT Press, Cambridge (2000)
40. Snyder, B.: Memory for music. In: Hallam, S., Cross, I., Thaut, M. (eds.) The Oxford Handbook of Music Psychology, pp. 107–117. Oxford University Press, Oxford, New York (2009)
41. Handel, S.: Listening: An Introduction to the Perception of Auditory Events. MIT Press, Cambridge, MA (1989)
42. Bregman, A.S.: Auditory Scene Analysis, The Perceptual Organization of Sound. MIT Press, Cambridge, MA (1990)
43. Fraisse, P.: The Psychology of Time. Harper & Row, New York (1978)
44. Longuet-Higgins, H.C.: Mental Processes: Studies in Cognitive Science. The MIT Press, Cambridge, MA (1987)
45. Longuet-Higgins, H.C.: A mechanism for the storage of temporal correlations. In: Durbin, R., Miall, C., Mitchison, G. (eds.) The Computing Neuron, pp. 99–104. Addison-Wesley, Wokingham, England (1989)
46. Thatcher, R.W., John, E.R.: Functional Neuroscience, Vol. I. Foundations of Cognitive Processes. Lawrence Erlbaum, Hillsdale, NJ (1977)
47. John, E.R., Bartlett, F., Shimokochi, M., Kleinman, D.: Neural readout from memory. J. Neurophysiol. **36**, 893–924 (1973)

48. Cariani, P.A., Delgutte, B.: Neural correlates of the pitch of complex tones. II. Pitch shift, pitch ambiguity, phase invariance, pitch circularity, rate pitch, and the dominance region for pitch. J. Neurophysiol. **76**, 1717–1734 (1996)
49. Cariani, P.: Temporal coding of periodicity pitch in the auditory system: an overview. Neural Plasticity **6**, 147–172 (1999)
50. Poeppel, D., Hickok, G.: Electromagnetic recording of the auditory system. Handb. Clin. Neurol. **129**, 245–255 (2015)
51. Will, U., Makeig, S.: EEG research methodology and brain entrainment. In: Berger, J., Turow, G. (eds.) Music, Science, and the Rhythmic Brain: Cultural and Clinical Implications, pp. xiv, 215 p. Routledge, New York (2011)
52. Arnal, L.H., Poeppel, D., Giraud, A.L.: Temporal coding in the auditory cortex. Handb. Clin. Neurol. **129**, 85–98 (2015)
53. Snyder, J.S., Large, E.W.: Gamma-band activity reflects the metric structure of rhythmic tone sequences. Brain Res. Cogn. Brain Res. **24**, 117–126 (2005)
54. Trainor, L.J., Zatorre, R.: The neurobiological basis of musical expectations. In: Hallam, S., Cross, I., Thaut, M. (eds.) The Oxford Handbook of Music Psychology, pp. 171–183. Oxford University Press, Oxford, New York (2009)
55. Koelsch, S.: Brain and Music. Wiley-Blackwell, Chichester (2012)
56. Malmierca, M.S., Sanchez-Vives, M.V., Escera, C., Bendixen, A.: Neuronal adaptation, novelty detection and regularity encoding in audition. Front Syst. Neurosci. **8**, 111 (2014)
57. John, E.R.: Electrophysiological studies of conditioning. In: Quarton, G.C., Melnechuk, T., Schmitt, F.O. (eds.) The Neurosciences: A Study Program, pp. 690–704. Rockefeller University Press, New York (1967)
58. Morrell, F.: Electrical signs of sensory coding. In: Quarton, G.C., Melnechuck, T., Schmitt, F.O. (eds.) The Neurosciences: A Study Program, pp. 452–469. Rockefeller University Press, New York (1967)
59. Schultz, W., Dickinson, A.: Neuronal coding of prediction errors. Annu. Rev. Neurosci. **23**, 473–500 (2000)
60. Schultz, S.R., Panzeri, S.: Temporal correlations and neural spike train entropy. Phys. Rev. Lett. **86**, 5823–5826 (2001)
61. Miller, R.R., Barnet, R.C.: The role of time in elementary associations. Curr. Dir. Psychol. Sci. **2**, 106–111 (1993)
62. Savastano, H.I., Miller, R.R.: Time as content in Pavlovian conditioning. Behav Processes. **44**, 147–192 (1998)
63. Arcediano, F., Miller, R.R.: Some constraints for models of timing: a temporal coding hypothesis perspective. Learn. Motiv. **33**, 105–123 (2002)
64. Tucci, V., Buhusi, C.V., Gallistel, R., Meck, W.H.: Towards an integrated understanding of the biology of timing. Philos. Trans. R. Soc. Lond. B Biol. Sci. **369**, 20120470 (2014)
65. Boring, E.G.: Sensation and Perception in the History of Experimental Psychology. Appleton-Century-Crofts, New York (1942)
66. Uttal, W.R. (ed.): Sensory Coding: Selected Readings. Little-Brown, Boston (1972)
67. Uttal, W.R.: The Psychobiology of Sensory Coding. Harper and Row, New York (1973)
68. Rieke, F., Warland, D., de Ruyter, R., Steveninck, V., Bialek, W.: Spikes: Exploring the Neural Code. MIT Press, Cambridge, MA (1997)
69. Cariani, P.: As if time really mattered: temporal strategies for neural coding of sensory information. Reprinted in: K Pribram, ed. Origins: Brain and Self-Organization, Hillsdale, NJ: Lawrence Erlbaum, 1994; 208–252. 12, 161–229 (Reprinted in: K Pribram, ed. Origins: Brain and Self-Organization, Hillsdale, NJ: Lawrence Erlbaum, 1994; 1208–1252) (1995)
70. MacKay, D.M.: Self-organization in the time domain. In: Yovitts, M.C., Jacobi, G.T., Goldstein, G.D. (eds.) Self-Organizing Systems 1962, pp. 37–48. Spartan Books, Washington, D.C. (1962)
71. Pratt, G.: Pulse computation. Department of Electrical Engineering and Computer Science, vol. Ph.D., pp. 214 leaves. Massachusetts Institute of Technology, Cambridge, MA (1989)

72. Braitenberg, V.: The neuroanatomy of time. In: Miller, R. (ed.) Time and the Brain, pp. 391–396. Harwood Academic Ppublishers, Australia (2000)
73. Abeles, M.: Synfire chains. In: Arbib, M.A. (ed.) The Handbook of Brain Theory and Neural Networks (2nd Ed.), pp. 1143–1146. MIT Press, Cambridge, MA (2003)
74. Izhikevich, E.M.: Polychronization: computation with spikes. Neural Comput. **18**, 245–282 (2006)
75. Cariani, P.: Temporal coding of periodicity pitch in the auditory system: an overview. Neural Plast. **6**, 147–172 (1999)
76. Cariani, P.: Temporal coding of sensory information in the brain. Acoust. Sci. Tech. **22**, 77–84 (2001)
77. Perkell, D.H., Bullock, T. H.: Neural coding. Neurosciences Research Program Bulletin. **6**,3 221–348 (1968)
78. Emmers, R.: Pain: A Spike-Interval Coded Message in the Brain. Raven Press, New York (1981)
79. Cariani, P.A., Delgutte, B.: Neural correlates of the pitch of complex tones. I. Pitch and pitch salience. II. Pitch shift, pitch ambiguity, phase-invariance, pitch circularity, and the dominance region for pitch. J. Neurophysiol. **76**, 1698–1734 (1996)
80. Orbach, J.: The Neuropsychological Theories of Lashley and Hebb. University Press of America, Lanham (1998)
81. Cariani, P., Micheyl, C.: Towards a theory of infomation processing in the auditory cortex. In: Poeppel, D., Overath, T., Popper, A. (eds.) Human Auditory Cortex: Springer Handbook of Auditory Research. Springer, New York (2012)
82. Zanto, T.P., Snyder, J.S., Large, E.W.: Neural correlates of rhythmic expectancy. Adv. Cogn. Psychol. **2**, 221–231 (2006)
83. Fujioka, T., Trainor, L.J., Large, E.W., Ross, B.: Internalized timing of isochronous sounds is represented in neuromagnetic beta oscillations. J. Neurosci. **32**, 1791–1802 (2012)
84. Nozaradan, S.: Exploring how musical rhythm entrains brain activity with electroencephalogram frequency-tagging. Philos. Trans. R. Soc. Lond. B Biol. Sci. **369** (2014)
85. Hickok, G., Poeppel, D.: Neural basis of speech perception. Handb. Clin. Neurol. **129**, 149–160 (2015)
86. Ding, N., Melloni, L., Zhang, H., Tian, X., Poeppel, D.: Cortical tracking of hierarchical linguistic structures in connected speech. Nat. Neurosci. (2015)
87. Thaut, M.: Rhythm, music, and the brain: scientific foundations and clinical applications. Routledge, New York (2005)
88. Large, E.W., Snyder, J.S.: Pulse and meter as neural resonance. Ann. N. Y. Acad. Sci. **1169**, 46–57 (2009)
89. Doelling, K.B., Poeppel, D.: Cortical entrainment to music and its modulation by expertise. Proc. Natl. Acad. Sci. USA **112**, E6233–E6242 (2015)
90. Cariani, P., Micheyl, C.: Towards a theory of information processing in the auditory cortex. In: Poeppel, D., Overath, T., Popper, A. (eds.) Human Auditory Cortex: Springer Handbook of Auditory Research, pp. 351–390. Springer, New York (2012)
91. Doelling, K.B., Arnal, L.H., Ghitza, O., Poeppel, D.: Acoustic landmarks drive delta-theta oscillations to enable speech comprehension by facilitating perceptual parsing. Neuroimage **85**(Pt 2), 761–768 (2014)
92. Farbood, M.M., Rowland, J., Marcus, G., Ghitza, O., Poeppel, D.: Decoding time for the identification of musical key. Atten. Percept. Psychophys. **77**, 28–35 (2015)
93. Swanson, L.W.: Brain Architecture: Understanding the Basic Plan. Oxford University Press, New York (2012)
94. Cariani, P.: Neural timing nets. Neural Netw. **14**, 737–753 (2001)
95. Cariani, P.: Temporal codes, timing nets, and music perception. J. New Music Res. **30**, 107–136 (2002)
96. Cariani, P.A.: Temporal codes and computations for sensory representation and scene analysis. IEEE Trans Neural Netw./Publication IEEE Neural Netw. Counc. **15**, 1100–1111 (2004)

97. Cariani, P.: Outline of a cybernetic theory of brain function based on neural timing nets. Kybernetes **44**, 1219–1232 (2015)
98. Sporns, O.: Networks of the brain. MIT Press, Cambridge, Mass (2011)
99. Abeles, M.: Local Cortical Circuits. An Electrophysiological Study. Springer, Berlin (1982)
100. Abeles, M.: Corticonics. Cambridge University Press, Cambridge (1990)
101. John, E.R.: Switchboard vs. statistical theories of learning and memory. Science **177**, 850–864 (1972)
102. Lorente de No, R.: Circulation of impulses and memory. In: Schmitt, F.O. (ed.) Macromolecular Specificity and Biological Memory, pp. 89–90. MIT Press, Cambridge, MA (1962)
103. Schmitt, F.O.: Biologically structured microfields and stochastic memory models. In: Schmitt, F.O. (ed.) Macromolecular Specificity and Biological Memory, pp. 7–17. MIT Press, Cambridge, MA (1962)
104. John, E.R.: Studies of memory. In: Schmitt, F.O. (ed.) Macromolecular Specificity and Biological Memory, pp. 80–85. MIT Press, Cambridge, MA (1962)
105. Cariani, P.: Symbols and dynamics in the brain. Biosystems **60**, 59–83 (2001)
106. Landry, C.D., Kandel, E.R., Rajasethupathy, P.: New mechanisms in memory storage: piRNAs and epigenetics. Trends Neurosci. **36**, 535–542 (2013)

Influence of the Cerebellum in Anticipation and Mental Disorders

Pascal Hilber

Abstract The cerebellum is involved in motor coordination and motor learning. Cerebellar plasticity can serve as a cellular basis of learning. Data obtained from humans and animals led to the supposition that this structure could be a comparator and play a crucial role in sensory anticipation and online sensorimotor control. Internal models aid in executing motion precisely and harmoniously with or without external sensory feedback. The analogy between control of body part motion and manipulation of mental representation suggests the cerebellum's possible involvement in non-motor mental functions. Moreover cerebellar-lesioned or mutant animals exhibit cognitive and emotional disturbances, especially high reactivity to environmental changes, behavioral disinhibition, and stereotyped behavior. All these results and theoretical approaches support the idea that the cerebellum and its role in anticipation could represent an interesting field of investigation in the pathophysiology and the treatment of neuropsychiatric disorders such as autism.

Keywords Anticipation · Cerebellum · Internal models · Psychiatric disorders

1 Introduction

In order to adopt adequate behaviors in their living environment, individuals must learn about it and anticipate the consequences of future motions and actions on their current state. Mental disorders, such as autism, are often associated with inadequate behaviors which can be regarded as a loss or lack of adaptive potential and the inability to anticipate environmental disturbances (see Brisson[1]). Studies in antic-

---

[1]Brisson, J., Sorin, A-L.: Anticipation and Child Development. In: Nadin, M.: (ed.) Anticipation and Medicine, pp. 188–199. Springer, Cham (2016).

P. Hilber (✉)
Laboratory of Psychology and Neurosciences of Cognition and Affectivity,
Rouen Normandy University, Mont Saint Aignan, France
e-mail: pascal.hilber@univ-rouen.fr

© Springer International Publishing Switzerland 2017
M. Nadin (ed.), *Anticipation and Medicine*, DOI 10.1007/978-3-319-45142-8_8

ipatory behaviors and their disorders bring together various areas of psychology, medicine, and neurosciences [1]. The mental disorders are sustained by neural activity disturbances and brain structure alteration. Among these structures, the hippocampus, the basal ganglia, and the cerebellum seem to be of more particular interest because of their complementary crucial roles in adaptive learning, memory, and anticipation [2]. Taking both Nadin's ideas [3] and Doya's proposition [4] into consideration, one can conclude that the basal ganglia and the reward system would be more involved in the expectation and the cerebellum in the internal model [5], prediction [6], and sensory anticipation [7]. The present article aims to highlight why this often neglected structure could be a good candidate for studying anticipation system alterations in mental disorders.

2   Cerebellar Functions, Pathologies, and Symptoms

In the French *Encyclopédie* (1751–1777), Albrecht von Haller wrote, "What is the specific function of the cerebellum? We do not know, just as we do not know the functions of many parts of the cerebrum. However, its function must be important since it is present in many species." In 1897, AndréThomas [8] (cited in [9]) held that "the cerebellum records peripheral excitations and central sensations and *reacts* to them […] it's an equilibrium reflex center." In 1824, Jean-Pierre Flourens [10] confirmed the cerebellum's motor role. Working on many animal species (e.g., frog, adder, cat, mole, duck, pigeon), he demonstrated that progressive and methodical ablation of the cerebellar layers provokes locomotor impairment but not intellectual alteration. Since that time, the cerebellum has long been considered as a structure involved only in motor control; its involvement in cognitive processes (including motor cognition) was greatly neglected. (For a historical review, please see [11].)

Three centuries later, Schmahmann and Caplan concluded, "the traditional teaching that the cerebellum is purely a motor control device no longer appears valid, if, indeed, it ever was" [12]. In fact, psychiatric disturbances were consistently observed in several forms of spinocerebellar ataxias [13–19], dementia, psychosis, depression, verbal and visual memory dysfunctions, and low intellectual performance—as well as deficiencies in mind processes. Cognitive and affective alterations have also been reported in numerous congenital and acquired lesions of the cerebellum [20, 21], such as disturbances in visuo-spatial organization, language, figural memory, speed of information processing and sequencing functions, and planning of daily activities. Lesions of the cerebellum can also disturb two types of motor learning: eye-blink conditioning [22] and acquisition of postural adjustment (see more on this below).

Structural and functional abnormalities were also clearly observed in subjects with mental disturbances [23], as revealed in cases of major depressive [24] and bipolar disorders, schizophrenia [25], attention-deficit hyperactivity [26], obsessive-compulsive disorders [27], developmental dyslexia [28, 29], and autism [30].

Schmahmann and Sherman [31] described the Cerebellar Cognitive Affective Syndrome (CCAS), which comprises executive function impairments (concerning attention shifting, abstract reasoning, and working memory), disruption of planning, visuo-spatial and spatial memory deficits, language problems, and affective changes characterized by disinhibition and maladaptive behaviors. The processing disruption leading to motor and/or non-motor dysmetria was named Universal Cerebellar Impairment (UCI) [12, 31–33].

3 Motor Cognition, Cerebellar Plasticity and Anticipation

Sensory motor learning and timing of muscle activation are classically impaired in cerebellar patients [32]. The cerebellum is a crucial actor in the acquisition of complex motor pattern during walking and equilibrium learned tasks [34]. Locomotor learning abilities were widely investigated in cerebellar rodents and the results are consensual: equilibrium motor learning is not completely abolished but strongly affected in cerebellar-lesioned animals. Furthermore, destruction of the inferior olive, the only source of the cerebellar climbing fibers, also leads to deficits (for a review see [35]).

Focused degeneration of the cerebellar cortex in mutant mice provokes strong alterations of complex sensorimotor learning capabilities. For instance, motor coordination and motor learning were thoroughly investigated in Lurcher mice (+/Lc) with early and complete loss of PC and olivocerebellar pathway [36–39]. When daily trained on the rotarod (a horizontal rotating pole on which rodents must learn to remain by adopting walking strategies), these mice remained able to improve their performance [38, 40]. The sensorimotor learning capabilities were also partially preserved in +/Lc in motor tests other than the rotarod: elevated unstable board, coat hanger, static beam, and rotating grid tests [39, 41–43]. Nevertheless, the data obtained clearly indicated a strong delay in a) improving their equilibrium performances (i.e., the latency before falling from the apparatus) and b) adapting behavior to the task (that is, to use a synchronous pace on the mast in the rotarod test) [40]. Lastly, the lack of cerebellar cortex in +/Lc not only altered their learning rate, but also impaired their maximal performance in comparison with control mice.

In consideration of all these results, we suggested that the proactive body adjustments (also named anticipatory body adjustments) elaborated during training solicited a high transcerebellar loop, implicating the cerebellar-thalamic-cortical pathway [40, 44], which was henceforth described as an activated network after motor training on treadmill running in rats [45].

Motor learning is associated with cerebellar cognitive processes [6, 46]. The involvement of the cerebellum in motor cognition (i.e., learning processes) is closely related to its role as a motor regulator and anticipatory system supporting internal models of the motor system [47]. In these models the cerebellum generates possible states regarding the forthcoming sensory (through a feed-forward model) and motor (through an inverse dynamic model) state of the organism [6, 47–52].

These models are then applied through an internal feedback process in order to correct the instructions to be delivered to the peripheral effectors. In the Marr-Albus model, the parallel fiber (PF)–Purkinje cell (PC) synapse strength is modulated through teaching signals provided by climbing fibers. This modulation serves as a basis for learning processes [53]. The functional plasticity of the cerebellum supports its role in motor learning. Long-Term Depression (LTD) is a modulation of the synaptic signal after training (a reduction response of the Purkinje cells), which is the biological support for motor learning. Indeed, it was demonstrated, first in rabbits and in many other species thereafter, that the PF-induced PC activity was diminished when PF and CF discharges occurred simultaneously [54].

From a pragmatic perspective, "motor learning" can be viewed as a cognitive function corresponding to the acquisition and enhancement of anticipatory capabilities during training sessions. In other words, it corresponds to the anticipation of potential learned errors resulting from an action. This entails acquiring efficient dynamic internal models of the motor system. Thus, in a supervised learning model and due to cellular plasticity, the cerebellum also learns and stocks the "error signal" in order to anticipate [55, 56].

4   Spatial Cognition, Cerebellum and Anticipation

The Morris Water Maze Test is the classical test used to investigate spatially mediated behaviors in rodents. Briefly: mice or rats are daily trained to find an escape platform in a circular pool filled of water. Available visual information shows that the animals create a cognitive map, and the time needed to find and to reach the platform decreases with training. These cognitive processes, highly hippocampal-dependent, were also investigated in many cerebellar animals consequent to the visuo-spatial disabilities observed in cerebellar patients. Spatial learning deficits were reported in the above-mentioned Lurcher mutants [57]; and total cerebellectomy induced alteration of spatial memory in mice [58]. Moreover, destruction of the olivocerebellar pathway with 3-acetylpyridine [59], or synchronous and rhythmical activation of the olivary neurons by harmaline administration [60], elicits visuo-motor, spatial learning, and spatial memory deficiencies in rats. Since activation and lesion of the olivocerebellar pathway have similar effects, the conclusion is that normal functioning of this pathway is required for achieving spatial learning. All the data obtained illustrates that the cerebellar cortex and its afferents are of global importance for performance of the spatial task.

The MWM was first used to show the hippocampal implication in spatial learning and memory [61, 62]. The cerebellar spatial impairments differ markedly from those observed in animals with hippocampal or cortical lesion [63–65]. In a complementary and elegant experimental approach with the star maze, Passot and his colleagues proposed a neurocomputational model of the cerebellar contribution to the spatial event, including internal models and anticipation [66]. The cerebellum would not directly contribute to the elaboration of cognitive maps, but would be essential as a "predictor" in selecting the appropriate motor procedure to explore the environment and to reach the anticipated position of the goal. From an integrative viewpoint, Gross et al. [67] proposed that the spatial functions of the cerebellum, whatever their precise nature, lie in its contribution to neuronal circuits, involving structures such as the fronto-parietal association cortex, the limbic system, the superior colliculus, or the basal ganglia. Neurocomputational models of sensori-motor perception were also proposed to explain the complementary roles of these multiple brain areas in solving a spatial task [67].

5   Emotional Reactivity, Cerebellum, and Anticipation

Anxiety is an emotion based on both anticipation and rumination. One can easily imagine that the role of the cerebellum is universal: in emotion, as in motor or spatial processes, it permits, first, learning about the environment that the animals live in, and, second, anticipating any changes in such environment in order to produce a shortened time-adapted response. Again, data obtained in Lurcher mutant mice revealed that cortico-cerebellar alteration provokes great disturbance: cerebellar mice are more stressed and less behaviorally inhibited than non-mutant mice when confronted with a stressful environment or an aversive stimulus [68, 69]. Further animal studies showed that the cerebellum, in particular the cerebellar cortex, was involved in fear conditioning [70–72]. Parvizi and his colleagues [73] considered that in human beings the cerebellum plays a modulatory role in some emotional behaviors, and that such action is the result of learning by pairing context with emotional response. Cerebellar activation was also recorded in anticipation of a task of aversive and noxious stimulus in humans [74–76].

6   Cerebellar Anticipation Deficits in Psychiatric Disorders: A Consensus?

The cerebellar participation of cognitive and emotional processes was confirmed through very many investigations that revealed cerebellar activation during emotional facial processing, orientation and maintenance of attention, imagined movements of objects, judgment of spatial orientation, and timing and anticipation [77–83]. As mentioned previously, patients and animals with cerebellar disturbances often exhibit cognitive and emotional disabilities that are not due to motor control alteration. Nowadays these clinical symptoms and behavioral deficits in animals are still an enigma, and no real consensus has emerged.

Here we propose that the ability to analyze a situation and to act properly in the environment is closely related to the ability to anticipate the consequences of our own actions on such environment. Thus we consider that the cerebellar contribution permits a) first to learn, in order to b) thereafter select and adopt oriented and adapted behaviors in the environment of the living. The discovery of cerebellar synaptic plasticity made concrete the biological support of such a complex mechanism. "Learning" emerges from experiences in the environment and allows for performance improvement—that is, for quickly adapting to a changing physical environment—and necessitates dynamic interactions between several neural networks organized in loop circuits, differentially involved in supervised, unsupervised, and reinforced learning [4]. The cerebellum participates in acquiring a mental representation of the world through its role in internal modeling, replacing external sensory feedback with internal feedback [50]. According to Ito [47, 50], the cerebellum would be the real active interface between the environment and its representation. As a whole, the studies accommodate the role of the cerebellum as a supervisor in the optimization of this cognitive process. According to the computational viewpoint, the internal representation would be learned and stored within the cerebellum, which allows for adapting behavior to "future-known" sensorimotor associations. From this point of view, we believe that disorders in anticipation seem to be a valid common denominator for shedding light on many psychiatric disorders displaying cerebellar deficits. Among these, Autism Spectrum Disorders (ASD) appear to be the most representative pathology, with complex aetiology associated with cerebellar disorders [84–89], potential internal model deficiencies [90–92], and, unfortunately, the absence of consensus. In fact, cerebellar degeneration is one of the most frequent physiological abnormalities observed in ASD [30, 89].

Because of their specific neural degeneration and behavioral disturbances, the above-mentioned cerebellar Lurcher mutant mice were proposed as a model animal model for the study of some biological and behavioral aspects of these mental disorders [93]. As illustrated in this article, several behavioral tests in animal models of human psychiatric disorders are based on anticipation (e.g., the Morris Water Maze test, the rotarod test); but hitherto, too little data in various reports has been interpreted in the light of anticipatory capabilities and/or disabilities. Serious consideration of anticipation and internal modeling [94] in mental disorders related to cerebellar abnormalities would be fruitful in both animal and humans studies. In the pathophysiology and treatment of mental pathologies with maladaptive behavior, such as autism, they would represent an interesting field of investigation.

References

1. Martin, G.B., Butz, G.B., Sigaud, O., Pezzulo, G.: Anticipations, brains, individual and social behavior: an introduction to anticipatory systems. In: Butz, M.V., Sigaud, O., Pezzulo, G., Baldassarre, G. (eds.) Anticipatory Behavior in Adaptive Learning Systems, pp. 1–18. Springer, Berlin (2007)
2. Fleischer, J.G.: Neural correlates of anticipation in Cerebellum, Basal Ganglia, and hippocampus. In: Butz, M.V., Sigaud, O., Pezzulo, G., Baldassarre, G. (eds.) Anticipatory behavior in adaptive learning systems, vol. 4520, pp. 19–34. Springer, Berlin (2007)

3. Nadin, M.: Can predictive computation reach the level of anticipatory computing? Int. J. Appl. Res. Inf. Technol. Comput. **5**(3), 171–200 (2014)
4. Doya, K.: Complementary roles of basal ganglia and cerebellum in learning and motor control Kenji Doya. Curr. Opin. Neurobiol. **10**, 732–739 (2000)
5. Imamizu, H., Miyauchi, S., Tamada, T., Sasaki, Y., Takino, R., Pütz, B., Yoshioka, T., Kawato, M.: Human cerebellar activity reflecting an acquired internal model of a new tool. Nature **403**(6766), 192–195 (2000)
6. Bastian, A.J.: Learning to predict the future: the cerebellum adapts feedforward movement control. Curr. Opin. Neurobiol. **16**(6), 645–649 (2006)
7. Serrien, D.J., Wiesendanger, M.: Role of the cerebellum in tuning anticipatory and reactive grip force responses. J. Cogn. Neurosci. **11**(6), 672–681 (1999)
8. Thomas, A.: Le cervelet – Etude anatomique, clinique et physiologique. Steinheil, Paris (1897)
9. Binet, A., Thomas, A., Henri, V.: Le cervelet. Revue **4**, 438–439 (1897)
10. Flourens, M. J. P.: Recherches expérimentales sur les propriétés et les fonctions du système nerveux dans les animaux vertébrés. Paris (1824)
11. Glickstein, M., Strata, P., Voogd, J.: Cerebellum: history. Neuroscience **162**(3), 549–559 (2009)
12. Schmahmann, J.D., Caplan, D.: Cognition, emotion and the cerebellum. Brain **129**(pt. 2), 290–292 (2006)
13. Bürk, K.: Cognition in hereditary ataxia. Cerebellum **6**(3), 280–286 (2007)
14. Cooper, F.E., Grube, M., Elsegood, K.J., Welch, J.L., Kelly, T.P., Chinnery, P.F., Griffiths, T. D.: The contribution of the cerebellum to cognition in Spinocerebellar Ataxia Type 6. Behav. Neurol. **23**(1–2), 3–15 (2010)
15. Bürk, K., Globas, C., Bösch, S., Klockgether, T., Zühlke, C., Daum, I., Dichgans, J.: Cognitive deficits in spinocerebellar ataxia type 1, 2, and 3. J. Neurol. **250**(2), 207–211 (2003)
16. Valis, M., Masopust, J., Bažant, J., Ríhová, Z., Kalnická, D., Urban, A., Zumrová, A., Hort, J.: Cognitive changes in spinocerebellar ataxia type 2. Neuro Endocrinol. Lett. **32**(3), 354–359 (2011)
17. Lilja, A., Hämäläinen, P., Kaitaranta, E., Rinne, R.: Cognitive impairment in spinocerebellar ataxia type 8. J. Neurol. Sci. **237**(1–2), 31–38 (2005)
18. Suenaga, M., Kawai, Y., Watanabe, H., Atsuta, N., Ito, M., Tanaka, F., Katsuno, M., Fukatsu, H., Naganawa, S., Sobue, G.: Cognitive impairment in spinocerebellar ataxia type 6. J. Neurol. Neurosurg. Psychiatry **79**(5), 496–499 (2008)
19. Garrard, P., Martin, N.H., Giunti, P., Cipolotti, L.: Cognitive and social cognitive functioning in spinocerebellar ataxia: a preliminary characterization. J. Neurol. **255**(3), 398–405 (2008)
20. Baillieux, H., De Smet, H.J., Dobbeleir, A., Paquier, P.F., De Deyn, P.P., Mariën, P.: Cognitive and affective disturbances following focal cerebellar damage in adults: a neuropsychological and SPECT study. Cortex **46**(7), 869–879 (2009)
21. Stoodley, C.J., Schmahmann, J.D.: The cerebellum and language: evidence from patients with cerebellar degeneration. Brain Lang. **110**(3), 149–153 (2009)
22. Zuchowski, M.L., Timmann, D., Gerwig, M.: Acquisition of conditioned eyeblink responses is modulated by cerebellar tDCS. Brain Stimul. **7**(4), 525–531 (2014)
23. Hoppenbrouwers, S.S., Schutter, D.J.L.G., Fitzgerald, P.B., Chen, R., Daskalakis, Z.J.: The role of the cerebellum in the pathophysiology and treatment of neuropsychiatric disorders: a review. Brain Res. Rev. **59**(1), 185–200 (2008)
24. Mohapatra, P.K., Misra, B.N., Patanaik, P., Sahoo, S.: Major depressive disorder—a co-morbid condition in a case of spino-cerebellar ataxia with writer's cramp. Indian J. Psychiatry **45**(4), 257 (2003)
25. Reyes, M., Gordon, A.: Cerebellar vermis in schizophrenia. Lancet **318**(8248), 700–701 (1981)
26. Perlov, E., Tebartz van Elst, L., Buechert, M., Maier, S., Matthies, S., Ebert, D., Hesslinger, B., Philipsen, A.: H1-MR-spectroscopy of cerebellum in adult attention deficit/hyperactivity disorder. J. Psychiatr. Res. **44**(14), 938–943 (2010)

27. Pujol, J., Soriano-Mas, C., Alonso, P., Cardoner, N., Menchón, J.M., Deus, J., Vallejo, J.: Mapping structural brain alterations in obsessive-compulsive disorder. Arch. Gen. Psychiatry **61**, 720–730 (2004)

28. Sun, Y., Lee, J., Kirby, R.: Brain imaging findings in dyslexia. Pediatr. Neonatol. **51**(2), 89–96 (2010)

29. Laycock, S.K., Wilkinson, I.D., Wallis, L.I., Darwent, G., Wonders, S.H., Fawcett, A.J., Griffiths, P.D., Nicolson, R.I.: Cerebellar volume and cerebellar metabolic characteristics in adults with dyslexia. Ann. N. Y. Acad. Sci. **236**, 222–236 (2008)

30. Allen, G.: Cerebellar contributions to autism spectrum disorders. Clin. Neurosci. Res. **6**(3–4), 195–207 (2006)

31. Schmahmann, J. D., Sherman, J. C.: The cerebellar cognitive affective syndrome. Brain 561–579 (1998)

32. Schmahmann, J.D.: Disorders of the cerebellum. J. Neuropsychiatr. **16**(3), 367–378 (2004)

33. Schmahmann, J.D.: Cognition, emotion and the cerebellum. Brain **129**(pt. 2), 190–192 (2006)

34. Topka, H., Massaquoi, S.G., Benda, N., Hallett, M.: Motor skill learning in patients with cerebellar degeneration. J. Neurol. Sci. **158**, 164–172 (1998)

35. Lorivel, P. H. T.: Animal models of cognitive and emotional functions of the cerebellum. In: Pombano, L. J., Evans, D. M. (eds.) Cerebellum anatomy, functions and disorders, pp. 31–58. Nova Publishers (2012)

36. Markvartová, V., Cendelín, J., Vozeh, F.: Changes of motor abilities during ontogenetic development in Lurcher mutant mice. Neuroscience **168**(3), 646–651 (2010)

37. Le Marec, N., Caston, J., Lalonde, R.: Impaired motor skills on static and mobile beams in lurcher mutant mice. Exp. Brain Res. **116**(1), 131–138 (1997)

38. Thifault, S., Girouard, N., Lalonde, R.: Climbing sensorimotor skills in Lurcher mutant mice. Brain Res. Bull. **41**(6), 385–390 (1996)

39. Porras-García, M.E., Ruiz, R., Pérez-Villegas, E.M., Armengol, J.Á.: Motor learning of mice lacking cerebellar Purkinje cells. Front. Neuroanat. **7**, 1–8 (2013)

40. Hilber, P., Caston, J.: Motor skills and motor learning in Lurcher mutant mice during aging. Neuroscience **102**(3), 615–623 (2001)

41. Lorivel, T., Hilber, P.: Motor effects of delta 9 THC in cerebellar Lurcher mutant mice. Behav. Brain Res. **181**(2), 248–253 (2007)

42. Lorivel, T., Hilber, P.: Effects of chlordiazepoxide on the emotional reactivity and motor capacities in the cerebellar Lurcher mutant mice. Behav. Brain Res. **173**(1), 122–128 (2006)

43. Cendelín, J., Korelusová, I., Vozeh, F.: The effect of repeated rotarod training on motor skills and spatial learning ability in Lurcher mutant mice. Behav. Brain Res. **189**(1), 65–74 (2008)

44. Hilber, P., Lalonde, R., Caston, J.: An unsteady platform test for measuring static equilibrium in mice. J. Neurosci. Methods **88**(2), 201–205 (1999)

45. Holschneider, D.P., Yang, J., Guo, Y., Maarek, J.I.: Reorganization of functional brain maps after exercise training: Importance of cerebellar—thalamic—cortical pathway. Brain Res. **1184**, 96–107 (2007)

46. Doyon, J., Benali, H.: Reorganization and plasticity in the adult brain during learning of motor skills. Curr. Opin. Neurobiol. **15**(2), 161–167 (2005)

47. Ito, M.: Bases and implications of learning in the cerebellum–adaptive control and internal model mechanism. Prog. Brain Res. **148**, 95–109 (2005)

48. Albus, J.S.: A model of computation and representation in the brain. Inf. Sci. (Ny) **180**(9), 1519–1554 (2010)

49. Doyon, J., Penhune, V., Ungerleider, L.G.: Distinct contribution of the cortico-striatal and cortico-cerebellar systems to motor skill learning. Neuropsychologia **41**(3), 252–262 (2003)

50. Ito, M.: Control of mental activities by internal models in the cerebellum. Nat. Rev. Neurosci. **9**(4), 304–313 (2008)

51. Ivry, R.: Exploring the role of the cerebellum in sensory anticipation and timing: commentary on Tesche and Karhu. Hum. Brain Mapp. **9**(3), 115–118 (2000)

52. Nixon, P.D., Passingham, R.E.: Predicting sensory events the role of the cerebellum in motor learning. Exp. Brain Res. **138**, 251–257 (2001)

53. Ito, M.: Cerebellar circuitry as a neuronal machine. Prog. Neurobiol. **78**(3–5), 272–303 (2006)
54. Ohyama, T., Nores, W.L., Murphy, M., Mauk, M.D.: What the cerebellum computes. Trends Neurosci. **26**(4), 222–227 (2003)
55. Wolpert, D.M., Miall, R.C., Kawato, M.: Internal models in the cerebellum. Trends Cogn. Sci. **2**(9), 338–347 (1998)
56. Miall, R.C., Weir, D.J., Wolpert, D.M., Stein, J.F.: Is the cerebellum a smith predictor? J. Mot. Behav. **25**(3), 203–216 (1993)
57. Tuma, J., Kolinko, Y., Vozeh, F., Cendelin, J.: Mutation-related differences in exploratory, spatial, and depressive-like behavior in PCD and Lurcher cerebellar mutant mice. Front. Behav. Neurosci. **9**(116), 1–19 (2015)
58. Hilber, J.C.P., Jouen, F., Delhaye-Bouchaud, N., Mariani, J.: Differential roles of the cerebellar cortex and deep cerebellar nuclei in learning and retention of a spatial task: Studies in intact and cerebellectomized Lurcher mutant mice. Behav. Genet. **28**(4), 299–308 (1998)
59. Gasbarri, A., Pompili, A., Pacitti, C., Cicirata, F.: Comparative effects of lesions of the ponto-cerebellar and olivo-cerebellar pathways on motor and spatial learning in the rat. Neuroscience **116**, 1131–1140 (2003)
60. Meignin, C., Hilber, P., Caston, J.: Influence of stimulation of the olivocerebellar pathway by harmaline on spatial learning in the rat. Brain Res. **824**(2), 277–283 (1999)
61. Garthe, A., Kempermann, G.: An old test for new neurons: refining the morris water maze to study the functional relevance of adult hippocampal neurogenesis. Front. Neurosci. **7**, 1–11 (2013)
62. D'Hooge, R., De Deyn, P.P.: Applications of the Morris water maze in the study of learning and memory. Brain Res. Brain Res. Rev. **36**(1), 60–90 (2001)
63. Mandolesi, L., Giuseppa, M., Spirito, F., Federico, F., Petrosini, L.: Is the cerebellum involved in the visuo-locomotor associative learning? Behav. Brain Res. **184**, 47–56 (2007)
64. Petrosini, L., Leggio, M. G., Molinari, M.: The cerebellum in the spatial problem solving: a co-star or a guest star? Prog. Neurobiol. **56**(98), 191–210 (1998)
65. Lalonde, R., Lamarre, Y., Smith, A.M.: Does the mutant mouse lurcher have deficits in spatially oriented behaviours? Brain Res. **455**(1), 24–30 (1988)
66. Passot, J.B., Sheynikhovich, D., Duvelle, É., Arleo, A.: Contribution of cerebellar sensorimotor adaptation to hippocampal spatial memory. PLoS One **7**(4), e32560 (2012)
67. Gross, H., Heinze, A., Seiler, T., Stephan, V.: Generative character of perception: a neural architecture for sensorimotor anticipation. Neural Netw. **12**, 1101–1129 (1999)
68. Hilber, P., Lorivel, T., Delarue, C., Caston, J.: Stress and anxious-related behaviors in Lurcher mutant mice. Brain Res. **1003**(1–2), 108–112 (2004)
69. Lorivel, T., Roy, V., Hilber, P.: Fear-related behaviors in Lurcher mutant mice exposed to a predator. Genes. Brain. Behav. **13**(8), 794–801 (2014)
70. Zhu, L., Scelfo, B., Tempia, F., Sacchetti, B., Strata, P.: Membrane excitability and fear conditioning in cerebellar Purkinje cell. Neuroscience **140**(3), 801–810 (2006)
71. Sacchetti, B., Scelfo, B., Tempia, F., Strata, P.: Long-term synaptic changes induced in the cerebellar cortex by fear conditioning. Neuron **42**(6), 973–982 (2004)
72. Sacchetti, B., Scelfo, B., Strata, P.: Cerebellum and emotional behavior. NSC **162**(3), 756–762 (2009)
73. Parvizi, J., Anderson, S.W., Martin, C.O., Damasio, H., Damasio, A.R.: Pathological laughter and crying: a link to the cerebellum. Brain **124**(pt. 9), 1708–1709 (2001)
74. Tillfors, M., Furmark, T., Marteinsdottir, I., Fredrikson, M.: Cerebral blood flow during anticipation of public speaking in social phobia: a PET study. Biol. Psychiatry **52**(11), 1113–1119 (2002)
75. Smith, K.A., Ploghaus, A., Cowen, P.J., Mccleery, J.M., Guy, M., Smith, S., Tracey, I., Matthews, P.M.: Cerebellar responses during anticipation of noxious stimuli in subjects recovered from depression: Functional magnetic resonance imaging study. Br. J. Psychiatry **181**, 411–415 (2002)

76. Moulton, E.A., Elman, I., Pendse, G., Schmahmann, J., Becerra, L., Borsook, D.: Aversion-related circuitry in the cerebellum: responses to noxious heat and unpleasant images. J. Neurosci. **31**(10), 3795–3804 (2011)
77. Durisko, C., Fiez, J.A.: Functional activation in the cerebellum during working memory and simple speech tasks. Cortex **46**(7), 896–906 (2010)
78. Balsters, J.H., Ramnani, N.: Symbolic representations of action in the human cerebellum. Neuroimage **43**(2), 388–398 (2008)
79. Balser, N., Lorey, B., Pilgramm, S., Naumann, T., Kindermann, S., Stark, R., Zentgraf, K., Williams, A.M., Munzert, J.: The influence of expertise on brain activation of the action observation network during anticipation of tennis and volleyball serves. Front. Hum. Neurosci. **8**, 568 (2014)
80. Lee, T.M.C., Liu, H., Hung, K.N., Pu, J., Ng, Y., Mak, A.K.Y., Gao, J., Chan, C.C.H.: The cerebellum's involvement in the judgment of spatial orientation: a functional magnetic resonance imaging study. Neuropsychologia **43**, 1870–1877 (2005)
81. Beaton, A., Mariën, P.: Language, cognition and the cerebellum: grappling with an enigma. Cortex **46**(7), 811–820 (2010)
82. Schweizer, T.A., Alexander, M.P., Cusimano, M., Stuss, D.T.: Fast and efficient visuotemporal attention requires the cerebellum. Neuropsychologia **45**, 3068–3074 (2007)
83. Kim, Y.T., Seo, J.H., Song, H.J., Yoo, D.S., Lee, H.J., Lee, J., Lee, G., Kwon, E., Kim, J.G., Chang, Y.: Neural correlates related to action observation in expert archers. Behav. Brain Res. **223**(2), 342–347 (2011)
84. Becker, E.B.E., Stoodley, C.J.: Autism spectrum disorder and the cerebellum. Int. Rev. Neurobiol. **113**, 1–34 (2013)
85. Gliga, T., Jones, E.J.H., Bedford, R., Charman, T., Johnson, M.H.: From early markers to neuro-developmental mechanisms of autism. Dev. Rev. **34**(3), 189–207 (2014)
86. Goines, P., Haapanen, L., Boyce, R., Duncanson, P., Braunschweig, D., Delwiche, L., Hansen, R., Hertz-Picciotto, I., Ashwood, P., Van de Water, J.: Autoantibodies to cerebellum in children with autism associate with behavior. Brain Behav. Immun. **25**(3), 514–523 (2011)
87. Kern, J.K.: Purkinje cell vulnerability and autism: a possible etiological connection. Brain Dev. **25**(6), 377–382 (2003)
88. Schroeder, J.H., Desrocher, M., Bebko, J.M., Cappadocia, M.C.: Research in Autism Spectrum Disorders The neurobiology of autism: theoretical applications. Res. Autism Spectr. Disord. **4**(4), 555–564 (2010)
89. Allen, G.: The cerebellum in autism. Clin. Neuropsychiatry **2**(6), 321–337 (2005)
90. Larson, J.C.G., Bastian, A.J., Donchin, O., Shadmehr, R., Mostofsky, S.H.: Acquisition of internal models of motor tasks in children with autism. Brain **131**(pt. 11), 2894–2903 (2008)
91. Stoit, A.M.B., van Schie, H.T., Riem, M., Meulenbroek, R.G.J., Newman-Norlund, R.D., Slaats-Willemse, D.I.E., Bekkering, H., Buitelaar, J.K.: Internal model deficits impair joint action in children and adolescents with autism spectrum disorders. Res. Autism Spectr. Disord. **5**(4), 1526–1537 (2011)
92. Haswell, C.C., Izawa, J., Dowell, L.R., Mostofsky, S.H., Shadmehr, R.: Representation of internal models of action in the autistic brain. Nat. Neurosci. **12**(8), 970–972 (2009)
93. Martin, L.A., Goldowitz, D., Mittleman, G.: Repetitive behavior and increased activity in mice with Purkinje cell loss: a model for understanding the role of cerebellar pathology in autism. Eur. J. Neurosci. **31**(3), 544–555 (2010)
94. Nadin. M.: Anticipation and the Brain. In: Nadin, M. (ed.): Anticipation and Medicine, pp. 135–162. Springer, Cham, CH. (2016)

Anticipation and the Brain

Mihai Nadin

Abstract We are our brains. The study argues for a theory of the brain based on the brain itself, not on theories generated to explain the world in some of its many aspects. Consequently, this study of the brain debunks those analogy assumptions, never confirmed by science, that still dominate brain science. The hypothesis advanced concerns the distributed nature of brain activity. It describes the role of interactions, the specific causality characteristic of brain related dynamics, and the broader questions of cognitive activity associated with awareness of change (usually subsumed as consciousness). Empirical evidence that brain processes are anticipatory in nature leads to the conclusion that reproducibility, as defined within the deterministic experimental method, cannot be expected. Significance for the performance of the living (reproduction, survival, evolutionary edge, etc.) informs brain processes and explains their non-deterministic nature.

Keywords Complementarity · Decidability · G-complexity · Interaction · Reactive brain · Anticipatory brain

1 Prediction

In the age of computation, consider X to be the next focus in science. In a time shorter then what it took for scientists and technologists to define, test, and deploy X, there will be a theory and an experimental proof generalizing from X to the living. The statement (known in a variety of formulations, from very crude to very subtle) that brains are computers is one example. It took less time than it did to define, test, and deploy computers, for McCulloch (1949) to come up with the formulation. Von Neumann, Newell, and Simon, etc. followed suit. The most recent

M. Nadin (✉)
anté—Institute for Research in Anticipatory Systems,
University of Texas at Dallas, Richardson, TX, USA
e-mail: nadin@utdallas.edu
URL: http://www.nadin.ws

© Springer International Publishing Switzerland 2017   147
M. Nadin (ed.), *Anticipation and Medicine*, DOI 10.1007/978-3-319-45142-8_9

such theory is the so-called "nature's brain." This "theory" sets forth that nature itself "learns and remembers past solutions just as our brains do" [1]. The novelty is that brains, in this theory, are now not von Neumann machines, but learning systems. This new theory goes on to ascertain: The "spontaneous emergence of... adaptive developmental constraints brings up the notion of foresight." The argument is straightforward: the ability to adapt is analogous to the ability of learning systems. Since Google DeepMind (i.e., deep learning) succeeded in winning over the world champion in the extremely complicated game called *Go*, nature must, of course, take the same path. The most successful accomplishment of recent science ("...consider X to be the next focus...." see above) becomes the new theory of how living nature behaves. The fact that foresight—a notion from which deterministic science distanced itself—is, in this new theory, associated with nature is in itself reason enough for further discussion of the anthropomorphic language in which nature's "brain" is described.

2 A Hypothesis

The circularity of such pronouncements has been repeatedly pointed out: take the construct "Theory of X" and examine something different from X from this theoretical perspective. The anthropic brain (based on Bolzmann's anthropic principle), the holonomic brain theory (Karl Pribam), and similar are good examples. "The brain is Bayesian" was stated many times by those ignoring the fact that the brain, of Bayes [2], came up with the Bayes formula. (Actually, it was Price [3], who edited Bayes' text.) Different brains developed the algorithm of Bayesian updating (i.e., inference): reduce the space of hypotheses by using new information as it becomes available.

However, the rose-colored glasses used to look at the world—in this case the Bayes formula—do not make the world rosy, but only color the perception of it: "Nature does deep learning." Many examples of similar generalizations are documented in the history of science: "Everything is a mechanism/machine," "Everything is a system," "Everything is a computation." And so on. "Everything is Bayesian" is only one more example. It can happen, though rarely, that a biological theory (such as that of the genetic code) is generalized to a domain different from that of the living (in this case, genetic computation).

This study argues for a perspective on the brain anchored *in the brain*, not in the theories that brains (actually human beings) generate in attempting to explain the world, that is, in addressing questions that transcend direct human experience. It also submits a hypothesis undergirded by such a perspective:

1. Brains are embodied in particular forms of living matter distributed in the entire body.
2. The brain is the locus of interactions that result in the perception of change and in the awareness of self-change.

3. Brains are always in anticipation of the actions through which they are expressed.
4. The dynamics of the living, in particular that of the brain, is of a broader causality than that of the entities in which it is embodied.
5. Brain processes take place in the domain of meaning.

To reduce the understanding of the brain and its functioning to the quantitative understanding expressed in the physics of its material embodiment (even of the most sophisticated expression) is to negate the brain's expression in the processes through which minds are constituted. In line with many scientists who realized that the brain is more than what meets the eyes examining it, Gelfand wondered whether "... neurons do not have, metaphorically speaking, a 'soul', but only electrical potentials" (Arshavsky [4]). We shall try to address his concerns. Moreover, we ascertain a model of the brain that corresponds to a *thinking body*. This model affirms the holistic nature of thinking, taking into account that the vast majority of neurons are not concentrated in the brain, but distributed in the entire body.

3 Imprinting—Beyond Lorentz's Ducks

The international nuclear fusion experiment, the international space station, the square kilometer array (of radio telescopes), the large hadron collider, are the highest priority for those institutions that fund the scientific community, with hundreds of billions of dollars allocated.

The human brain, approached by the European HBP Project (112 partners from 24 countries), the USA-based Brain Activity Map (BAM), Japan's Brain/MINDS project are funded in the range of 10 billion US dollars (Figs. 1 and 2).

Inference from the level of funding to the relevance or importance of the subject could help in recognizing motivations: the "Why?" question. It could even suggest an awareness of what is possible: to understand the physics of the universe (a high-order task) or to obtain some insight into how human beings live, think, act, how they formulate their questions and advance answers to them. The disparity in allocated resources does not disclose the degree of adequacy of the effort. Even the most critical voices will not abstain from acknowledging the progress made in the study of nuclear fusion, in the research of outer space (to which radio astronomy contributes broadly), or in the study of matter and energy. Physics—up to the 19th century practiced as natural philosophy—underlies the scientific revolution (to which Galileo largely contributed). The deterministic foundation came from Descartes (in respect to causality) and Newton, who made natural law the expression of human knowledge of the universe. It is not the purpose of this study to recount the history of physics and the various steps it took to reach the impressive current state of knowledge of physical phenomena. There is nothing comparable on record concerning the evolution of knowledge about the living, and even less about the brain. This is not surprising since beyond Descartes—focused

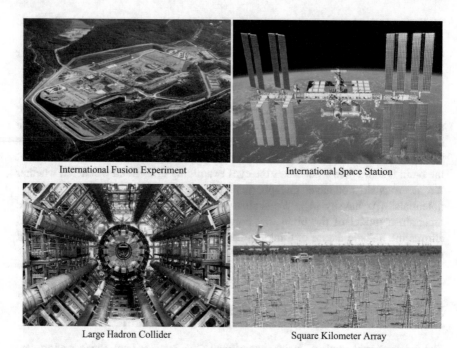

Fig. 1 First-tier science: high priority research nuclear fusion, space exploration, particle physics

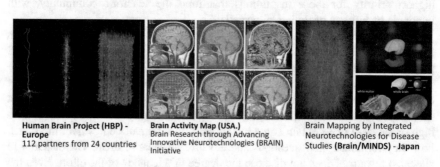

Fig. 2 Second tier research: the brain

on the "pineal gland"—and Newton, all that there is, alive or not, has been considered as a subject of physics—including the brain.

The living and the physical share at least in the matter from which all that exists is embodied, in the energy involved in their respective dynamics. Nevertheless, in seeking knowledge about the living, researchers of all tendencies have acknowledged differences. For the same reason, the history of physics should not be recounted here; neither will the discussion on the fundamental difference between the physical and the living be detailed at this time. (Concerning the living, the reader can consult Rosen [5], Nadin [6], Elsasser [7].) Albeit, it should be noted that the brain has been continuously explained in terms of what physics has ascertained in its successive steps from a naïve understanding of the world to its current explanatory models (such as quantum mechanics). The brain has been considered in relation to the early sources of hydraulic power (fountains, pumps, water clocks). It was also put in relation to clockwork mechanisms, to steam engines, to telegraphic networks, to relay circuits, and—surprise! surprise!—to today's newest machine: the computer. After quoting MacCormac [8] ("Theory requires metaphors to be both hypothetical and intelligible"), Daugman makes a striking observation: "But perhaps like the goslings of Conrad Lorenz…we are too easily imprinted with the spectacle of the day, the 'duckiest' device in sight" [9, p. 33]. (Imprinting describes the instinctive bonding with the immediate moving objects that birds see soon after hatching.)

Whoever examines brain research as a major endeavor of our time cannot avoid a simple observation: "Big Neuroscience" (the name under which brain research is aggregated) is in reality another name for computer science, or computer-based experimental physics relying on neuroimaging as its main source of data. In the issue dedicated to 20 years of MRI, *Neuroimage* uses the expression "workhorse method" to describe MRI's role in cognitive neuroscience [10]. For as long as looking at the human brain was not possible (unless the subject was dead, i.e., a post-mortem brain), the focus was on what could be seen from outside: the shape and morphology of the skull. Shakespeare's skull was presumably stolen from the grave with this intention. Phrenology (as study of the shape of the brain was called) gave way to examining patients affected by the loss of some abilities. The inability of two of Broca's [11] patients to talk inspired his desire to establish which parts of the brain were responsible for this condition. With electricity came the focus on brains' electrical signals and electroencephalography (EEG) emerged. It was initially invasive to the extent of actually not measuring but triggering a variety of processes. X-rays became the visualization medium in the computerized tomography (CT) scanner. The entire scanning technology became a subject of its own. It extends from CT to CAT (computerized axial tomography) and to PET (positron emission tomography), which detects the gamma rays emitted from the radioactive chemical injected. From here to functional magnetic resonance imaging (fMRI), the progression includes computer visualization and the associated programming that turns changes in the blood flow in different parts of the brain into images supposed to represent brain processes. A large number of studies question the method [12].

Yet again, this is not an account of technologies, rather a suggestive sequence of steps documenting how scientists and doctors have tried to "get into the brain" in order to understand it. Available methods (the "glasses" through which we look at the brain) dictated the outcome. With the advent of molecular biology, the focus was placed on what makes up the grey mass one sees when the skull is opened or when post-mortem brains are examined. Brain cells continue to be analyzed in detail; the chemistry and the genetics of the brain are among the most advanced fields of research. "A brief history of the brain" [13] covers billions of years leading to the human brain as it is known today. It also reports on the variety of angles from which the brain was considered. From cells to nerve cells, to neurons, from a distributed configuration to a central brain, this is an evolution-anchored narration impressive in what it ascertains, but also in what it leaves out—or does not yet understand. Willis [14] suggested something that we can call "the functional organization of the brain." Galvani [15] discovered the role of electric signals; Broca (as mentioned) cortical localization; Ramon y Cajal [16] came up with the neuron perspective; Loewy and Dale [17] discovered neurotransmitters, etc. (Finger [18] among others, gives a decent account for such contributions). All these eminent men were consequential in guiding brain researchers. One example: Barlow's "neuron doctrine" [19] (as it is called) states that functions of individual neurons can be extrapolated in order to explain the function of the brain as a whole [20]. To stick to the imprinting metaphor, various "ducks" lead to various conceptions. Nevertheless, the controversial aspects of the technology deployed for measuring the brain are of significance to our purpose, i.e., lending credence to the hypothesis advanced. There is no denying that research of the brain has resulted in impressive accomplishments. Still, we know very little about the brain. The major reason for this is that while the physics of the brain is important, to the extent that the physics and the chemistry of the living are, they remain a partial description only.

4 The Need for a New Framework

To explain why previous descriptions, anchored in the deterministic paradigm, are deficient, and to provide a broader context for my hypothesis, we shall proceed in two directions:

1. Niels Bohr's notion of complementarity, i.e., impossibility of a clear-cut distinction between the behavior of a system and its interaction with the measuring instruments. The particle-wave duality is one of his examples; the distinction between organic and inorganic chemistry is another.
2. Gödel's notion of decidability, i.e., ability to distinguish between what is stated and what is described in the statement.

This suggests that a full non-contradictory description of a complex system is not possible.

We shall examine each.

4.1 Complementarity

Complementary characteristics cannot be observed or evaluated at all at the same time. According to this view, we can say that the reactive brain—which is the subject of physics—and the anticipatory brain—which is the subject of an understanding of the living built around the knowledge domain of meaning—are complementary. The reactive brain is deterministic to the extent that all physical processes are: there is a cause (or a multitude of causes) in the past, and there are effects of the cause, extending from the present, where they can be immediately observed, to the future, where inferences from a state of the matter to the cause of the state noticed can be made. (This view was also defined as mechanistic [20].)

The reactive brain admits fractionation, i.e., the reduction to components whose behavior is aggregated in the behavior of the whole. (For more on fractionation, see Rosen [21] and Louie [22].) One such example was already given: the single neuron in Barlow's doctrine. Further examples are those of molecules, genetic associations with cortex functions or within the structure of the brain; attempts to manipulate, pharmacologically or otherwise, particular neurotransmitters or receptors. Cells were also singled out in direct brain stimulation of surgical patients, or in recordings of such patients (Libet [23], who identified the anticipation component, was active in this domain.) Populations and networks were studied with the aim of seeking correlations between cognitive aspects and structural morphometry, and functional connectivity. The understanding of the reactive (mechanistic) brain within the reductionist approach should not be discarded. Rather, it should be considered in its unity with the understanding of the anticipatory brain. The anticipatory brain, as a particular embodiment of the living, is not fractionatable. The complexity of the whole and that of any part naturally defined—thalamus, cortex, amygdala, etc.—or artificially obtained—e.g., individual neurons extracted from the brain, receptors, neurotransmitters, etc.—is the same, that is, like the complexity of any living entity. However, none is identical with another. Most important: the anticipatory brain, which can be understood only as holistic given the infinite diversity of its ever changing constituents, displays a broader causal dynamics, involving past, present, and future.

From a system perspective, this is expressed as

$$x(t) = f(\text{past states, current state, possible future states}) \tag{1}$$

in which various states of the system would be defined in some manner (not a trivial task). If, for reasons contrary to the holistic nature of the living, someone were to "extract" the brain from the individual (who, in turn, was "extracted" from interaction with others; what von Uexküll [24] defined as *Umwelt*), the following would ensue:

$$(\text{previous state of the brain}), \mathcal{R}_{pp}(\text{current state of the world}), \\ \mathcal{R}_{pf}(\text{possible future states of the world}) \tag{2}$$

In both of the relational connections, $\mathcal{R}$ concerns a future state of the brain (the present is a future for what was the past state)

$$\mathcal{R}_{pp}—\text{relates past and present states—perceptual relation} \tag{3}$$

$$\mathcal{R}_{pf}—\text{relates present and future—goal directedness relation} \tag{4}$$

The functional aspect (cf. 1) of this state transition in a system and the relational aspect [cf. (2)–(4)] are Bohr-complementary. This is an open system.

It should be noted that the various brain metaphors and explanatory models assume a closed system whose parts (wheels, pipes, bolts, levers, etc.) define its functioning. Those aware of the irreducible nature of the living realize that the brain continuously remakes itself, just as the living to which it belongs (is part of) interacts with a changing world, more precisely within its *Umwelt*.

4.2 Decidability

Let us extend Gödel's view of decidability to knowledge about the brain. A paraphrase will help: any effectively generated theory capable of expressing knowledge of the brain and its dynamics cannot be both consistent and complete. In particular, for any consistent, effectively generated formal theory that proves certain basic brain functions, there is a statement about the brain that is true, but not provable in the theory.

For those educated in the spirit of logic, this is an easily recognizable paraphrase of Gödel's Theorem VI of the original formulation [25], which served as inspiration (and reference). In a different context (G-complexity) [26], I advanced the view according to which the distinction between the living, of which anticipation is definitory, and the nonliving (the physical) is that between undecidable complexity and decidable complication. Once again, for those inclined to extract the brain from the whole to which it belongs—i.e., the thinking body—and which it ultimately characterizes, the results can only be fragmentary. A complete description is an illusory task, since the brain is in continuous change. But if *ad absurdum* such a description would be performed, the description would lack consistency. We can continue to describe how seeing, hearing, smelling, etc. take place. The effort is at times impressive, but what is obtained from such descriptions is mechanism, not the understanding of the integrated nature of cognitive processes.

Empirical evidence from the study of brains is overwhelming in respect to the contradictory nature of all there is to it. Brain injury (TBi) and vocational reha-bilitation brought up questions regarding factors that affect the outcome. Contradictory evidence regarding gender, occupational background, race, etc. has been reported [27]. But just as overwhelming are the findings regarding the inner workings of the brain, the various time scales at which it works, the nature of its connectivity. If the brain were a computer (or any machine), descriptions of its functioning would not be contradictory.

It is impossible to ignore that the heroic effort of the *mechanicists* to understand what the brain is and how it works is undermined by the premise they adopt. For inherited aspects of cognitive expression (what used to be called psychological functions), they have searched genetic underpinnings of differences from one person to another. Twins or family members were chosen as subjects. But in order to "put their eyes on" what genes contribute, the mechanicists assumed some associations between biological processes related to blood oxygen level dependent signals (neuroimaging's BOLD technology) and gene expression in peripheral tissues. Experiments could not quantify and verify this assumption.

In general, the brain-mapping obsession with localization (Where do mental process take place? How can they be triggered?) never explained the "Why?" and "How?" of cognitive activity [28]. Machine learning (mentioned in the Introduction of this text [1]) is more and more deployed in order to connect patterns of activation in the brain (or of the brain) and cognitive outcome: the multi-voxel pattern analysis (MVPA) method. Representational similarity analysis (RSA) pursues the opposite path, with the same degree of ambiguity, in associating stimuli and responses. Mechanicists go even further, undisturbed by the fact that their experiments are not reproducible. They do not understand that integrating neuroimaging and computational modeling will not make living processes submit to the expectation of reproducibility. The fact that the brain, in its condition of G-complexity, cannot be algorithmic escapes their understanding. Such methods produce not only a skewed notion of the brain, but also one of no predictive use.

5   The Brain Is What We Do

Nikolai A. Bernstein (but not only), who, in order to examine the brain (at a time when visualization techniques did not exist), suggested that examining human actions (the motoric system) would provide access to cortical processes. It is no accident that Bernstein is mentioned here. "Anticipatory excitation"—under which label Bernstein understood what others called "neuromuscular tone" or "orienting reaction"—was used to "characterize the physiological premises for the creation by the brain of directing models of required future," [29]. Veresov [30] submitted an acceptable summary of Bernstein's view: from reactiveness to activeness, from mechanism to organism.

The process leading to the foundation of these empirically derived models extends from criticism of Pavlov's physiology of reactiveness to Bernstein's attempt to understand motoric expression and its relation to the brain. The principle of sensory adjustments that he advanced, and documented through the cyclogram, is in fact an anticipation-based theory of motoric activity. The model (which informed Pavlov's work) ascertained a mechanism of movement represented by the reflex arc stimulus-brain processing-motor reaction. Bernstein took note of the fact that the stimulus is but only one factor; many other control processes reflect the status of the motoric systems, and actually introduce deviations. The reflex arc is replaced by a reflex ring (usually understood as a form of biological feedback). For all practical purposes, Bernstein was right in deriving knowledge about how the brain works from the study of motion. It is through motion that space and time are cognitively constituted. In his view, the brain does not only send commands to the muscles, but also receives signals from the peripheral sensory terminals. It is a process in which adjustments are continuously performed. At this juncture, the "models of the required future"—state of preparedness—appears, together with the suggestion of an anticipated result. Reactions are triggered by stimuli. An active act, such as lifting one's arms, turning one's head, or moving one's foot is the expression of something not existent or something to come about as the action predates a cause. In Bernstein's words (which I am not trying to force into aligning with my understanding of anticipation):

> "We may assert that at the moment when a movement begins, the entire aggregate of engrams needed to bring this movement to completion is already present in the central nervous system. The existence of such engrams is proved by the very fact of the existence of motor skills and of movements that have become automatic" [31, p. 62]

We would say "autonomous," i.e., expression of a distributed cognitive activity, and not automatic.

The goal-driven activity is fully confirmed by empirical evidence. This brings up a question: how should we understand how the goal to be accomplished in action can be the cause of the action? Since the goal defines the processes corresponding to the goal-directedness, the action through which the goal is attained (avoiding some danger, for example) can only be the expression of both the brain and the body. Veresov [30, p. 9] brings up Spinoza's "thinking body" (a construct not really present in his writings) or Kant's "creative capacity of the soul" in order to explain how goal-directedness is reached. What counts here are not such references (providing the usual shield of authority), but rather what prompted their explanation of how the brain works, and even what the brain is. Indeed, the external world and the internal reality of integrated mental and physiological activity go into what is expressed in the anticipatory action—a realization from the large space of the possible.

A "reflex" movement, corresponding to the action-reaction aspect of the physically embodied living, can acquire, in a given context, new significance: hitting one's toe against a table corner prompts a reaction. Avoiding the accident can be the result of a voluntary act or of an anticipatory action. Learning—which I claim is the explanation of many anticipatory characteristics (for example, the grip and the anticipation of gauging it as a harder grip on a heavy object, a light grip on a fragile, light object [32]—explains how volitional acts (which the living initiates) and non-volitional acts (such as reactions) can fluidly turn into each other. Between unconditioned reflexes (no sensory trigger), such as blinking, and conditioned reflexes (Pavlov's salivating dog), there is a continuum of expression in action: in a subject's reaction to a blow of the fist, either by avoiding more or by striking back, there is a wide space of possibilities: "alternative programs" as Bernstein saw them. The selection of the response is part of the process and can be either the result of an evaluation or of spontaneity. Moreover, in the continuum of active expression, the majority of the acts are triggered from within (and usually defined as volitional). In Bernstein's view, rest is movement at zero speed, the unconditioned reflex is a reaction with a zero degree of initiative ("activeness"). The model of the future is probably the aggregate of perceptional information (related to the context) and of *information generated* by the mind itself [33, 34]. Centralized and distributed processes are integrated in a coherent action—this is the expression of the thinking body.

In the jargon of those seeking functional connectivity through neuroimaging (fMRI is still the preferred instrument), the resting brain is never quiescent. We still do not really know what the fluctuation in the BOLD signals are. But we know that the anterior and midline regions, as well as the temporoparietal cortex, change their pattern of activity depending upon tasks performed. This and other arguments of the same nature prompt our suggestion to consider spectrum disease as affecting not only the brain (e.g., Alzheimer's, with its signature amyloid deposition and reduced metabolic performance), but the whole body. Progress has been made in documenting the role of physical exercise (in addressing autism, Alzheimer's, comorbidities of epilepsy, etc.). In the AnticipationScope, the expression of anticipation in action documents the hypothesis discussed here [35] (Fig. 3).

"The cardinal premise" for Bernstein's concept "is the ability to make prognoses and model the future" [37]. We don't know what the model of the future informing the anticipatory expression is, as we don't know why in spectrum disease anticipation at work is either too high or too low. Some have questioned the possibility of getting 2D visual perception converted to a realistic 3D model. From my perspective, this is a *non sequitur* matter since in a holistic perspective, all senses—on the continuum of sense, not just the five identified so far—contribute to a coherent model. Moreover, the whole body is mapped to the brain, as much as the brain extends to the body.

Let us consider one more important aspect: stimuli, which sensors translate into data. Exposed to the world, or better yet, integrating with the world, the living takes

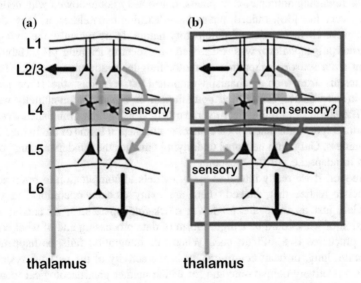

Fig. 3 The thalamus is not only a stopover for sensory data, but also a locus for data generation [36]

it in through stimuli. These are of a very large variety and of different energy levels. Albeit, what ultimately shapes behavior, i.e., expression in action, are not the stimuli, but their "significance for the individual" [38, p. 4], "internal purposefulness." Meaning does not quantify. It is constituted in the context. The same stimulus, or combination of stimuli, can have a different meaning in various circumstances. Just for the sake of example, a snake's threatening hiss may mean different things in different contexts: in the zoo, on a recording, in the desert or jungle, etc. Using arguments different from those of Uexküll, I have argued in favor of an understanding of the living and of living processes in the meaning domain, not in the data domain [39] for two reasons:

1. Anticipatory processes are the outcome of interpretations, i.e., perception of consequences.
2. The dynamics of the living is the expression of interactivity.

To be successful (in the evolutionary sense), action (through which anticipation is expressed) has to predate, not to follow, changing circumstances. Ergo: Evolution is predicated upon successful anticipatory action driven by the meaning of change in the world, not by quantitative variations. No living entity measures—it interprets. Measurement is a human activity associated with the construct called *numbers*, which define quantities. Interpretations, in a variety of forms, are part of the dynamics of life. It should be noted that Uexküll (*The Theory of Meaning*) believes "that life can only be understood when one has acknowledged the importance of meaning" [24].

Those who study the brain, human or that of other species, might not realize that brainless living entities display anticipatory characteristics as well. The body of evidence regarding anticipation in plants, notoriously not endowed with neurons, is growing very fast [40]. Indeed, plants have learning capabilities, as they also display adaptive characteristics of anticipatory nature. More recently, the living slime (Physarum polycephalum) was researched in respect to learning [41]. Habituation, different from sensory adaptation or motor fatigue, is stimulus specific. The evidence results from non-mechanistic empirical observations: the slime samples, grown in petri dishes, found their nourishment on a path that eventually was polluted. The slime "showed a clear aversive behavior," which diminished over time, as it learned (got "habituated"). Of course, in non-neural forms of the living, there is no awareness. Only with neuronal underlying emerges the "understanding" of what it takes to adapt, i.e., of the meaning.

Moreover, if we really intend to apply models of computation to brain activity, we'd better realize that, indeed, there are many clocked operations, at various timescales, just as there are no fewer clock-free operations, of analog nature. Integration of what could be a digital form of data processing and of what could be analog processes is a difficult task. When this integration fails, as happens with some brain, lung, or heart conditions, or in the activity of the nervous system, the result is a relatively unique outcome for which neither pharmacological means nor others (e.g., surgery, prosthetics) seem adequate. We have good deterministic theories regarding biological clock processes, but still don't know how to "fix" them, because we are obsessed with the matter, not with the function.

In our days, there are many languages used to describe the dynamics of the living: our own so-called natural language; the large variety of mathematical languages (some would call them jargon); the language of chemistry (actually two languages, one for organic chemistry and another for inorganic); the language of genetics, the language of molecular biology. There are formal languages and there are programming languages, which algorithmic computation relies on. There are interaction languages characteristic of interaction design, visual languages (to which diagrams and animations belong), the languages of sounds. (NB: sonification is a sound representation of phenomena not related to sound.) But there is no specific language that corresponds to the nature of the brain. The neural network is a mathematical construct corresponding to the neuron construct, but not to the variety of neurons making up the brain. Moreover, while everything else in the world is described (better yet, *represented*) and eventually comprehended by our brains, to understand the brain, we use our own brain and the language associated with it.

This raises a simple question: which language? So-called natural language (in which this text is written) is the outcome of interactions that cover direct, indirect, and mediated forms of human self-constitution through what human beings do: labor, leisure, feeding and eating, sexual encounters, combat, etc., etc. They can be immediate or delayed, concrete or more general in nature, or abstract. Parallel to natural language, and corresponding to a variety of practical experiences (of survival, for instance), other forms of expression have emerged as well: images, sounds, rhythms, sculpted and shaped objects. Numbers and associated operations, eventually leading to the language of mathematics, and of physics, chemistry, botany, zoology, etc., facilitated brain expression focused on quantity, as well as the description of the brain in some of these languages. With or without our intention, this creates an unavoidable open-ended spiral: brain → knowledge about brain with the knowledge of the brain → further knowledge about the brain, and so on and on (Fig. 4a, b).

Only recently have we become aware of the relation between the language that describes something and the described. Indeed, representation, including those of the brain, depends on the means of representation. From among those making an explicit argument for what is called an "adequate language," I highlighted Gelfand's contribution [42]. In a nutshell, he argued that formulations of biological subjects had better come from within biology.

> ...the point is not to apply mathematics to biology from outside, but to create new "biological" chapters of mathematics originating from inside, from the very existence of the problems pertaining to the science about life [42, p. 66].

This epistemological prerequisite condition is evinced by the fact that we are our brains; that is, there is no way to distinguish between the whole human being (integrated system of systems) and the brain (separated from the whole to which it physically extends via the so-called nervous system).

(a)

(b)

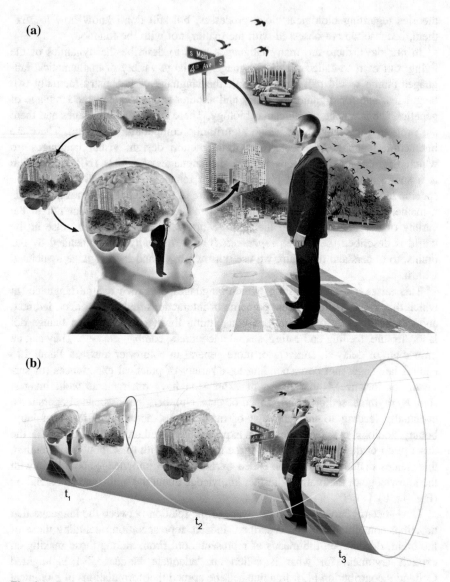

Fig. 4 a Each interaction refreshes the state of the thinking body. **b** The infinite spiral of self-constitutive interactions

The whole of which the brain is constitutive is engaged in quite a number of different representations of the future (Fig. 5). These are internal representations expressed in a great variety of physical and chemical processes, but also in the associated meaning, realized at the molecule, cell, neuron, etc. level. In the interaction among organisms, these representations become perceptions. Finally, in

Fig. 5 Representations of the future

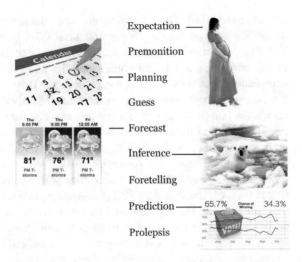

Expectation —
Premonition
— Planning
Guess
— Forecast
Inference —
Foretelling
Prediction —
Prolepsis

sharing, i.e., communication, they are expressed in language (everyday language, mathematical language, etc.) as:

Prediction → inference based on probability
Forecast → inference from past data-based predictions to the future under involvement of self-generated data
Expectation → evaluation of outcome based on incomplete knowledge

(For more details on these distinctions, see Nadin [43].)

While such distinctions are rather of theoretic relevance within a study of anticipation and the brain, those who study the brain (mostly from a physics-based perspective) often face practical questions. What we know from the classic type of behavioral experiment is that a certain action ("Press button") appears to correlate in some ways with the trigger. Neuromarketing took full advantage of the conditioning implicit in such experiments. The brain is placed in a closed-system context, where various messages (advertising, marketing tactics, product shape and color) are associated with motivation or reward activity. "Optimism bias," i.e., test a hypothesis with the group where it was generated, is probably the least disturbing shortcoming of such "self-fulfilling" prophecies. Unfortunately, they are used also in "engineering" diets, educational methodology, legal applications. Indeed, the machine part of the living (in this case, the human being) can be driven by all kinds of stimuli. Neuroimaging does not explain the brain in such applications, but rather applies physics in order to describe what conditions behaviors. The outcome of such so-called experiments is not knowledge, but conditioning. Bem [44] went as far as to generate experimental evidence for anomalous retroactive influences on cognition and affect. The scientific community did not necessarily accept the results, despite the strict protocol accompanying them. What is strange with such experiments is that they are carried out against the understanding of the nature of the phenomena—in this case, prescient action ("guessing" the trigger). If this were an anticipatory expression, experiments could not capture them. They would be irreproducible (as is the case with most experiments on the brain). By their condition, anticipatory processes are non-deterministic. To document a non-deterministic event in the deterministic setting of the experiment is a contradiction in terms.

In the case of predictive hearing, the results are well aligned with the anticipation characteristic of sound (in particular music) perception. The physics of hearing brought Hodgkin and Huxley a Nobel Prize [45]. Theirs was a very inspired electric model (Fig. 6), tested and applied (for instance, in the hearing aid). The "machine" that Hodgkin and Huxley identified captures all sounds as data. But it can be called an "autistic" device: it cannot capture or provide for meaning. Yes, the data could match the pattern of a bird's song or of a symphony, but could not realize their respective meanings. Generalizing from the simple neural making of a squid (known to have a large axon) to the human is in itself questionable.

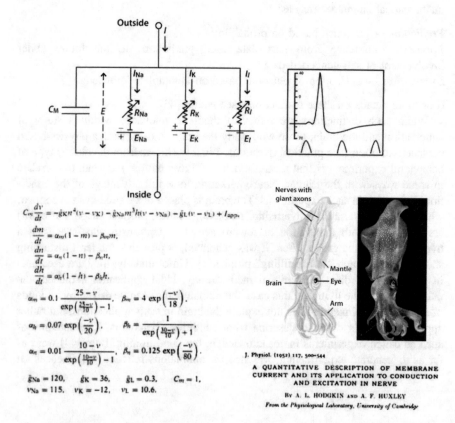

Fig. 6 The notion of circuitry associated with cognitive processes infested brain science after Hodgkin and Huxley advanced their theory

It is surprising therefore that Jeff Hawkins, a dedicated technologist, the man who designed the first handwriting recognition-based personal assistant (the Palm), built upon this understanding. The following, inspired by his research on the brain, expresses his views: "We should find cells in all areas of the cortex, including the primary sensory cortex, that show enhanced activity in anticipation of a sensory event, as opposed to in reaction to a sensory event" [46]. A daring scientific hypothesis—anticipation of a sensory event, turned into prediction: "We should find cells." At least, Hawkins is looking for biological evidence, not for circuitry of physico-chemical representations. Regardless, my concern is less with Hawkins's approach (which still owes to a mechanistic viewpoint) and more with a perspective co-substantial with the subject. Hawkins's handwriting recognition operated on the syntactic level. It was not meant to realize the meaning of words or their pragmatic function.

The model of the *thinking body* (a concept erroneously attributed to Spinoza [47]) ascertains that the brain is distributed throughout the body. In other words, it affirms the holistic nature of thinking, taking into account that the vast majority of neurons are not concentrated in the brain, but distributed in the entire body.

5.1   A Central yet Distributed Activity

The spectacular deployment of neural networks (i.e., connected artificial neurons, which, as already mentioned, are mathematical constructs) in the theory and the applications of machine learning created an interesting epistemological illusion. It has been assumed that since they were inspired by the brain, they could model brain processes. The single neuron in the neural network assumed to be made up of identical entities is connected to all others and participates in the aggregate computational behavior that training the network brings about. In the brain, no single neuron is identical to any other. Neurons are not machines, with a defined input, threshold values of data, processing, output. Rather, they are extremely individualized entities, known to be self-sustained oscillators. Intensity (wave amplitude) and frequency depend upon excitability, synaptic activity (degree of integration in the living, ever changing network), and the glial cells. The chemistry of the process is extremely complicated. Here are only a few details. Neurotransmission affects the coupling of individual neurons. Potassium or magnesium can raise or lower the excitability. Calcium stimulates synchronization (as demonstrated in experiments with a living network in vitro [48]).

Without going into more details—since neither the neuron as such nor the neuronal living network is the subject—we can suggest that the central, yet distributed, brain activity is the result of a dynamic process of aggregation resulting from synaptic interactions. Actually, many expressions of the thinking body do not rely on a central command. The extreme of this distributed cognitive activity is illustrated by the octopus, whose arms are, for all practical purposes, almost autonomous. The human being does not exhibit this extreme of the thinking distribution. The principle of minimum energy expenditure explains the richness of physiological activity and motoric expression. It also explains why centralism and distributed processing together define the entire mobility of the living, in degrees that vary from species to species.

Of course, a living entity endowed with a reduced number of neurons is capable of limited movement expression and of lower adaptive capability. Awareness emerges once the autonomic expression of anticipation is accompanied by learning. Eventually it constitutes consciousness. Those aware of the fact that to observe neurons is to affect their condition have no difficulty in understanding that under conditions of experiment, anticipation is rendered impossible. Multielectrode arrays (MEAs), for instance, are deployed to connect the living neuron to some measuring circuitry. Neuron oscillation or neuron firing (the synapses) observed under these conditions at best lead to a record of the intrusion, not one of organic functioning. Techniques such as fluorescent calcium imaging, intended to facilitate the visualization of larger sets of connected neurons, is also intrusive. Therefore, it is not surprising that the purpose of the exercise is less the knowledge gained through experiment and more the preparation for the hippocampal prosthesis [49]. Dedicated to the notion that the brain, like everything else associated with life, is yet another subject of physics, this type of approach will eventually go for a whole brain machine.

6   Laws and Records

If biology, and implicitly all disciplines pertinent to the living, wants to be like physics, it chases a chimera. The physical is defined through a limited dynamics that can be described through laws. To reach back to Windelband's distinction of knowledge domains, this is the realm of the nomothetic.[1] In contradistinction, the realm defined as idiographic[2] is that of Gestalt, knowledge expressed as the record of change. The knowledge gained through such descriptions defines dynamics as itself subject to change: dynamics of dynamics. Laws and records are of a different condition. This is quite evident when we consider brain activity. No two brains are the same; no two epileptic seizures are the same, and strokes are as original (i.e., unique) as artworks are expected to be. As exciting and promising as the Human Connectome Project is (probably completed before this study is published), it chases after the wrong goal: to find how cognitive disorders or impairments (such as dyslexia, amnesia, etc.) might be the consequence of failed connection. To model functional connectivity is a high-order goal; to hope that fMRI data might help is slightly misleading, given the arguments within the community of researchers.

---

[1]Of or relating to the study or discovery of general scientific laws.

[2]Of or relating to the study or discovery of particular scientific facts and processes, as distinct from general laws.

Worse is to expect regularities similar to those captured in the laws of physics. The convergence of data in studies focused on connectivity, or on other brain characteristics, is as relevant as numerology.

In the examples to follow, we shall see how particular physics-driven research, or research generalizing across the scale of brain manifestations, falls prey to, or avoids the pitfall of, nomothetic obsession.

6.1 What Do We Want to Know?

How does the brain work? How do technology and the science expressed in it perform? These are two different goals, unfortunately most of the time confused. We shall examine some examples exactly in order to make sure that our main thesis —seek knowledge about the brain in the brain, not in the means for describing it— is understood in its specific articulation, and not as a general call.

Research of the brain, in its *vivo* condition, post mortem, or on computer models, is often carried out on animals or on other simpler living entities. To which extent science can generalize across species and across scales (from the miniscule brain of an insect, to the brain of a mouse, or to the human brain) will continue to be debated. It is doubtful that inferring from the functioning of the over 70 million neurons of the mouse (and ca. 10^{11} synapses) to the functioning of almost 90 billion neurons of the human being is defendable. (It can be done, but with awareness of what can be expected, not what one wishes to find out.) The fact that much can be learned from the 302 neurons of the nematode *Caenorhabdites elegans* is undisputable. But learning is by no means generalizable across the scale. Neuronal recordings of a variety of non-human forms of the living contributed much to our understanding of variety, but by no means to causal necessity. Reversible inhibition or excitation applied to various examples (rats are preferred, for some reason) could not, even at the low end of the scale, explain the statistical distribution of the multivariate data generated in cognitive studies.

It is worth mentioning that efforts are under way to study nothing other than consciousness of insects and other invertebrates—with the implicit assumption that they have consciousness. Bees are a choice, given that in their *Umwelt* they exhibit a very rich motoric behavior easily associated with what are called conscious choices. The fact that the human mid-brain is involved in awareness seems a settled matter. (This is not the place to argue otherwise.) It remains to be seen if the "first person perspective" of the bee and that of the human being have the anticipatory dimension in common [50]. This could add to the arguments that this study has offered so far.

Less open to argument remains the perspective adopted. For the reductionists, localization remains a valid question. "To fight or not to fight," to take a recent example [51], is associated with the deep brain structure (the *habenula*) of the zebrafish. According to the research at the RIKEN Brain Science Institute, the *habenula* contains two neuronal circuits. Local field potentials (electric current) in win/lose fish were measured; fluorescence signals linked to neuronal brain activity in brain slices were corroborated with the electric signal. The so-called interpenduncular nucleus (IPN) evaluates the habenular inputs and informs on the fight behavior: "It makes sense" to fight or not. After that, fish were "prepared"—the circuits in question were silenced with a nerve toxin—to win. Transgenic fish (the "prepared" fish) had no learning: winning did not affect a successive choice.

For the record: this is an example from among many extremely well conceived experiments. It was carried out using the most recent technology to localize activity at the neuronal level. I chose it exactly because, within the paradigm it illustrates, it is exemplary. The conclusion: "The same circuits exist in all vertebrates, including humans, with possibly the same bistable mechanisms." The authors further claim to predict fight outcomes purely from neural data. What the research suggests would generalize the machine model, i.e., a nomothetic achievement, as illusory as the premise upon which the research was carried out. This is different from the attempts to deal with consciousness from the perspective of the pragmatics (in the case of the bees, the mobility aspects, i.e., the large space of possible future states from among which one becomes awareness of finding nectar or getting back to the hive).

The reductionist-deterministic view in the RIKEN experiment might be nothing more than yet another example of how the premise is echoed in the experiment: circular analysis, not different from that in neuroimaging research. A holistic perspective would entail considering the brain in connection with the whole (in this case, the entire zebrafish). It would also require placing the subject in its *Umwelt*, or at least in the social context. Aggression, as documented, is an individual expression only under conditions of pathology. Otherwise, it corresponds to social interaction and is associated with social behavior. Obviously, the transgenic fish is a fabricated pathological case, and the correlation to neural activity only qualifies the measured electric activity as characteristic of the pathological. Scientists coming from the biochemical tradition would not measure electricity but identify which substances are involved, which ions, etc., they would deliver specific data (and publish the results, inviting others to validate their findings). It should be at least mentioned here that such and similar attempts illustrating different perspectives of the problem end up delivering valid data, about a partial aspect of the processes researched, without really explaining them.

The *complementary* anticipatory view, not simply an alternative perspective, could be further exemplified. Mapping the bees' nervous system—around one million neurons and a very small brain—could lead to hasty computational models, or to non-deterministic descriptions that reflect the holistic nature of anticipatory behavior. The hypothesis advanced in this study suggests that consciousness is the outcome of interactions always missed by those who attempt to localize a function or explain a certain performance in the living. Obviously, the same discussion on how a particular knowledge perspective affects the outcome of research can be extended from the focus on the brain to the state of the body in general—bee or human being—to what those involved in healthcare practice can derive from it. We are proceeding in this direction.

6.2 Creativity and Engineering

Biology (with some exceptions such as mathematical and computational biology) is associated with hands-on activity. The lab, in some form or another, is where biologists observe, measure, and manipulate their subject—and realize (hopefully) the uniqueness of each. It is about the living, from the scale of large communities to molecular biology and genetics. The Cartesian revolution, and subsequently Darwin's evolution theory, foisted upon biology a particular scientific underpinning, rarely questioned. Biologists rejected the vitalist view of distinguishing between the living and the nonliving on account of entelechy, the hormic schema, *l'élan vital*, *sentiment intérieur*, and the like. Vitalists undermined their own work. Hans Driesch correctly understood that his experiments (dividing cells of sea urchin embryos) would not confirm mechanistic theories of ontogeny. Nuclear cytoplasmic interaction and nuclear equivalence became part of the conversation in biology long after the reported experimental biologist that Driesch was sought explanations for his failed mechanistic hypothesis in entelechy (life force). Even Bergson's attempt, or some years later Schrödinger [52] and Elsasser [7], or more recently, the notion of self-organization (Varela and Maturana [53]) have not helped the discussion of whether the living and the nonliving are the same. Vitalism, a demeaning label always suspect of mysticism or religious infiltration, made the controversy difficult. The reductionist-deterministic view (in reaction to which vitalist explanations were advanced) had the advantage of bridging to the successful path of practicality. The making of things, of machines in particular, of assembling them from parts, and of fixing them (often by taking them apart) shaped the human being more than thinking about their own condition did. The future biologists felt like craftsmen (artists and engineers, we would say today) performing grafts, tissue culture, genetic modification, staying away from mathematics. The farm became a factory, and the farmer an engineer. The same holds true for physicians: they want to be more like mechanics.

Whether in the Indian, Chinese, Babylonian, or similar culture, medicine, as a form of applied biology, involved the arts of dealing with herbs, minerals, and animals and body parts. The earliest known prosthesis (the Egyptian Greville Chester Great Toe, ca. 1295-664 BCE), and the Roman Capua leg (ca. 300 BCE) embodied reductionism and determinism well before it was formulated. The 16th-century Ambroise Paré, who performed amputation surgery, also designed a hinged prosthetic arm. There was no room, and no real need, for philosophical discussion. The healer's art and skills prevailed, plus quite a bit of mysticism, later religious devotion since the outcome seemed to depend on immortal forces that transcended the mortal doctors (some qualified through dedicated study involving empirical knowledge, others charlatans of the like encountered in our days).

The mechanistic view facilitated what biologists were after: closeness to the real thing. To ask *What?* the real thing was, was more attractive than asking *Why?*—as to why does the living behave differently from the nonliving; and why (again) is the human being not reducible to animals. The more important *Why?* of disease, suffering, and death were avoided or turned over to the church. The mechanistic perspective informed the short path; the long path, the Lamarckian tradition, was time consuming. Behavioral, genetic, or even anatomical changes take time and are not, like mechanistic procedures, guaranteed to succeed. The opposite of determinism—non-determinism—reflects the fact that in the living some processes related to what is defined as illness take a path difficult to understand (not to say accept). And even more difficult to reproduce. In the final analysis, they are always unique.

Generating from the unique—the idiographic realm—has a certain attraction to it: the promise of finding not the cure for one person, but for all those who experience some shared symptoms. Individualized care is similar to individualized life. The individual lives the disease. There is no one description for all individual lives. Creativity comes to mind in this connection, as an original expression. Indeed, biological processes are by their nature an expression of *sui generis* creativity. There is no homogeneity in the living, as there is no repetition.

6.3   The Holistic View

The originator of the notion of holism, Jan Smuts, would not distinguish between the physical and the living. In this respect, holism was actually even less meaningful than Spinoza's totality (contradicting the Cartesian mind/body distinction). When awareness of needs for parts for one another is realized, love for each other follows immanently. (The metaphor of the holistic body eventually underlined some of his ethical considerations.)

Regardless, the physics reduction is omnipresent: "The ultimate aim of the modern movement in biology is in fact to explain biology in terms of physics and chemistry" [54]. Time and again physics- and chemistry-based assertions proved, to put it carefully, at best to be incomplete descriptions of living phenomena. Still, this has not triggered the questioning of reductionism, not to say reformulation of physical or chemical principles.

While Jan Smuts's holism was actually of no consequence to biology, the consubstantial notion of *Umwelt* deserves to be considered. Uexkühl described it [24] with much detail in respect to the organism's functions. The example of the tick can help in understanding the thought. Given that the purpose of the living is reproduction, the tick seeks the medium for planting eggs. It can take a long time—up to 18 years (each form of life has its own time scale)—until the coat of a furry animal passing by becomes available. The blind tick registers butyric acid (in the sweat), body warmth, texture. *Umwelt* becomes accessible to the bug through the receptor and effector system. For another animal, the meaning would be different. This is not a matter of physics, nor one of chemistry; it is a matter of significance, i.e., what counts are not numbers to describe quantities, but meanings. The tick does not measure how much butyric acid, how warm the body, how dense the fur. The meaning is: medium for multiplication. Reproduction is the fundamental characteristic of the living (the purpose aspect, i.e., the teleology of existence).

The human being accesses the world through representations, which means reality re-presented, made into a virtual reality. For this to happen, the whole of the body—all the systems we are aware of—is engaged. It is a generic act of autonomic interpretation of the world. The molecules of the neurotransmitters do not know what numbers are, or even what a threshold is. They interface with reality through sensors, and are interfaces with various cells eventually constituted in dynamic cell populations and neural networks. Together they make up the entire body. Therefore, when Bernstein suggested accessing the brain through motoric system expression, he was not using a metaphor. For him, the course of life consists not in reaching "homeostasis with the environment," but actually in overcoming the environment. As we know, homeostasis was derived from the dynamics of machines, and, if properly captured, would describe how much of the human being's functioning is truly analogous to a machine's functioning. Its final goal would be "to maintain uniformity" [55]. Rheostasis includes a change in set-point, a reactive mechanism of adaptation to different circumstances. Allostasis, "cephalic anticipatory adaptation" [56] would transcend the machine model in favor of self-adjusting processes.

"The concept of Allostasis was introduced to take account of the physiology of change and adaptation...to the behavioral and physiological anticipation of future events" [57]. The potential for cephalic anticipatory adaptation is characteristic of the human brain. The type of control associated with allostasis is "not just in reaction (to change), but in anticipation of it" [58, 59].

From the mechanistic perspective, movement control means that the brain commands a muscle and the joints execute movement. A simple examination of the spine (separate vertebrae connected by flexible joints), of the neck (involved in very complicated head movements), of the glossopharyngeal (that supports the ability to speak), of eye movements (coordinated by 24 eye muscles) gave Bernstein, and many others, reason to doubt that the machine reduction is an effective way to understand the motoric system. He realized that the human body has quite a number of different types of mobility. The variable number of degrees of freedom renders implausible the understanding of choice as a central function. Bernstein's *Atlas des Ganges und Laufes des Menschen* (Atlas of the walking and running gait of the human being [translation mine]) is a detailed account of human movement [60]. The manuscript, dated 1929, brings irrefutable evidence of the non-deterministic aspects of leg movement.

All this is relevant to the suggestion already made: to understand centralism and task distribution, sequentiality and parallelism, homogeneity and heterogeneity together. Maybe one more example for explaining what is meant: the octopus (of course not related to the human being) was described as having "eight brains" (in the eight arms). Peter Godfrey-Smith, a philosopher interested in animal intelligence, remarks, "Perhaps in octopus we see intelligence without a centralized self [61, p. 16]. An octopus's arms have a high degree of autonomy. With the nerve connecting the arm to the brain cut, stimulating the skin resulted in independent behavior (including the acts of reaching and grabbing food). An octopus can even tear its own arm. The unity of central and distributed activity leading to coherent behavior is what I try to exemplify. In their *Umwelt*, octopuses display what for an observer would qualify as intelligent behavior [62].

Imagine, for a change, an observer of the human being. Motoric expression, i.e., how humans move, how they perform certain activities (from sexual expression to what we call intellectual activity and artistic performance), together with patterns of nourishment, could easily be catalogued. Mood, emotion, affectivity would probably require further examination. Individuality, as the identifier of a selected specimen within the social context, would also become clear after a longer period of observation and comparative effort. The fundamental distinction between the observer and the observed could concern consciousness, i.e., the extent to which the observer could infer from a current state to a future state, without affecting the transition from one to another. When humans observe their own consciousness, they cannot avoid changing it (see Fig. 4). The recursive loop makes every observation falsifiable. But yet again, the observer would take note of the creative nature of human dynamics. From insemination to birth, adolescence, maturity, and death, creativity is omnipresent. What humans qualified as illness, regardless of its nature, would appear as well as a unique expression. Sometimes it is overcome without any outside intervention. Most of the time, it is subject to a particular form of activity (called healthcare), performed within a pattern that does not distinguish between the living and the nonliving.

The "thinking body"—a conscious effort to describe that of which the observer is part—is actually the human being in its complementary condition: brain and movement (through which both time and space are constituted). Variability is achieved in light of the fact that a great deal of movement, or better yet, brain and change in its broadest sense, is locally initiated. The multitude of joints and muscles, of various degrees of freedom, continuously check each other. Anticipation is the outcome of holistic processes expressed in action: attraction, avoidance, propagation, intervention. The future state is one of significance for survival and for pursuing goals (which are part of the possible future).

In this holistic view, it is not possible to ignore the fact that the human being has to be understood in its unity, not only with the extended world, but also with its microbiome—all the microorganisms that coexist (one can say "co-think" and co-act). Microbiology estimates that the 30 trillion human cells are complemented by fungi, archaea, viruses, and microbes. At this time, we probably do not know enough about what all this means, but we can easily realize that the undecidable nature of the living, in particular in its realization as human being, might have something to do with it.

7   Relevance

Within the *Seneludens* research project [63] dedicated to various aspects of aging, attention was given to capturing anticipatory expression in some form. The *AnticipationScope* [35] provided both a record (data along a timescale of activities) and a kinematic visual expression (the film of the activities). Time reference, in the sense of interval, is essential. Duration informs regarding preparation, onset, performance, outset. During the *Seneludens* experiments, a hypothesis, informed by the presence of a large set of subjects of spectrum disease affected individuals was formulated: Autism is the expression of skewed anticipation.

Almost eight years later, I can be quite satisfied that the hypothesis was proven right. A patient (John Elder Robison [64]) and a neurologist at Harvard University (Alvaro Pascual-Leone [65])[3] shared an experience involving transcranial magnetic stimulation (TMS)[4] applied to a well-documented case of autism, manifested as "missing emotional language." TMS was used for diagnostic purposes (localization of the brain areas involved) and therapy. The frontal lobe was targeted with the purpose of identifying possible association with empathy. For the evaluation of the ability to induce change, motor areas were stimulated. The patient, a professional with an excellent record of engineering sound systems, had an almost perfect physical processing of sounds. Missing was the ability to interpret, to live the melody, to experience emotions. In other words, data was processed with the precision of a machine, but there was no meaning associated with it. Thus was answered Gelfand's question whether the neuron was only a substratum for electrical signals, or something else made it as important as it is in neural networks for the human perception of the world.

For all practical purposes, I do not assume that my hypothesis, or for that matter Gelfand's question, was known to the group involved in the TMS application. Whether they were aware of my hypothesis or of Gelfand's formulation is irrelevant. What really counts is the realization that on a different path, which I am not necessarily enthusiastic about (for reasons evident to those who are reading this study), the scientific hypothesis points to a perspective different from that guiding deterministic-reductionist brain science. With this in mind, I would suggest that the following is of practical relevance:

1. Consider the uniqueness of each form of cognitive disorder.
2. Consider the whole within which it became symptomatic.
3. Consider the pre-symptomatic phase (for instance, in Parkinson's, this extends to over six years).
4. Consider the interactive dimension, and identify which are the factors influencing it.
5. Consider the immune system, as part of the larger "thinking body," as the immediate ally in addressing disorders, no matter of which nature.
6. Identify the way in which the anticipatory expression in the state called *health* (as relative as the term is) is affected and seek alternatives consubstantial with the living.

---

[3]Interview on Fresh Air, National Public Radio, April 21, 2016.

[4]In the jargon of science: Repetitive transcranial magnetic stimulation induces long-lasting changes in protein expression and histone acetylation.

The degenerative path of the species will not be reversed by reprogramming the genetic code or by retraining neuronal networks via transcranial electric or magnetic stimulation (TES or TMS), or through the so-called deep brain stimulation (DBS). Evolution, from the embrace of which humans liberated themselves to a large extent, suggests the patient path of engaging biological factors, not succumbing to the seduction of becoming machine-like. The trans-species condition claimed by Neil Haubisson (featuring an implanted antenna and several sensors that make him experience the world in a way different from how a human would) is not exactly what this study argues for, although by no means is it an argument against seeking knowledge beyond the traditional boundaries of science.

Acknowledgments The Russian Academy of Sciences, N.P. Bechtereva Institute of the Human Brain (St. Petersburg, Russia) challenged me to present my understanding of the brain and the role it plays in anticipatory processes. The keynote address at the conference Modern Trends in Human Brain Neuroscience (September 24–26, 2015) offered the opportunity to engage scientists of reputation in a discussion of the ideas presented. I am indebted to Dr. Valentina Ilyukhina and Dr. Elena Nikolaeva for initiating the invitation, and to the anté–Institute for Research in Anticipatory Systems for supporting the research. Hanse Wissenschaftskolleg made possible the revision of this manuscript and preparation for publishing. I am grateful to Dr. Matthew Goldberg (UT-Southwestern Medical Center) for suggestions regarding pre-symptomatic diagnostics of Parkinson's, and to Dr. Michael Devous and the study group on fMRI for helping me clarify the potential and the limits of this technology. Kalevi Kull and Andres Kurismaa provided useful references to the work of F.S. Rothschild. The Deutsche Forschungsgemeinschaft supported some of the research.

References

1. Kouvaris, K., Clune, J., Kounios, L. Brede, M., Watson, R.A.: How evolution learns to generalise: principles of under-fitting, over-fitting and induction in the evolution of developmental organization (2015). https://arxiv.org/abs/1508.06854. See also https://www.newscientist.com/article/mg22930660-100-evolution-learn-natural-selection/
2. Bayes, T.: An Essay Towards Solving a Problem in the Doctrine of Chances, pp. 370–418. Philosophical Society, London (1763)
3. Price, R.: A demonstration of the second rule in the essay towards the solution of a problem in the doctrine of chances, published in the Philosophical Transactions, vol. LIII. Communicated by the Rev. Mr. Richard Price, in a Letter to Mr. John Canton, M. A. F. R. S. Philosophical Transactions of the Royal Society, **54**, 296–325
4. Arshavsky, YI.: Gelfand on mathematics and neurophysiology. http://www.israelmgelfand.com/bio_work/arshavsky_biomed.pdf
5. Rosen, R.: Life Itself: A Comprehensive Inquiry Into the Nature, Origin, and Fabrication of Life (Complexity in Ecological Systems). Columbia University Press, New York (1991)
6. Nadin, M.: Anticipation: a spooky computation. In: Dubois, D. (ed.) CASYS, International Journal of Computing Anticipatory Systems, Partial Proceedings of CASYS 99, vol. 6, pp. 3–47. CHAOS, Liege (1999)
7. Elsasser, W.: Reflections on the Theory of Organisms. Holism in Biology. Johns Hopkins University Press, Baltimore (1998) (Originally published in 1987 by ORBIS Publishing, Frelighsburg, Quebec)
8. MacCormac, E.R.: A cognitive theory of metaphor. MIT press, Cambridge (1985)

9. Daugman, J.G.: Brain metaphor and brain theory (Chap. 2). In: Bechtel, W., Mandik, P., Mundale, J., Stufflebeam, R. (eds.).Philosophy and the Neurosciences. A Reader. Wiley-Blackwell, Oxford (2001)

10. Bandettini, P.A.: Twenty years of functional MRI: the science and the stories. Neuroimage **62** (2), 575–588 (2012)

11. Broca, P.P.: Localisations des fonctions cérébrales. Siège de la faculté du langage articulé. Bulletin de la Société d'Anthropologie **4**, 200–208 (1863). See also http://www.oxfordscholarship.com/view/10.1093/acprof:oso/9780195181821.001.0001/acprof-9780195181821-chapter-10

12. Shifferman, E.: More than meets the fMRI: the unethical apotheosis of neuroimages. J. Cognition Neuroethics **3**(2), 57–116 (2015)

13. Robson, D.: A brief history of the brain. New Scientist (Feature) 21 September 2011. https://www.newscientist.com/article/mg21128311.800-a-brief-history-of-the-brain/

14. Molnar, Z.: Thomas Willis (1621–1675), the founder of clinical neuroscience. Rev. Neurosci. **5**, 329–335 (2004). See also: Willis, T.: The functional organization of the brain. http://www.oxfordscholarship.com/view/10.1093/acprof:oso/9780195181821.001.0001/acprof-9780195181821-chapter-7

15. Galvani, L.: De viribus electricitatis. University of Bologna, Bologna (1791). See also http://www.oxfordscholarship.com/view/10.1093/acprof:oso/9780195181821.001.0001/acprof-9780195181821-chapter-8

16. Ramon y Cajal, S.: Histologia del system nervioso del homre y de los vertebrados. Ministerio de Sanidad y Consumo, Madrid (2007). See also http://www.oxfordscholarship.com/view/10.1093/acprof:oso/9780195181821.001.0001/acprof-9780195181821-chapter-13

17. Loewi, O., Dale, H.: The discovery of neurotransmitters. http://www.oxfordscholarship.com/view/10.1093/acprof:oso/9780195181821.001.0001/acprof-9780195181821-chapter-16

18. Finger, S.: Mind Behind the BRAIN. A History of the Pioneers and Their Discoveries. Oxford University Press, Cambridge (2000)

19. Barlow, H.: Single units and sensation. A neuron doctrine for perceptual psychology. Perception **1**(4), 371–394 (1972)

20. Poldrack, R.A., Farah, M.J.: Progress and challenges in probing the human brain. Nature **526**, 371–379 (2015)

21. Rosen, R.: Fundamentals of measurement and representation of natural systems, pp. xiii, 88–90, 114–120. Elsevier, Amsterdam (1978)

22. Louie, A.H.: More Than Life Itself, pp. 217–221. Ontos Verlag, Frankfurt (2009)

23. Libet, B.: Neural destiny: does the brain have a mind of its own? Sciences **29**(2), 32–35 (1989)

24. von Uexküll, J.: A Foray Into the Worlds of Animals and Humans: with a Theory of Meaning (O'Neil, D.J.: Trans.) University of Minnesota Press, Minneapolis (2010) (Originally appeared as Streifzüge durch die Umwelten von Tieren und Menschen. Verlag Julius Springer, Berlin, 1934)

25. Gödel, K.: Über formal unenscheidbare Sätze der Principa Mathematica und verwandte Systeme. Monatshefte für Mathematik und Physik, **38**, 173–198. The first incompleteness theorem originally appeared as Theorem VI. (1931)

26. Nadin, M.: G-complexity, quantum computation and anticipatory processes. Comput. Commun. Collab **2**(1), 16–34 (2014)

27. West, M., Target, P., Yasuda, S., Wehman, P.: Return to work after TBi. Brain injury medicine. In: Zasler, N.D., Katz, D.I., Zafonte, R.D. (eds.) Principles and Practice pp. 791–813. Demos Medical Publishing, New York (2007)

28. Poldrack, R.A., Yarkoni, T.: From brain maps to cognitive ontologies: informatics and the search for mental structure. Annu. Rev. Psychol. **67**, 587–612 (2016)

29. Latash, M. (ed.): Progress in motor control. Bernstein's Traditions in Movement Studies, vol 1. Human Kinetics, Champaign, IL (1998)

30. Veresov, N.: Guest editor's introduction Nikolai Bernstein: the physiology of activeness and the physiology of action. J. Russ. East Eur. Psychol. **44**(2), 3–11 (2006)

31. Bernstein, N.A.: Ocherki po fiziologii dvizheniy i fiziologii aktivnosti [Очерки по физиологии движений и физиологии активности] (In English: Outline of the Physiology of Movements and the Physiology of Activity). Medizina, Moscow (1966)
32. Nowak, D.: Different modes of grip force control: voluntary and externally guided arm movements with a hand-held load. Clin. Neurophysiol. **115**(4), 839–848 (2004)
33. Nadin, M.: Mind—anticipation and Chaos. Belser Verlag, Stuttgart (1991)
34. Nadin, M.: Antecapere ergo sum: what price knowledge? AI in life and society 25th anniversary volume: a faustian exchange: what is "to be human" in the era of ubiquitous computing? pp. 39–50. Springer, London (2013)
35. Nadin, M.: Quantifying anticipatory characteristics. The AnticipationScope and the anticipatory profile. In: Iantovics, B., Kountchev, R. (eds.) Advanced Intelligent Computational Technologies and Decision Support Systems, Studies in Computational Intelligence, vol. 486, pp. 143–160. Springer, New York (2013)
36. Constantinople, C.M., Bruno, R.M.: Deep cortical layers are activated directly by thalamus. Science **340**(6140), 1591–1594 (2013)
37. Nadin, M.: Variability by another name: "repetition without repetition." In: Nadin, M. (ed.) Learning from the Past. Early Soviet/Russian Contributions to a Science of Anticipation. Cognitive Science Monographs, vol. 25, pp. 329–340. Springer, Cham (2015)
38. Meijer, O.G., Bongaardt, R.: Bernstein's last paper: the immediate tasks of neurophysiology in the light of the modern theory of biological activity. Motor Control **2**(1), 3–9 (1998)
39. Nadin, M.: Semiotics is fundamental science. In: Jennex, M.E. (ed.) Knowledge Discovery, Transfer, and Management in the Information Age, pp. 76–125. IGI Global, Hershey (2014)
40. Novoplansky, A.: Future perception in plants. In: Nadin, M. (ed.) Anticipation across disciplines, pp. 57–70. Springer, Cham (2016)
41. Boisseau, R.P., Vogel, D., Dussutour, A.: Habituation in non-neural organisms: evidence from slime moulds. Proc. R. Soc. B: Biol. Sci. **283**(1829) (2016). See also http://datadryad.org/resource/. doi:10.5061/dryad.51j89
42. Gelfand, I.M., Tsetlin, M.L.: Mathematical modeling of mechanisms of the central nervous system. Models of the Structural-Functional Organization of Certain Biological Systems (Gelfand, I.M. (ed.), Beard, C.R., Trans.). Cambridge, MIT Press (1971)
43. Nadin, M.: Anticipatory and predictive computation. In: Laplante, P. (ed.) Encyclopedia of Computer Science and Technology. Taylor & Francis, London (2016)
44. Bem, D.J.: Feeling the future: experimental evidence for anomalous retroactive influences on cognition and affect. J. Pers. Soc. Psychol. **100**(3), 407–425 (2011)
45. Hodgkin, A.L., Huxley, A.F.: A quantitative description of membrane current and its application to conductive and excitation in nerve. J. Physiol. **117**(4), 500–544 (1952)
46. Hawkins, J., Blackeslee, S.: On Intelligence. St. Martin's Griffin, New York (2005)
47. Ilyenkov, E.V.: The ideal in human activity. Dialectical Logic, Essays on its History and Theory (Creighton, C., Trans.) Erythros Press and Media, Kettering OH (1977)
48. Penn, Y., Segal, M., Moses, E.: Network synchronization in hippocampal neurons. Proc. Nat. Acad. Sci. **113**(12), 3341–3346 (2016)
49. Berger, T.W., Hampson, R.E., Song, D., Goonawardena, A., Marmarelis, V.Z., Deadwyler, S.A.: A cortical neural prosthesis for restoring and enhancing memory. J Neural Eng. **8**(4) (2011). http://www.ncbi.nlm.nih.gov/pubmed/21677369
50. Barron, A.B., Klein, C.: What insects can tell us about the origins of consciousness. Proc. Natl. Acad. Sci. **113**(18), 4900–4908 (2016)
51. Chou, M-Y., et al.: Social conflict resolution regulated by two dorsal habenular subregions in zebrafish. Science **352**(6281), 87–90 (2016)
52. Schrödinger, E.: What Is Life?. Cambridge University Press, Oxford (1944)
53. Varela, F., Maturana, H.R.: Autopoiesis and Cognition: The Realization of the Living. Reidel Publishing, Dordrecht (1980)
54. Monod, J.: Chance and Necessity: An Essay on the Natural Philosophy of Modern Biology. Knopf, New York (1971)

55. Cannon, W.B.: Organization for physiological homeostasis. Physiol. Rev. **9**(3), 399–431 (1929)
56. Schulkin, J.: Allostasis: a neural behavioral perspective. Horm. Behav **43**, 21–27 (2003)
57. Schulkin, J.: Social allostasis: anticipatory regulation of the internal milieu. Frontiers in Evolutionary Neuroscience, January 31 (2011). http://journal.frontiersin.org/article/10.3389/fnevo.2010.00111/full
58. Moore-Ede M.C.: Physiology of the circadian timing system: predictive versus reactive homeostasis. Am. J. Physiol. **250**(5) (pt. 2), 737–756 (1986)
59. Bauman, D.E.: Regulation of nutrient partitioning during lactation: homeostasis and homeorhesis revisited. In: Cronje, P.J. (ed.) Ruminant Physiology, pp. 311–328. CAB Publishing, NY (2000)
60. Bernstein, N.A.: Atlas des Ganges und Laufes des Menschen. Deutsche Gesellschaft für die Geschichte der Sportwissenschaft, Dortmund (1929). See also http://archive.thulb.uni-jena.de/hisbest/receive/HisBest_cbu_00030812
61. Godfrey-Smith, P., Matthew, L.: Long term-high density occupation of a site by Octopus tetricus and possible site modification due to foraging behavior. Mar. Freshw. Behav. Physiol **45**(4), 261–268 (2012)
62. Montgomery, S.: The Soul of an Octopus. Atria Books, New York (2015)
63. Project Seneludens: http://seneludens.utdallas.edu/
64. Robison, J.E.: http://www.johnrobison.com/
65. Pascual-Leone, A.: http://tmslab.org/aboutus-faculty-pascual-leone.php

Part IV
Anticipation and Medical Data Processing

The Role of MHealth and Wearables for Anticipation in Medicine

Alice Ferng, Vishal Punwani and Shiv Gaglani

Abstract As the market for health-tracking wearable devices continues to expand, there is an emerging niche for healthcare applications, and data acquisition and usage. Within this area exists a wealth of clinically relevant data already collected from wearers, including physiological and lifestyle data. This information allows us to not only optimize current medical treatments and health planning, but also to expand preventive medicine by applying *anticipation* to medicine. We propose that much of the data collected through these wearable devices can be used to inform both patient and clinician of long-term physiological trends, and to anticipate potential onset of illnesses with a view to stemming their progression, or even mitigating their occurrence altogether. This paper highlights important issues within the health-wearable paradigm and presents upcoming applications of wearable technologies in medicine.

Keywords Mhealth · Tracking · Wearables · Technology · Anticipation · Prevention

1 Introduction

Health-tracking wearable devices have seen a tremendous rise over the last several years. Previously, the majority of "wearables" have been focused on the realms of activity and exercise-tracking through step-counting and heart rate measurement.

A. Ferng (✉)
University of Arizona College of Medicine, Tucson, AZ, USA
e-mail: aferng@email.arizona.edu

V. Punwani (✉)
University of Melbourne School of Medicine, Melbourne, Australia
e-mail: vishal.punwani@gmail.com

S. Gaglani
Johns Hopkins School of Medicine, Baltimore, MD, USA
e-mail: shiv@osmosis.org

© Springer International Publishing Switzerland 2017   179
M. Nadin (ed.), *Anticipation and Medicine*, DOI 10.1007/978-3-319-45142-8_10

We are now seeing a new wave of devices that capture physiological data, including more sensitive biochemical measurements such as blood glucose testing and urine analysis [1]. For example, "Smart Bands" for the Apple Watch, intended for release in 2016 will include specialized sensors that may collect this and other similarly sensitive data. New data collection devices compatible with existing smartphones and other electronics will also help drive a shift toward enabling the recording of consumer physiological data through adjunct technologies.

Mobile apps will soon begin to share the spotlight with clinical medicine in health monitoring. In June 2013, WellDoc, a Baltimore-based healthcare technology company, received FDA approval to sell the first prescription-only smartphone app [2]. WellDoc persuaded insurance companies to offer reimbursements for their app, which is designed specifically for type 2 diabetes management. The billion-dollar question is whether this development has opened the floodgates for physicians prescribing apps in addition to, or in lieu of pills.

2 Anticipation in Medicine

2.1 Mobile Health (MHealth) Advantages for Patient and Clinician

There is significant, untapped potential of these devices and apps to inform clinical monitoring and planning. Patients with both acute and chronic conditions may benefit. The acutely ill patient may be able to detect a fever using a smartphone-enabled thermometer, and even perform a urinalysis to help diagnose a urinary tract infection. In this way, acutely ill persons may be able to easily and reliably self-diagnose simple conditions [3].

A patient with a chronic medical condition may also benefit from enhanced, regular monitoring. In the currently diagnostic model, decisions regarding patient care are often made based on incomplete information. For example, when a patient visits her family physician, management is often guided by her presentation at that specific point in time. While the physician may have a complete medical history, he or she is being provided with a snapshot of the patient's health at any given consult. With health-tracking wearables, longer-term physiological pictures can be painted, thus better positioning patients and doctors toward more appropriate management options, which may improve outcomes [4].

Wearables also open the door for a type of 'shared care' between doctor and patient [3]. For example, if a diabetic patient regularly checks his blood glucose using smartphone attachments and an app, he might also permit his family doctor or primary care physician to view his glucose trends. Instead of diabetic wellness check-ups every three months, a clinician might request to see the patient earlier or later based on his personal blood glucose tracking, with an objective of minimizing the destructive complications of diabetes. This is an example of how healthcare may become more proactive as opposed to reactive.

Similarly, consider the patient who regularly checks her blood pressure. A rising trend in BP levels over weeks or months might inform the patient and their doctor approximately how long from any given time point a diagnosis of hypertension can be projected to occur without lifestyle modification or medical intervention. A new layer is therefore added to the model of disease prevention, or at the least, the lead-time for diagnosis can be increased.

2.2 How Valuable Is Anticipation? Can It Improve Medical Care?

The ability to collect a wealth of physiological and lifestyle data introduces the concept of 'anticipation' in medicine. Anticipation is not unlike prediction, but is slightly nuanced—it is the strong prediction of an outcome, however it is made with the backing of clinical evidence collected over time [5]. Medical data is already collected through widespread use of wearables and apps. We suggest that this data be used to inform patient and clinician of long-term physiological trends, and perhaps more importantly, to *anticipate* potential onset of illness with the ultimate goal of prevention. We highlight in this paper a number of important issues with this paradigm and upcoming applications of technologies in medicine.

One way to determine the benefit of any intervention is to look at a relevant endpoint [6]. Wearables can be thought of as interventions if they are used for anticipatory care. For example, the presumed purpose of continually monitored blood glucose is to reduce the development of impaired glucose tolerance and diabetes. If use of a medical wearable reduces patient progression to any adverse endpoint, then that suggests that there is value in that wearable device.

Decreasing progression to undesired health endpoints could have further positive implications other than better patient outcomes. Insurance providers may opt to cover the cost of a wearable for risky patient-groups in order to lower the combined risk of a patient developing an expensive illness. John Hancock Insurance has already begun providing complimentary Fitbits, and offering reduced cost insurance plans for Fitbit wearers [7].

Beyond benefiting patients and insurance companies, wearables offer unique opportunities for medical researchers. First, participants would have options to allow or deny collection of individual metrics as they see fit. Second, provided that the relevant data is accurately measured by the device being used, or is at least reliable enough for correlations to be made, the continuous stream of data could benefit clinical trials. The FDA would have more clinical data with which to consider relevant regulatory decisions, and the increase in clinical applications of new technology will inform the improvement and use of even more technologies to improve evidence-based medicine [8].

3 Users of MHealth Technology

Two broad groups of users benefit most from wearables. The first general group includes young, healthy, tech-savvy persons who primarily use these devices for activity and exercise tracking. Companies such as Under Armour and Adidas have acquired mobile application companies such as MyFitnessPal and RunKeeper to program and provide a more complete digital fitness/data collection platform. As another example, researchers have used data from Apple's HealthKit package to obtain helpful de-identified data that can be used to study a disease, condition, device, or sensor [9, 10]. Thus, while the most apt (and likely) to use these digital fitness applications, for the most part this younger cohort has less use for metrics associated with possible pathology—glucose, cholesterol, blood pressure, etc. This group primarily benefits from wearables through activity-tracking, and the resulting lower risk of disease conferred by their healthier lifestyles.

The second broad group consists of more elderly and chronically ill patients. This population is more likely to be diabetic, arrhythmic, and/or rife with cardio-vascular problems, such as those arising from high cholesterol or long-term smoking. Ironically, this group could stand to benefit from health-monitoring wearables the most, but there is often a hesitancy to adopt new tech in this group for reasons including high financial costs, privacy concerns, and misconceptions around ease of use [11, 12]. This group primarily benefits from wearables by anticipatory means—predicting adverse outcomes with a view to mitigation. For example, loss of balance upon rising from bed or a chair is due to failing antici-pation in the body's ability to prepare for the change in position.[1] By tracking the effects of aging and recording them on an individual's *Anticipatory Profile*™ the intention is to develop means for maintaining anticipation in the aging.

Usage and de-identified physiological data collected from different age groups and subpopulations will provide useful stratified data that not only allows for better applications of the technologies, but also allows for better marketing to target consumer groups.

4 FDA Regulation of Medical Devices, Wearables, and Applications

The Food and Drug Administration (FDA), which has regulatory control over medical devices, categorizes medical devices into three classes [13]. A medical device is defined by the FDA as any product or equipment used to diagnose a

---

[1]By tracking the effects of aging and recording them on an individual's *Anticipatory Profile*™ the intention is to develop means for maintaining anticipation in the aging. See: The anticipatory profile. An attempt to describe anticipation as process, http://www.tandfonline.com/doi/abs/10.1080/03081079.2011.622093.

disease or other conditions, to cure, to treat or to prevent disease. Class I devices are simple in design and have no potential risk. Examples include tongue depressors and Band-aids. These devices must be registered, exhibit proper branding and labeling, and be produced using proper manufacturing techniques. Class II devices are more complicated in design but have minimal risk. Examples are X-ray machines, powered wheelchairs, and surgical or acupuncture needles. Class III devices are intricate in design and have the strictest guidelines because they pose the greatest risk. Examples include implanted pacemakers and prosthetic or artificial heart valves. However, in general, the FDA does not regulate most health-related wearable devices as long as they are low-risk for consumers to use.

4.1 FDA Guidelines

To provide oversight and clarity to the development of mobile health (mHealth) applications, the FDA recently released their guidelines on regulations involving mobile medical apps (February 2015). By some definitions, a mobile app can itself be considered a medical device, leading to stricter regulations. This can be the case when a mobile app transforms a mobile platform into a regulated device with the addition of a sensor that is used in the diagnosis of disease or other conditions. An ECG sensor attached to a mobile phone would transform the mobile app into a device since the readouts become useable patient data. These rules are nevertheless malleable and will vary depending on application. The FDA will not regulate but will "exercise enforcement discretion" over apps that help users self-manage, track, or monitor their disease or conditions without providing medical advice or suggestions. Fitness wearables and mobile sensors and trackers for diet, exercise, sleep and mood that bring control to the user are examples, and therefore are not under regulation. Additionally, aiding patient communication with physicians via video-conferencing or telemedicine portals is not strictly regulated.

The line is drawn at the point where apps help transform a mobile device into a medical device, for example as a blood pressure cuff, otoscope, sphygmo-manometer, pulse oximeter, spirometer, or ophthalmoscope. The argument is that the non-digital version of these medical devices are already FDA-regulated and if they were used incorrectly or malfunctioned, they may pose a risk to patients. By the FDA's definition, a 'mobile medical app' is a mobile app intended to either (1) be used as an accessory to a regulated medical device, or (2) transform a mobile platform into a regulated medical device. What is a regulated medical device? The FDA guidance states:

> When the intended use of a mobile app is for the diagnosis of disease or other conditions, or the cure, mitigation, treatment, or prevention of disease, or is intended to affect the structure or any function of the body of man [or woman], the mobile app is a device.

4.2 Data Security

On the patient or consumer's side, one of the first concerns regarding a mobile application is the security of their personal data. Many devices allow for background monitoring of activity, such as a heart rate-sensing mirror or a car seat that measures weight. Notably, many mobile phones are also able to log keystrokes or GPS location data in addition to number of steps taken. Furthermore, commercial wet labs such as Theranos, 23andMe, and Cleveland Heart Lab that are offering their services to the public are making genetics, proteomics, and epigenetics increasingly available. Used laboratory equipment can also be easily purchased at auctions or on websites such as eBay. Easy access to equipment and in-depth personal data makes it possible for consumers to bring the lab into their homes and personal health analyses to themselves without having to go to an institution, company, or outside location. There is a perhaps unintended consequence of these devices and technologies whereby they are providing health care capabilities to the consumer/patient. As the average person lacks a medical education with which to put collected data into context, it is important that clinicians help to provide the platform, knowledge, and experience for these shifts to take place. Undoubtedly, data security will become a critical component to consider, as it could be disastrous for certain devices, for example, a cardiac pacemaker, to become "hackable" and controlled by an unauthorized third-party.

4.3 Electronic Medical Records

Physicians often dictate patient charts on their mobile devices or computers, take images on their mobile phones, and check medical records remotely. The major healthcare corporations in the United States have already begun multi-million dollar shifts from paper to electronic records to allow healthcare professionals to have structured methods of keeping patient data. On the other end, patient portals are beginning to be built into these systems whereby patients are slowly being given the option of checking their own health data remotely. This gives patients a chance to read their diagnoses and check their medications list. There is public health potential here, where through Internet resources, patients can monitor side effects of drugs, or monitor their symptoms even more closely using new devices and technology.

A potential danger is that patients risk becoming afflicted with "cyberchondria," a state where patients become hyper-aware of symptoms, and believe they may have diseases and syndromes that they do not actually have. There are likely to be many false positives in certain diagnostic devices and false diagnoses in general as these technologies are developed.

5 Current Devices, Mobile Applications, and Commercial Companies

Mainstream fitness wearable devices from companies such as Fitbit, Jawbone, Garmin, Misfit, Moov, and Apple currently provide users with personal fitness data that includes heart rate, steps taken, stairs climbed, calories consumed, body mass index (BMI), and sleep activity. The appearance of these devices has created a psychological shift in the mindset of users, toward being more proactive with their lifestyle and health. Studies have demonstrated that just being aware of the number of steps taken per day and wearing a device that measures these steps can be powerful enough to motivate the wearer to walk the longer route to a destination, although more extensive studies involving behavior change techniques implemented in technologies need to be performed [14]. Moreover, the ability to analyze health trends and data on an individualized basis can encourage the user to be more self-conscious about health and fitness.

5.1 Gamification and Interactivity

"Gamification" of fitness milestones has become one method of encouraging users to link their peers to their accounts and make their data viewable. In this way they can compete with each other in tasks such as walking 20,000 steps a day, or climbing the most number of stairs in a week. Similarly on the mobile app software end, push notifications are another way to remind or encourage the user to interact with the app interface more often and frequently. The Fogg behavior model describes 3 main elements that must converge in order for there to be a change in behavior: motivation, ability, and trigger [15]. In the case of a fitness wearable device, the motivation is often a desire for better health, and a trigger to act may be facilitated by friends who also own the same wearable. Ability is defined by simplicity factors that include time, money, and physical effort, where the idea is to make a behavior simpler to accomplish.

For many fitness technology companies, much of the success of their devices is owed to the user experience and interface (UX/UI) of their mHealth app and fitness tracker, as well as the culture that has developed around fitness wearables. Ease-of-use, portability, convenience, and devices that require very little time to use are favored by consumers. Additional appeal comes from functions beyond passive data tracking, which includes nudges for time-sensitive events or activities, reminders, or options for personalized programmable prompts.

5.2 Usability

Devices that require the least amount of work to maintain and that allow for the maximum interpretable data collected are therefore ideal, e.g., long battery life or wireless syncing. A major design flaw would be the requirement of a device to be physically connected to a computer in order to sync data, and its marketability would be significantly affected even if the physical design was aesthetically desirable. Devices must excel on both the hardware and software levels. The hardware must be attractive and streamlined, such that the consumer feels good about wearing and using a device in public, and the software must be intuitive and easy to use and follow, offering simple-to-understand analytical data outputs. Admittedly, a portion of user attraction to a product is often due to creative marketing strategies and intelligent branding.

6 New Commercial Opportunities

Bridging into the more specialized mHealth app arena are health trackers such as iRhythm for the detection of cardiac arrhythmias, Neumitra for measuring the autonomic nervous system, or BodyMedia armbands that can additionally measure skin temperature and heat flux. These health trackers can potentially offer clinicians data that can be integrated into more advanced levels of healthcare. An example use case would be using skin temperature and heat flux to predict onset of a hot-and-cold flash during menopause, or to predict the onset of heat stroke. Interpretation and utilization of these collected data offers a unique challenge in the years to come, as our increasingly advanced sensors and technology will allow us to identify and monitor parameters we have not previously considered.

Technology giants such as Google X, Google's semi-secretive facility dedicated to major technological advancements, have continued to research and release various health-related products. For example, Google Glass was created in this facility and has been used by physicians to help with their daily clinical work in ways that include telemedicine, accessing patient records and charts, and during surgery to aid in recognizing anatomical structures. Other interesting innovations include a contact lens capable of detecting glucose levels in tear fluid for diabetic patients, and a project known as Baseline Study. Baseline Study analyzes medical information and uses genomics to define what a healthy human body actually is, or rather, gives a 'baseline' such that deviations from that genomic "norm" will allow a change to be detected that may suggest predisposition to a certain disease state. This data is determined through an aggregate of data from a population of anonymized individuals. The clinician would have the ability to more easily predict the onset of a major disease or condition and provide prophylactic options.

In an effort to combine many of the new mHealth apps and technology, the Smartphone Physical was curated and implemented in 2012 [16, 17]. The

Smartphone Physical includes the use of nine devices total, categorized into 3 broad groups each aimed at a unique demographic. The first group of devices, intended for everyday use by consumers, includes a scale for weight, a blood pressure cuff, and an oxygen saturation monitor. The second group, intended to reach specific patient-groups, includes an ECG, spirometer, and otoscope for ear examination. The third group, for medical providers, includes more advanced devices requiring professional training to interpret, and includes the stethoscope, ophthalmoscope, and ultrasound.

7 Data Management, Analysis, Application

Utility of collected data from a medically approved wearable will be maximized if the output can be easily reviewed and interpreted by the patient's family physician or other clinicians. Application data scientists will be key in properly parsing out large data sets such that usable conclusions, or at least strong correlations can be made between trends and diagnoses. While some sensors are able to offer relatively objective data, e.g. a glucose meter, other sensors that record spirometry, skin tone coloring for stages of bruising, or blood flow may offer more subjective correlations, and large data sets would likely have to be analyzed for statistical significance.

Preventice, as a standout example of how to do things in an informed manner, has created a remotely monitored arrhythmia tracker [18]. Physicians are given access to a 'dashboard' of data on which displays their patient's heart rhythm and other key biometric data. This type of connectivity represents a seamless adaptation of wearable-collected data into clinical decision-making. This type of data usage, however, does not involve more complex embedded algorithms that have thresholds set to help interpret the data. Instead, it continues to require the clinician's expertise to make decisions and diagnoses.

Algorithms that perform more complex analyses of collected data will become more prevalent over the next few years. Clinician input will be most critical when setting the relevant thresholds and clinical significance of individual patient data. AliveCor's FDA-approved atrial fibrillation diagnosis algorithm is a good example of point-of-collection algorithms that are used at the time of device usage to provide ECG data about the heart's current state [19, 20]. Collected information can then be shared with a healthcare professional if any unusual heart rhythms are recorded.

7.1 Processing Big Data

There will be no shortage of data collected by current and future devices, but there will be a lag phase wherein our processing of this large set of data will need to catch up to hardware development. Scientists and clinicians will need to determine how to interpret and utilize vast amounts of collected information. Going forward, for algorithm purposes, monitoring for each disease or health condition will require the creation of 'thresholds' to distinguish between good and ill health. Subsequent clinical trials may need to be conducted to test the thresholds that are set. Given the complexity of the human body, co-morbidities on top of lifestyle choices will make this task of parsing out data even more complex. Machine learning in the form of using patient data to train an electronic ecosystem on what is significant will become widespread practice in mHealth app development. Hypothetically, machine learning could be used to identify the group of healthy individuals that have increasing blood pressures over 4.5 years with a given slope, and determine that this trend may result in a diagnosis of hypertension at the 5-year mark. Similar analyses could be performed for other health conditions.

Within a generation we will witness an effort to replace the need for a live, real-time clinician with a virtual "artificial intelligence" clinician for a general checkup. The ability to perform experiments such as mobile polymerase chain reaction (PCR) will give individuals the ability to self-diagnose, given that the correct primers for a disease biomarker or gene are provided. Advanced technologies could allow for complex microarrays or multivalent ligand targeting to be performed remotely and on a mobile device—potentially useful in diagnosing genetic conditions or infectious diseases. Biopsies of tumors could potentially be performed in the future, where advanced imaging technologies are able to reconcile pathological states and offer a tentative or definitive diagnosis immediately (e.g. suspected squamous or basal cell carcinoma). However, while we are making fast progress from science fiction to reality, a trained pathologist's eye will still be the preferred method of diagnosis for the coming years.

8 Conclusions

Health-tracking wearables have become far more sophisticated over the past few years. Previously only able to measure rudimentary data such as heart rate and steps taken, they are now able to directly measure and/or estimate metrics such as blood glucose, heart rhythm, and brain activity [1].

8.1 Anticipation

Anticipation as a concept within medicine refers to prediction of a health outcome that is informed by collected physiological data. If disease or complication-onset can be anticipated, clinician and patient alike are empowered to redouble their effort to delay or prevent morbidity [5]. There may be an increased dependence by insurance companies on collecting personalized health information that may begin to drive medicine towards a world involving 'genetic discrimination' (not unlike in the science fiction film Gattaca) [21]. It will be critical that clinicians understand that prediction does not always result in onset, just as a healthy lifestyle and behavior do not guarantee a disease-free state. We will want to stray away from examining humans as machines using machines, but instead use these technologies in ways that best augment current healthcare. Since human beings are each unique biological systems, we need to consider each individual as a unique case and not as only part of an aggregate population. We also need to be careful not to further discriminate in healthcare against those who have a predisposition for certain congenital or familial conditions, and to appreciate what makes us organic and distinctive as humans. Part of the human condition involves illnesses and diseases, and new technologies should first aim to alleviate these health conditions before considering the potential creation of super-humans or cyborgs.

8.2 Anticipating New Challenges

Despite the benefits that medical wearables have, there are still major challenges to overcome. Given the diverse depth of collected information, where will wearables fit in FDA guidelines? How will data be resolved and interpreted? How will we determine patient-specific thresholds for various diagnoses and conditions, and can they be generalized to the population? How can we ensure we reach the demographic who will benefit the most from wearable technology? Finally, devices will need to remain portable and simple to use in order to drive widespread market adoption.

Regardless of the challenges on the horizon, medical wearables and mHealth applications objectively provide unique, long-term insight into a patient's health, and will undoubtedly continue to improve patient outcomes as they become more and more sophisticated. The key lies in balancing the use of newer technologies between modern and traditional healthcare. It is important to recognize that the use of technologies will fall at different ends of the spectrum according to their applications and utilization in different medical fields. Some technologies will only be used to further verify a diagnosis or aid in creating one, while other technologies may provide the gold standard for diagnosis or treatment. Certain clinical occupations may be eventually replaced by advanced computer programs and devices— but there is a balance there still. Consider medical imaging diagnostics: someday a machine may be better able to discern minor differences on imaging than a human clinician, but a physician's skill in parsing subtle clues from a patient's history with findings on imaging to weave a diagnosis cannot easily be replaced. Nevertheless, new technologies can augment and improve the lives of patients and general consumers, and have the potential to allow for better, cheaper, and more accurate diagnoses.

References

1. Gaglani, S.M., Topol, E.J.: iMedEd: the role of mobile health technologies in medical education. Acad. Med. **89**(9), 1207–1209 (2014). doi:10.1097/ACM.0000000000000361
2. Quinn, C.C., Clough, S.S., Minor, J.M. et al.: WellDoc™ mobile diabetes management randomized controlled trial: change in clinical and behavioral outcomes and patient and physician satisfaction (2008)
3. Steinhubl, S.R., Muse, E.D., Topol, E.J.: Can mobile health technologies transform health care? JAMA **310**, 2395–2396 (2013). doi:10.1001/jama.2013.281078
4. Hooge, A. (ed.): Five predictions for the future of wearables. Smashingboxes.com (2015)
5. Nadin, M.: Can predictive computation reach the level of anticipatory computing? Int. J. Appl. Res. Inf. Technol. Comput. **5**, 171–200 (2014). doi:10.5958/0975-8089.2014.00011.6
6. Sullivan, J.E.: Clinical trial endpoints. FDA Clinical Investigator Training Course (2012)
7. John Hancock Introduces as Whole New Approach to Life Insurance in the U.S. That Rewards Customers for Healthy Living. PR Newswire (2015)
8. Lecklider, T. (ed.): Powering medical haute couture. Eval. Eng. (2015)
9. Marbury, D. (ed.): 10 Apple HealthKit mobile apps for physicians and their patients. Med. Econ. (2014)
10. Lang, M.B. (ed.): What if Apple were an ACO? HealthCare Purchasing News (2015)
11. Peek, S.T.M., Wouters, E.J.M., van Hoof, J., et al.: Factors influencing acceptance of technology for aging in place: a systematic review. Int. J. Med. Inf. **83**, 235–248 (2014). doi:10.1016/j.ijmedinf.2014.01.004
12. Dehzad, F., Hilhorst, C., de Bie, C., Claassen, E.: Adopting health apps, what's hindering doctors and patients? Health (2014). doi:10.4236/health.2014.616256
13. What Does it Mean for FDA to "Classify" a Medical Device? Center for Devices, Health R
14. Lyons, E.J., Lewis, Z.H., Mayrsohn, B.G., Rowland, J.L.: Behavior change techniques implemented in electronic lifestyle activity monitors: a systematic content analysis. J. Med. Internet Res. **16**(e192) (2014). doi:10.2196/jmir.3469
15. Fogg, B.J. (ed.): What causes behavior change? (2015)
16. Gaglani, S.M. (ed.): A glimpse into the Smartphone physical. MedGadget (2013)
17. Gaglani, S.M., Batista, M.A.: The future of Smartphones in health care. Virtual Mentor **15**, 947 (2013). doi:10.1001/virtualmentor.2013.15.11.stas1-1311
18. Preventice announces commercial availability of BodyGuardian remote patient monitoring system. J. Innovations Card. Rhythm Manage. (2015)
19. Muhlestein, J.B.: QTC intervals can be assessed with the AliveCor heart monitor in patients on Dofetilide for atrial fibrillation. J. Electrocardiol. **48**, 10–11 (2015). doi:10.1016/j.jelectrocard.2014.11.007
20. Tarakji, K.G., Wazni, O.M., Callahan, T., et al.: Using a novel wireless system for monitoring patients after the atrial fibrillation ablation procedure: the iTransmit study. Heart Rhythm **12**, 554–559 (2015). doi:10.1016/j.hrthm.2014.11.015
21. Jabr, F.: Are we too close to making Gattaca a reality? In: Are We Too Close to Making Gattaca a Reality? (2013) http://blogs.scientificamerican.com/brainwaves/are-we-too-close-to-making-gattaca-a-reality/. Accessed 6 Sept 2015

Coaching of Body Awareness Through an App-Guide: The HealthNavigator

Wilko Heuten, Hermie Hermens, Jan-Dirk Hoffmann,
Monique Tabak, Janko Timmermann, Detlev Willemsen,
Anke Workowski and Johannes Technau

Abstract The article describes the process of the development of an app to support people in lifelong secondary prevention, e.g. after a heart attack or with chronic diseases. The goal of this app-guide, called the HealthNavigator, is to coach people walking on leisure paths through guidance, motivation, and better understanding of physical abilities (body awareness). There were three development cycles, each one followed by a demonstrator test with health professionals and patients. The outcomes of these tests were taken into account for the following development cycle. Focus was placed on how to motivate people to exercise regularly and in a healthy manner. The evaluations showed that HealthNavigator was found to be usable tool for teaching body awareness and making users feel safe, while motivating them to take walks in the open air.

W. Heuten (✉) · J. Timmermann
OFFIS Institut Für Informatik, Oldenburg, Germany
e-mail: Wilko.Heuten@offis.de

J. Timmermann
e-mail: Janko.Timmerman@offis.de

H. Hermens · M. Tabak
Roessingh Research and Development, Enschede, The Netherlands
e-mail: h.hermens@rrd.nl

M. Tabak
e-mail: m.tabak@rrd.nl

J.-D. Hoffmann · D. Willemsen · A. Workowski
Schüchtermann-Klinik, Bad Rothenfelde, Germany
e-mail: JHoffmann@schuechtermann-klinik.de

D. Willemsen
e-mail: DWillemsen@schuechtermann-klinik.de

A. Workowski
e-mail: AWorkowski@schuechtermann-klinik.de

J. Technau
GewiNet—Kompetenzzentrum Gesundheitswirtschaft, Osnabrück, Germany
e-mail: johannes.technau@gmail.com

© Springer International Publishing Switzerland 2017
M. Nadin (ed.), *Anticipation and Medicine*, DOI 10.1007/978-3-319-45142-8_11

Keywords Physical activity · Rehabilitation · Health-app · Motivation · User-centered design

1   Introduction

Regular physical activity has nothing but positive effects on the human body. There is one main premise for this: the activity must be exercised individually within an optimal range and at optimal intensity in order to avoid overexertion, injury, and physical problems, such as circulatory or coronary disorders. A moderate aerobic activity of 30 min per day reduces the risk of coronary heart disease [1], and, carried out regularly, it enhances the subjective quality of life [2]. An ideal endurance activity in a moderate range can be performed through taking walks—hiking, trekking, strolling—which can be done by anyone and anywhere in the open air. Controlling intensity while performing endurance activity in an outdoor setting can be difficult, especially for people with little or no experience in this. Control requires objective parameters, such as heart rate, and subjective parameters, such as self-assessment in order for the walker to be coached regarding individual optimal walking intensity. To this end, an app-guide, called HealthNavigator, was developed to coach people during walking on leisure pathways (e.g., streets with sites of historic or cultural interest, paved or unpaved paths in parks, country roads, forest paths). The app provides route guidance, motivation stimulus, and real-time body awareness. The HealthNavigator is an app-guide for people who want to improve their physical condition by hiking outdoors in nature; it can also be used, for example, by persons with cardiovascular diseases, who do not take any medication (e.g. ß-blockers) to regulate heart rate. In that sense, potential users are persons in the phase of lifelong secondary prevention or in the final phase of rehabilitation. *Caveat*: HealthNavigator is part of a research project, not a medical device in the legal sense.

HealthNavigator consists of an app on a smartphone, an app on a smartwatch, and a belt with an integrated ECG. The system should encourage users to optimize their performance and impart confidence in respect to physical capabilities by displaying physical parameters, monitoring self-estimated exertion by the Borg-Scale [3], and providing individual feedback. Due to simultaneous detection of objective and subjective conditions of the user, the HealthNavigator is able to recommend user-tailored load profiles and suggest optimal walking routes. To motivate the user to go for a walk, HealthNavigator provides information on points of interest (such as historical buildings, memorials, and museums in towns or villages), rest areas (such as restaurants or sitting areas with attractive panoramas).

An iterative approach was followed in developing the HealthNavigator, with several cycles building on each other. There were three main iterations, each one consisting of a requirements analysis, system-development, and evaluation. These will be discussed in the sections to follow. Comments from users and professionals helped in adapting the system to achieve steady improvement towards the final demonstration model.

2 Requirements

Although the number of applications and telemedicine systems is rapidly evolving, many of such initiatives do not reach an operational phase. Insufficient technical performance and low usability are considered to be among the major barriers for successful implementation [4]. Involvement of patients and professionals in the requirements analysis and the design process is crucial for adjusting its utilization in the user's daily routine and for successful implementation. Therefore, for developing the requirements of HealthNavigator, we chose an iterative, user-centered design approach. It considers the user as the basis for the design and involves him/her in the evaluation of design choices [5].

A scenario was developed following the PACT approach (People, Activity, Context and Technology) related to patients using a technology in their daily life within a certain (medical) context. Incorporating the principles of evidence-based medicine into PACT scenario development provides starting points for more effective and efficient design of such applications [6]. We studied the literature documenting state-of-the-art research, and we assessed the user's medical knowledge and needs by means of PACT tables and expert interviews. By means of open interviews, medical experts were asked about the target users. The developers provided feedback by proposing FICS extensions to the scenario. (FICS stands for Function and events, Interactions and usability issues, Content and structure, Style and aesthetics, related to system use.) The scenario was updated, containing PACT elements combined with FICS elements (Fig. 1).

After elicitation of the requirements in focus groups with medical experts and patients, the functional requirements were prioritized following MoSCoW (Must have, Should have, Could have, Wouldn't have). Table 1 shows the Must haves (M) and the Should haves (S). Mock-ups were created and evaluated with the patients and specialists. In addition, a field study was performed with the users, to evaluate simple navigation techniques on a smartphone.

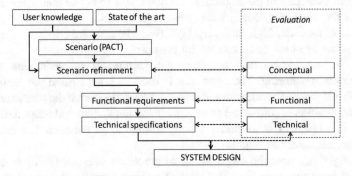

Fig. 1 The HealthNavigator's user-centered design approach (reproduced with permission from [4])

3 HealthNavigator

3.1 Concept

The HealthNavigator application combines two main support systems. First, there is
health status monitoring, which is realized by measuring heart rate, but also by
letting the users reflect their perceived exertion. The information measured by and
entered into the system is then used to give users advice for walking faster or slower
according to their goals. Second is the navigation system, which includes points of
interest. It helps users to navigate on previously defined routes and to adapt the
route according to user needs. It also supports navigation back to the starting point
(e.g., a healthcare facility) using the shortest route (which may be different from the
predefined routes).

While both systems work together, we carefully separated them by using different devices. The navigation system works on the smartphone, since it is not
necessary all the time, because walking routes tend to be straight for a longer time,
and the smartphone provides a big enough screen with high resolution for getting a
clear overview of the map for the route. Health status is constantly monitored.
Interaction with the user should be easy, fast, and not distracting. Thus, we decided
to use a smartwatch for this interaction.

3.2 Health Status Monitoring

The navigation of the HealthNavigator also focuses on the user's exertion level.
Health status monitoring is supposed to keep the exertion of users within appropriate limits. In the HealthNavigator, this is achieved by measuring heart rate and
querying the perceived user exertion. The perceived exertion can be measured using
the Borg-RPE-Scale [4], which allows the rating from 6 (no exertion at all) to 20
(maximum exertion). In the beginning, the users have to define their target exertion
level. In the HealthNavigator, users were able to choose between four different
levels: low, medium, high and very high (see Fig. 2a). Additionally, the fitness
level and the age have to be entered by the users.

The HealthNavigator uses the perceived exertion as the main parameter for
health status monitoring. The heart rate is used as a supportive parameter. This
means that if the system measures exertion different from what the user entered, the
perceived exertion is considered to be correct and the system will adapt to that. The
goal of this behavior is to cultivate body awareness in the user. We created an
algorithm, visualized in Fig. 3.

The algorithm uses the target rating of perceived exertion (RPE) to compute
estimated heart rate corridors. The HealthNavigator uses five heart rate corridors:

Table 1 Requirements of HealthNavigator, showing the Must haves (M) and Should have (S) requirements

1. Start-up and settings		
1.1	For first time use, the system must create a user profile based on responses regarding height, medication, preferences, etc.	M
1.2	At start-up, the system should ask only a subset of questions, relevant for that moment, for the user profile	S
1.3	At start-up, the system must propose suitable routes to the patient, based on the user profile	M
1.4	The app should allow for adjusting the following settings: User profile, Walking or Biking, Routes, Display Parameters	S
1.5	The system can be used with default settings	M
2. Monitoring and sensing		
2.1	The system must specify the user's location in order to provide directions	M
2.2	The system must measure the user's activity in order to show the level of activity and set an activity goal	M
2.3	Activity must also be accurate (i.e., correlates to METs) when the user is cycling	M
2.4	The system can determine the amount of activity the user should aim for, which can be adapted by the healthcare professional	S
2.5	The system should measure physical parameters through additional sensors Heart rate should always be monitored (mandatory) Saturation and breathing should be monitored when appropriate (optional)	S
2.6	Additional information should be calculated from the gathered data: location, time, speed, duration etc.	S
2.7	The system should acquire information about surroundings: weather, points of interest, etc.	S
3. Adaptive routing		
3.1	The system proposes the most suitable route to the patient, which corresponds in difficulty and length to the physical abilities and wishes of the patient, based on the user's profile, history of walks, and past performance	M
3.2	Besides the most suitable route, the system must also provide alternative routes	M
3.3	The user must be able to decide which route to walk/bike	M
3.4	For each route the system must show the difficulty/intensity, prospective arrival time (based on user's profile and most suitable physical performance), and points of interest	M
3.5	On route, the system must provide route navigation	M
3.6	On route, the system must provide information about the surroundings	M
3.7	In case the user gets off the route, the system should be able to guide the user back	S
3.8	The device should support the user in finding the shortest way back	S
4. Display parameters		
4.1	During walking/cycling, the system must be able to display heart rate, time, location, estimated duration	M
4.2	The system should be able to display altitude, slope, oxygen, breathing rate, calories burned. The patient can choose these parameters in the settings	S

(continued)

Table 1 (continued)

	5. Feedback on performance	
5.1	The system must provide feedback about optimal physical performance (e.g., "Slow your pace")	M
5.2	The patient must be alerted when measured physical parameters are beyond the patient's personal thresholds	M
5.3	The system should provide feedback in the form of advice, not instruction	S
5.4	The feedback should be tailored and personal: the system should automatically adapt the feedback to the patient's personal preferences, performance, and available information about the environment	S
5.5	Feedback must be provided in such a manner that it should not be necessary to hold the device in the hand the whole time	M
	6. Reports	
6.1	The system should document the measured parameters and route in a report	M
6.2	The user should be able to make print-outs of the report	M
6.3	Effect of the training should be perceptible in order to increase motivation	S
6.4	Report should be sent in a secure manner	S
6.5	The patient needs to log into gain access to the data	S

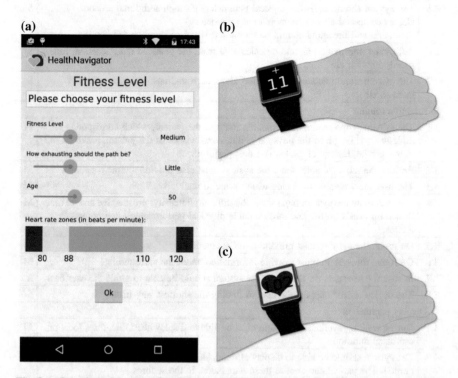

Fig. 2 a Selection of load level and smart watch interaction of the Health Status Monitoring. **b** Borg-RPE-Scale input and **c** heart rate display

Fig. 3 Algorithm for the
health status monitoring using
the perceived exertion as main
parameter

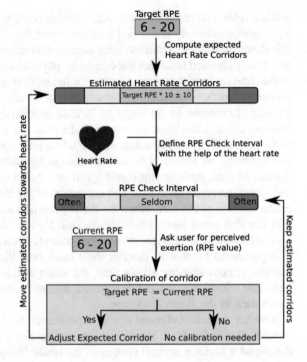

one optimal corridor, two warning corridors, and two alarming corridors. The corridors are estimated based on the user's the age, the target exertion level, and the maximum heart rate of the user, all of which is estimated by the popular formula 210–age. The HealthNavigator then utilizes the measured heart rate to estimate whether the users are within their desired load limits. If the heart rate is within the estimated optimal heart rate corridor, the HealthNavigator assumes that the user has obtained the desired load and only asks about perceived exertion every ten minutes. If the heart rate lies outside the optimal corridor, the system will shorten this interval. If the deviation is only small (low and/or high corridors), the interval is lowered to three minutes. If the deviation is high (too low and/or too high corridors), the interval is lowered to one minute. This way, the users won't be asked for perceived exertion too often if it is very likely that they are within their desired limits.

Just as the correlation between the heart rate and the RPE can be high for everyone, since each person reacts differently to physical activity, the interpretation of the RPE levels is also very individual and can differ, too. To address these issues, the algorithm uses the regularly user-entered RPE values to adjust the estimated heart rate corridors, and in conjunction, the interval of user queries for the RPE. This calibration is made if the actual RPE that the user entered equals the target RPE that the user entered at the beginning. In this case, the user has obviously reached his/her defined target intensity—which is the overall goal. If the estimated heart rate corridors do not reflect this, they must be adjusted. The algorithm does this slowly in order to avoid too great adjustments caused by temporal effects, such as a briefly elevated heart rate not caused by physical activity or by short elevations of the exertion level due to geography (for instance, slopes). We chose to adjust the estimated heart rate corridors by a maximum of 5 beats per minute per adjustment. By way of example: let the target be RPE 12 and the estimated heart rate corridor between 110 and 130 beats per minute. If the heart rate is 100 and the user enters 12 as the current RPE, the algorithm will adjust the estimated heart rate corridor to 105 and 125 beats per minute. This will not change the query interval the first time; but the second time, the heart rate corridor will be reduced to 100 and 120, respectively, and match the actual heart rate. The navigation of the HealthNavigator also focuses on the exertion of the user. When the desired exertion level is selected (Fig. 2a), the user can also select how exhausting the path for the hike should be.

The requirements analysis showed that patients in cardiac rehabilitation consider it important to be able to monitor their heart rate. Since entering the perceived exertion is not necessary all the time, the smart watch shows feedback about the heart rate when no user input is required. Figure 2b and c show both smartwatch views used in the HealthNavigator.

Another important element of the requirements analysis is that the system can provide feedback about the user's optimal performance. Therefore, we have developed a decision support system so the HealthNavigator can provide feedback to help the user achieve the desired exertion level: sending the user messages about walking pace (for example, "Decrease your pace," when, based on the system input, exertion becomes too high). This feature of HealthNavigator is under development and has not yet been fully evaluated.

3.3   Touristic Navigation

In addition to selecting the desired exertion level (Fig. 2), the user can also select how exhausting the path for the walk should be. The HealthNavigator then locates them and downloads available routes from Open Street Maps. These routes were entered beforehand by experts and contain a necessary rating of the route and possible shortcuts in case users feel over-exerted during the hike. The system then shows available paths to the user and colors them according to the previously entered desired exertion level. Routes that fully meet this requirement are marked green, and routes that are too easy are colored yellow. Routes that are too difficult are colored red and display an additional warning symbol to alert the user against choosing a route that is too hard (Fig. 4).

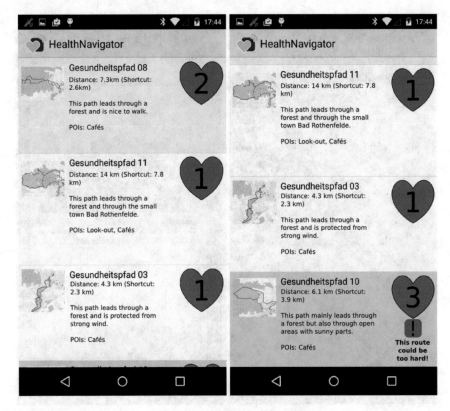

Fig. 4 Examples of route lists indicating suitable routes for the user

After choosing a route, the HealthNavigator displays the navigation view (Fig. 5). It consists of a self-rendered map based on Open Street Map data. This data is downloaded as soon as the route is known. During the walk, this allows the HealthNavigator to always guide the user back to the starting point through the shortest path, even if the mobile data connection fails. The navigation itself is realized using a highlighted blue path displayed on the map. The user can zoom in and out to get a better overview of where the path leads. A dialog for showing the expected amount of time to arrival can be opened and closed by the user.

Every route has a predefined shortcut. When it is reached, the HealthNavigator opens a dialog informing the user about both possible ways to continue. The user can then decide whether he/she would like to walk the original route as planned or take the shortcut (due to overexertion, time planning, or other reasons). The user can always activate the navigation back to the starting point by choosing this option from the application menu.

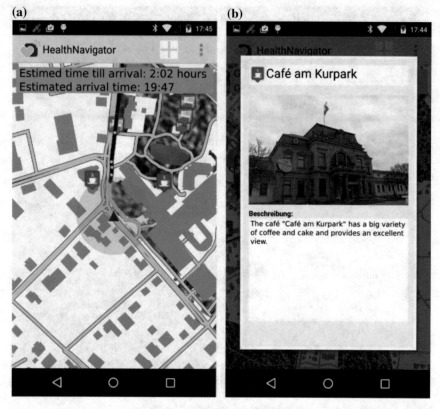

Fig. 5 The navigation view of the HealthNavigator. **a** The navigation map and **b** a detailed display of a point of interest

4 Evaluation

Evaluations have been conducted annually, which allows for incorporating the respective partial results into each iteration of the HealthNavigator. Similarly, the usability of the HealthNavigator has been evaluated by various user tests in several separate studies. The design of the HealthNavigator was set based on these results. Final evaluation of the HealthNavigator was performed under real conditions, such as the handling and using of the device by the test participants. A standardized protocol was used at two testing stagess, consisting of a pre-interview, a field study, and a post-interview.

Participating in the study was a total of 12 subjects (5 in group 1 and 7 in group 2),with a mean age of 56.3 ± 8.7 years (group 1: 48.6 ± 6.9 years; group 2: 61.9 ± 4.7 years) with a range of 42 years to 68 years (group 1: 42–57 years; group 2: 56–68 years). Group 1 consisted of healthy people (four of whom were healthcare professionals) and group 2 consisted of persons undergoing physio-therapy. The subgroups differed significantly in age.

HealthNavigator received a good usability rating (SUS score: 71.0 ± 22.3) based on [7, p. 118]. The majority of the participants (83.3 %) believed that their body awareness will increase by using the HealthNavigator system frequently (group 1: 80 %; group 2: 85.7 %). Healthcare professionals explained that this is the case because dyspnea (i.e., difficult breathing) and other symptoms could be diagnosed. The user will become more aware of his/her body condition and learn about his/her capacity. Nine of all participants (75 %) felt safe and confident with the HealthNavigator system (group 1: 60 %; group 2: 85.7 %); and six participants (50 %) thought that the HealthNavigator system can remove anxiety over exer-cising (group 1: 60 %; group 2: 42.9 %). Four of the seven patients voiced their intention to use HealthNavigator once it becomes available.

The health status monitoring (entering the Borg) was rated the most important current feature of the HealthNavigator, together with the feedback provided. This may be due to the fact that the HealthNavigator is a device with new features, in contrast to the sport navigation devices currently on the market.

The results regarding the features of the HealthNavigator show that the (heart rate) feedback, the tourist information, and the selectable routes are the main motivational functions. The results show that the users are intrinsically motivated to use HealthNavigator to improve their physical condition, to explore the surround-ings, and to learn something about the surroundings, while they find the opinions of others regarding the device of less importance.

5 Discussion

With the HealthNavigator, a new method was implemented to help patients regarding lifelong secondary prevention and for healthy people who want to increase their physical condition. The HealthNavigator combines tourist navigation, health status monitoring, and feedback in order to increase body awareness and to coach users in walking on scenic and tourist routes. Evaluations show that the HealthNavigator is user friendly, can teach body awareness, and makes users feel safe, while motivating them to walk and hike outdoors.

The need for people in general and especially for patients to maintain an active and healthy lifestyle was the original impetus behind the project and was closely related to experiences of project partners with patients in current healthcare. The result—the combination of the measured exertion and perceived exertion—is from a technical and medical point of view, new and opens broad prospects for follow up.

We designed a technical system that the target groups find usable and under-standable. To achieve ease of usability, we ourselves designed the routing, which allowed us to take the individual preferences of older people into account. There are, of course, improvements to be made for further development of the HealthNavigator. For example, to explain the system to a larger audience, a manual is needed. Also more effort needs to be put into the user interface and the integration of sensors in order to obtain better view of the patient while walking.

With the automatic determination of heart rate corridors and the algorithmic verification of heart rate matching the rating of perceived exertion (RPE), we offer the user a unique possibility to control him- or herself and to learn what his or her individual thresholds are. Very interesting would be to further develop the feedback of the HealthNavigator system for increased guidance regarding navigation and health status monitoring. The decision support that provides feedback and coaches the user to acquire the best walking pace is largely developed, but should be precisely evaluated regarding safety, effects on walking behavior, and motivational influence. In respect to the navigation, currently the user receives a message when an alternative route is available; a future system should automatically coach the user to alternative routes when the input data requires doing so. In light of the recent unobtrusive sensors that are becoming more and more available, together with available context information, the system could be extended through more input data (e.g., physical activity, weather), to make the automatic routing and decision support feedback as accurate and tailored to the user as possible.

The testing with two groups (healthcare professionals and potential users in secondary prevention) delivered two different viewpoints: one from the actual user and another from the persons who are likely to recommend and explain the system to the user. Both views are important when you think in terms of market introduction and operational conditions. We received valuable feedback from professionals and patients, who were mainly positive. Physiotherapists from the Netherlands were willing to test the system and prompted its usefulness for their daily work, as well as patients and practitioners from Germany, who were assured that the HealthNavigator would be of great benefit. However, first we need to perform larger scale trials in order to investigate the effects on the walking behavior and quality of life of the users, and the added value for daily practice. And in terms of market introduction, a closer look at the needs of practitioners and general rehabilitation processes must be taken so that HealthNavigator can be incorporated into regular (care) programs.

Acknowledgments HealthNavigator is a research project funded by the INTERREG IV A Programme D/NL.

References

1. Balady, G., et al.: Core components of cardiac rehabilitation/secondary preventionprograms: 2007 update. Circulation **115**(20v), 2675–2682 (2007)
2. Bjarnason-Wehrens, B., et al.: Leitlinie körperliche Aktivität zur Sekundärprävention und Therapie kardiovaskulärer Erkrankungen (2009)
3. Borg, G.: Anstrengungsempfinden und körperliche Aktivität. Deutsches Ärzteblatt **101**(15), 1016–1021 (2004)

4. Broens, T.H.F., et al.: Determinants of successful telemedicine implementations: a literature study. J. Telemedicine Telecare **13**(6), 303–309 (2007)
5. op den Akker, et al.: Development and evaluation of a sensor-based system for remote monitoring and treatment of chronic diseases—the continuous care & coaching platform. In: Proceedings of the 6th International Symposium on eHealth Services and Technologies, EHST 2012, 3–4 July 2012, Geneva, Switzerland. pp. 19–27 (2012)
6. Huis in't Veld R.M.H.A., et al.: A scenario guideline for designing new teletreatments: a multidisciplinary approach. J. Telemedicine Telecare **16**(6), 302–307 (2010)
7. Bangor, A., Kortum, P.T., Miller, J.T.: Determining what individual SUS scores mean: Adding an adjective rating scale. J. Usability Stud. **4**(3), 114–123 (2009)

Part V
Anticipation and Psychological Aspects of Patient Treatments

Anticipation and Child Development

Julie Brisson and Anne-Laure Sorin

Abstract Anticipation serves as an interesting clue in the field of developmental psychology since it can be assessed at a behavioral level during interaction within both social and physical environments. Assessing anticipatory behaviors leads to a better understanding of child development, providing clues to the development of a child's cognitive skills. On a theoretical level, several questions arise. Does a child who anticipates necessarily understand the situation he is in, or does he merely make a perceptive association learned through conditioning? Can it be stated with certainty that a child who does not (or cannot) anticipate does not understand? In the case of a child's typical development, anticipation is also one means for understanding atypical development, such as autism spectrum disorder. On a practical level, whatever the underlying explanation, anticipation deficit could be one interesting screening procedure for detecting developmental disorders as early as possible.

Keywords Anticipation · Autism · Behavior · Development

1 Introduction

In the field of developmental psychology, researchers are attempting to delve into the minds of infants: what they already know at birth, what they learn and how they relate to their environment, how they further develop. Children come to the world within a social environment and progressively learn from observing this environment. They observe and experience several repeated situations, for instance, routines like meal times, bath or care periods, play times, and interactions with parents. If they understand these situations and the repetitive patterns are assimilated, children adapt and

J. Brisson (✉) · A.-L. Sorin
Laboratoire de Psychologie et Neurosciences de la Cognition et de l'Affectivité,
Université de Rouen, Rouen, France
e-mail: julie.brisson@univ-rouen.fr

A.-L. Sorin
e-mail: anne-laure.sorin@univ-rouen.fr

© Springer International Publishing Switzerland 2017
M. Nadin (ed.), *Anticipation and Medicine*, DOI 10.1007/978-3-319-45142-8_12

respond quickly to these various repeated solicitations. That is, they learn. (Learning is a defining characteristic of the living [1]). The information they assimilate goes into building their anticipatory functioning; it becomes part of their anticipation repertory. Seen in this light, a child's actions can reveal clues to the development of anticipation. Thus, observing a child "in action," researchers can track not only a child's understanding of the given situation, but also the role anticipation plays.

Our research attempted to answer two main questions:

1. How typically developing infants understand their world and what can help them to understand it more quickly?
2. Do infants later diagnosed with autism display a lack of anticipation, implying that they do not understand as quickly as typically developing infants do?

1.1 Observing Anticipatory Behavior

As researchers in developmental psychology attempt to enter the minds of infants, they have elaborated sophisticated paradigms: the habituation paradigm [2] and the violation of expectation method [3]. Anticipatory behavior has recently been taken into consideration. It is indeed an interesting clue for researchers because it can be observed, measured, and interpreted in terms of the child's abilities. An anticipatory behavior could be considered a sign that a child recognizes a specific situation, might have understood it, and adapts his behavior. We maintain that there is a link between motor anticipation and cognitive anticipation.

From among the several methods used in developmental psychology, observation is generally employed to assess motoric behavior [4]. Software that allows frame-by-frame viewing is used to assess, for instance, whether the child readies his arms in order to protect his body from a fall. One specific motoric behavior is oculomotor behavior. Eye-tracking technology facilitates new perspectives because it can track the child-subject's visual exploration path. The child's gaze shows whether he anticipates the aim of a behavior as anticipatory saccades are generated toward a target ahead of an ongoing observed action [5]. When a child sees a person eating soup, for example, his anticipatory behavior could be to look at the mouth area before the spoon touches it. This could lead to the conclusion that the child knows the aim of this activity. Anticipation can also be observed in postural changes, especially anticipated postural adjustments. Every motoric action is due to coordination between movement and posture [6, 7]. Posture is usually assessed through kinematics and electromyography recordings [8].

When can anticipatory behavior be observed? What does anticipation reveal about the processes that give rise to it? Are infants able to anticipate from birth? Is anticipation dependent on the context? These questions will be addressed in the first part of this text. The second part will deal with the difficulties of people with autism spectrum disorder (ASD) to anticipate, and the possibility to use such difficulty as a screening criterion.

2 Anticipation—A Working Definition

It is important to define what anticipation means in the context of this research. Anticipation seems to be a familiar concept. People use the word *anticipation* (and its various forms) without giving it much thought. Therefore, it is important to define anticipation, a complex process, in the scientific framework. An anticipatory process is a process "in which the current state results not only from a past state, but especially from a possible future state" [9]. It is expressed in action. It implies understanding the consequences of one's action on oneself and the environment in which the action takes place [10]. Thus one needs not only to be aware of what happens, but also to calculate the possible outcome that will prepare the corresponding action and to carry it out [11]. For example, when a bottle falls from a table and you want to catch it, you need to determine its trajectory in order to prepare your reach and act. In order for a person to conjure this project of action (which takes place within milliseconds [12]), information must be "processed." The perceptions induced by the situation, such as visual or proprioceptive information, are analyzed. The information perceived in the present has to be connected to personal knowledge. Indeed, anticipation requires using a representation of one's self and of the environment, as well as a representation of the interplay between the two. These are not present from birth; they will be accumulated, beginning with birth, through infancy [10] and through a certain point in adulthood [13].

Furthermore, the situation is analyzed in comparison with similar observed or experienced situations that have been stored in memory. Although they are from different theoretical branches in psychology, several authors have included these memorized patterns in their theories. For instance, Piaget [14] defined his action schemes as blocks of knowledge about organized actions that can be generalized or used in similar situations. These knowledge units are constantly updated and utilized for obtaining another major component: the goal of the future action.

The goal is an important concept in anticipation, as is the preparation needed to attain it, taking into account the parameters of the situation, or inferring someone else's goal in a social situation. Preparation—the readiness potential according to Libet [12]—leads to executing the action. Possible outcomes are linked to the goal and correspond to the ideal result of the action before its execution. Finally, the result of the action is compared to this internal model while executing it (see Hilber [15] on this issue). Learning takes place and the anticipation is reinforced—or rejected if it turns out to be wrong.

This process could be effective for analyzing what underlies the actions of adults. But what about children? They progressively conjure or construct a world that is more and more coherent and predictable, leading to the development of their anticipatory capabilities. In the next part, we shall describe how anticipation is utilized to assess the infant's and child's abilities, from the simplest motoric action to more and more complex social situations, and what can help the child to analyze a situation and adapt to it.

3 Typical Child Development

Studies in different domains of developmental psychology show that healthy infants quickly acquire anticipatory capabilities. When it comes to following a regular moving object, infants 18–36 weeks old anticipate the linear path of a rapidly moving object [16]. They are able to assess distances and velocity in space. In a smooth pursuit task, 5-month-old infants will form expectations of the position and time of the target, even if the path is not sinusoidal, but triangular [17]. The same anticipation capability was evinced in children at 2 and 3 months of age when they were shown pictures in a predictable sequence. The infants' eyes moved to where the next picture would appear just before it was shown [18–20].

3.1 Evidence of Anticipation

Postural Adjustment. When a voluntary motion begins, the movement of body parts tends toward the center of gravity and creates a disequilibrium. The posture instantly adjusts so that the body maintains balance while executing the motion. With children, this adjustment is at first retroactive and gradually becomes proactive with experience [7]. At around 15–18 months of age, babies can adapt the posture of their trunk and neck muscles in anticipation in order to reach an object with the arm without losing balance [21, 22]. Between the ages of 18 and 30 months, they progressively learn how to adjust in anticipation the body's center of gravity while taking their first steps [23].

To assess anticipatory postural control at an older age, the bimanual load-lifting paradigm was elaborated [24]. When a tray carried by a participant is either loaded or unloaded, the carrier anticipates the weight transfer. Children between ages 21 and 40 months are more able to stabilize the tray while loading it themselves than when the experimenter loads it [7]. The unloading condition has been studied with older children between 4 and 8 years of age. In these tasks, children learn to integrate the object's weight and center of gravity as they anticipate loading or unloading. The anticipatory postural adjustments (APA) are at first immature and lack synchronization with the movement itself. It improves with maturation and experience and allows the child to perform a greater variety of actions.

3.1.1 Reaching for an Object

When it comes to reaching for an object, most children from 5 months on adapt their grasp to the target by starting to close their hand around the object before touching it [25]. This action is executed in anticipation of, not in reaction to, touching the target's shape. Nevertheless, it is only from 13 months on that a child's grasp is as accurate as an adult's when it comes to smoothness of the movement and accuracy of hand aperture. When two pieces of information about the target have to be processed, 10–12-month-old children adapt their grasp first to the orientation of the object they want to grasp and then to its size [26]. Whereas 12 month-old children are able to simultaneously adapt their hand configuration to both orientation and size of the object they want to grasp, it is more difficult for 10-month old children, who have more of a problem in anticipating the object's size. When only one aspect of the object is present, the younger children can adapt to the object's size; but adding another dimension entails complication. They need advanced preparation abilities (readiness potential) in order to integrate varied information about the object.

As children advance in age, their ability to construct a motor program improves. Proaction becomes more efficient and priority is given to proaction. This was demonstrated with bimanual coordination. Before the age of 11 months, children try to reach for big objects with one hand; they use the second hand after receiving haptic feedback. From 11 months on, children start their reach using both hands, implying better anticipation of and preparation for the final result [27]. From the age of 3 years, they can anticipate a more complex action, pursuing not the easiest path, but the path that will lead them to the most comfortable end-state—although both paths could lead to the same goal. Children not only perform the task, but also anticipate the path. This ability increases with age, and as a function of the task difficulty and children's' expertise in accomplishing the task [28].

3.1.2 Anticipation in Social Situations

Children already use anticipatory behaviors in situations that require no social knowledge. The social environment is more complicated to analyze than the physical environment partly because it is less predictable. In this context children have to deal with their own goals and also with the supposed goals of others. How are actions understood in a social context? When do infants become able to anticipate the goal of a perceived action in a social context?

Children constantly experience social interactions or observe interactive situations [29]. Multiple experiences help them understand the situation as well as predict the goal of the people around them [30]. Experience plays an important role as they learn to adapt their own behaviors in social situations. Any observed action as simple as observing another person transferring a toy into a bucket can imply anticipation. Whereas 12-month-old children and adults anticipate the action by looking ahead of the toy and toward the goal container, 6-month-old infants do not [31].

Elsner et al. [29] showed that the properties of the goal of the social action can help a child to understand a gesture and attribute a meaning. Twelve month-old infants observed an interactive scene of give-and-take in which a hand or an affordant U-shaped object was utilized. The infant's look shifts to the goal more quickly when the receiving hand is presented with the U-shape than when the palm of the hand faces the ground. The infants are also faster when the hand is replaced by a U-shaped object. This proactive gaze-shift to a long-trajectory reach has also been noticed in 6-, 8-, and 10-month-old, but not 4-month-old infants [32].

Dealing with more complicated social actions, 12-month-old infants anticipate the goal of a feeding action while observing as a third party. At 12 months of age, they look particularly at the adult who should receive the food [33].

We could assess that goal anticipation would be displayed much earlier in situations in which the child is directly involved. Some authors support the idea that anticipating the goal of an action is correlated to being able to produce the action (e.g., [5, 34, 35]). Another facilitator to anticipation could also be goal saliency [36]. The multiplicity and heterogeneity of the tasks remind us that anticipation is action specific. Indeed, it implies awareness of the other's goal within an action. This awareness is linked to the knowledge about a situation or an action and potentially to the ability to perform an action or a sequence of actions.

Knowing the child's abilities is important to better understand child development. It is also a prerequisite for discovering differences in developmental disorders. The next part deals with anticipation ability in the case of autism spectrum disorders.

4 Development of Infants with Autism

The fifth edition of the Diagnosis Statistical Manual (DSM-5 [37]) defines autism spectrum disorders as resulting in impairments in everyday life and particularly in social relations, and as having an early onset. The main criteria are a deficit in communication and social interaction, together with repetitive and restricted behaviors, interests, and activities. Anticipation deficit is not included among the main criteria, but it has appeared in several aspects of development.

Children with autism display postural adjustment failure [38]. When they perform a reaching motion, the ability to adapt their speed or the body segments position to their target is delayed. In a bimanual unloading task, they do not present anticipatory contraction of the muscles that stabilize the arm position [39]. They do not display anticipatory postural adjustments, but rather use a feedback control mode. In other words, instead of anticipating, they react [8].

4.1 Autism and Social Interaction

In order to test for social cognition, 3-year-old children were presented with a movie asserting a false belief. A puppet places an object in one of the two available boxes. An actress is watching the scene. While she turns away, the puppet returns, removes the object from the box and leaves. The actress turns back to the scene and wants to grab the object. In this situation, children who can assess false beliefs are able to understand that the actress did not see the puppet removing the object and therefore she believes that the object is still in the box. An eye-tracking device was used to measure whether the child participant anticipates the actress's goal by looking longer and in advance at the box in which the object is believed to be. Results showed that children at risk for autism displayed difficulties in understanding a mental state [40].

4.2   Autism and Timely Diagnosis

Although the scientific and clinical communities are attempting to diagnose a child's autism at the earliest age possible, it is rarely made before the child reaches the age of 3 years. Several studies have been performed with children already diagnosed in order to assess their anticipation capabilities. To learn more about children's early years and their specific development, researchers have two main options: follow infants at high risk for autism (because an older sibling was already diagnosed), and/or study family home movies. Parents recorded these movies before they learned that the child was diagnosed as autistic. Parents agreed to lend a copy so that researchers could analyze their child's behavior in comparison with a group of typically developing infants. Although the parents did not produce this material for research purposes, it does provide a lot of information upon cautious and close analysis. Both types of observation provide information about atypical development.

Several studies using the observation method pointed out the lack of anticipation in young children with autism. In contrast to typically developing children, children with autism do not raise their arms in anticipation of being picked up by an adult [41, 42]. When they are seated and about to fall down, they do not protect themselves by holding out their arms. They may not even realize the consequences of falling [43–45]. Trevarthen and Daniel [46] observed an interaction between a father and his 11-month-old twin daughters. One was diagnosed with autism, the other was not. The typically developing sister displayed anticipatory emotional behaviors. As the interactive climax approached, she showed more joy and pleasure behaviors, such as smiling, shouting, and excitement, until the sequence ran its course. The other sister did not seem to have any expectation concerning the father's behavior and did not show any anticipatory interest.

When researchers observed family home movies of a feeding situation, they noticed that infants—as young as 4 months of age—later diagnosed with autism do not anticipate the spoon's approach. In contrast to typically developing children, they succeeded less in opening their mouth in anticipation when the parent presented the spoon to feed them [4].

Researchers observed infants during their first year of life in order to assess their production of anticipatory smiles during initiation of joint attention in groups of infants with high or low risk for autism [47]. When they want to share their interest in an object, infants usually shift their gaze from the partner to the object of interest. The act is accompanied with smiling. When the smile occurs before the child turns to the adult, as the latter is looking at the object, it is called anticipatory smiling [48]. It typically emerges around the age of 6–12 months [49]. Gangi et al. [47] showed that the sibling with high risk for autism displayed fewer anticipatory smiles than do low-risk children. Nevertheless, they did not find any association between the anticipatory smiling and the severity of the ASD symptoms. These results could be interpreted as signs calling for a broader autism phenotype, one showing early differences in communication. It seems that, depending on the complexity of the task, at-risk children are able to anticipate. Nevertheless, when they learn the behavior in a specific situation, they need more time or more repetition of the situation to be able to perform as well as typically developing children.

All the signs discussed above could be part of a screening tool for targeting early signs of autism at a specific age.

4.3 Open Questions

There are numerous explanations for this lack of anticipation. There are many areas in which shortcomings are acknowledged. Information about perception is treated in some other way, either because of a weakness of central coherence [50], enhanced perceptual functioning [51], or problems with motion vision [52]. Slowing down the visual or auditory stimuli seems to have a positive impact on the subject-child's behavior [53–55]. In regard to their knowledge, self-awareness [56], understanding of others' emotions [57], or the creation of social routines, there is some alteration. It has also been shown that children between ages 6 and 36 months cannot easily detect the contingency within a situation [58]. The lack of anticipation could be due to a lack of interest in social stimuli (e.g. [59]). On the other hand, it could also be due to a motor (dyspraxia-like) disorder [60], or more likely to difficulties in the executive functions, such as readying a sequence of actions or executing the sequence without any mistake [38, 61–63].

Assessing anticipatory capabilities can provide insight into the development of children's cognitive skills. On a theoretical level, some questions remain. Does a child who anticipates necessarily understand the situation he is in, or does he learn perceptive associations through conditioning? Can we say that a child who does not anticipate does not understand? Or might the understanding be good but the readiness potential or execution is altered?

5 Practical Considerations: How Does This Information Help?

In comparison to typical development, anticipation is also a means for understanding atypical development, such as autism spectrum disorder. On a practical level, whatever the underlying explanation, anticipation deficit could be one interesting screening item among others for detecting developmental disorders as early as possible. Early screening could lead to closely following the infants who have an atypical development and offer them playful activities to stimulate their attention, imitation, and communication abilities, etc. The earlier the intervention, the better the development [64–70]. Because of a better development of intellectual abilities, language, and social behavior, early intervention reduces the total cost for society (round £148,000 per person, or 6 % saving, [71]). Early screening also reduces parental distress and anxiety due to diagnosis wandering. Parents notice quite early that their child development is atypical, even during the first year of life [72, 73]. They try to understand their child and his unusual behavior. From this perspective, early screening could lead to more information for the parents and possible invitation to parental support groups.

This goal will be attained in the long term, after performing a series of studies to confirm the trend and assess the differences across diagnoses.

References

1. Nadin, M.: Anticipation—The End Is Where We Start From. Müller Verlag, Basel (2003)
2. Colombo, J., Mitchell, D. W.: Infant visual habituation. Neurobiol. Learn. Mem. **92**(2), 225–234 (2009) http://doi.org/10.1016/j.nlm.2008.06.002
3. Baillargeon, R., Spelke, E., Wasserman, S.: Object permanence in five-month-old infants. Cognition **20**(3), 191–208 (1985)
4. Brisson, J., Warreyn, P., Serres, J., Foussier, S., Adrien, J.-L.: Motor anticipation failure in infants with autism: a retrospective analysis of feeding situations. Autism **16**(4), 420–429. (2011) http://doi.org/10.1177/1362361311423385 (2011)
5. Gredebäck, G., Kochukhova, O.: Goal anticipation during action observation is influenced by synonymous action capabilities, a puzzling developmental study. Exp. Brain Res., **202**(2), 493–497 (2010) http://doi.org/10.1007/s00221-009-2138-1
6. Jover, M., Schmitz, C., Bosdure, E., Chabrol, B., Assaiante, C.: Anticipatory postural adjustments in a bimanual load-lifting task in children with Duchenne muscular dystrophy. Neurosci. Lett. **403**(3), 271–275 (2006) http://doi.org/10.1016/j.neulet.2006.04.054
7. Jover, M., Schmitz, C., Bosdure, E., Chabrol, B., Assaiante, C.: Développement de l'anticipation posturale chez l'enfant sain et pathologique: revue de travaux. Approche Neuropsychologique des Apprentissages chez l'Enfant - A.N.A.E., **81**, 66–75 (2005)
8. Schmitz, C., Martineau, J., Barthélémy, C., Assaiante, C.: Motor control and children with autism: deficit of anticipatory function? Neurosci. Lett. **348**(1), 17–20 (2003) http://doi.org/10.1016/S0304-3940(03)00644-X
9. Nadin, M.: What speaks in favor of an inquiry into anticipatory processes? Prolegomena to the 2nd edition of anticipatory systems, by Robert Rosen. In: Klir, G (ed.) International Book Series on Systems Science and Systems Engineering, Springer, London, xv–lx (2012)
10. Schmitz, C., Forssberg, H.: Atteinte de la motricité dans l'autisme de l'enfant. In: Andres, C., Barthélémy, C., Berthoz, A., Massion, J., Rogé, B. (eds.) L'autisme, De la recherche à la pratique, 227–248. Odile Jacob, Paris (2005)
11. Buser, P.: Où et quand le cerveau anticipe-t-il? In: Berthoz, A., Debru, C. (eds.) Anticipation et prediction. Odile Jacob, Paris (2015)
12. Libet, B.: Unconscious cerebral initiative and the role of conscious Will in voluntary action. Behav. Brain Sci. **8**, 529–566 (1985)
13. Nadin, M.: Project Seneludens. http://seneludens.utdallas.edu/
14. Piaget, J.: La naissance de l'intelligence chez l'enfant. Delachaux et Niestlé, Neuchâtel (1936)
15. Hilber, P.: Influence of the cerebellum in anticipation and mental disorders. In: Nadin, M. (ed.) Anticipation and Medicine, pp. 124–134. Springer, Cham (2016)
16. Von Hofsten, C.: Predictive reaching for moving objects by human infants. J. Exp. Child Psychol. **30**(3), 369–382 (1980) http://doi.org/10.1016/0022-0965(80)90043-0
17. Von Hofsten, C., Rosander, K.: Development of smooth pursuit tracking in young infants. Vis. Res. **37**(13), 1799–1810 (1997) http://doi.org/10.1016/S0042-6989(96)00332-X
18. Haith, M.M., Hazan, C., Goodman, G.S.: Expectation and anticipation of dynamic visual events by 3.5-month-old babies. Child Dev. **59**, 467–479 (1988)

19. Canfield, R.L., Haith, M.M.: Young infants' visual expectations for symmetric and asymmetric stimulus sequences. Dev. Psychol. **27**(2), 198–208 (1991)
20. Wentworth, N., Haith, M. M., Hood, R.: Spatiotemporal regularity and interevent contingencies as information for infants' visual expectations. Infancy **3**(3), 303–321 (2002) http://doi.org/10.1207/S15327078IN0303_2 (2002)
21. Van Der Fits, I. B. M., Hadders-Algra, M.: The development of postural response patterns during reaching in healthy infants. Neurosci. Biobehav. Rev. **22**(4), 521–526 (1998) http://doi.org/10.1016/S0149-7634(97)00039-0
22. Van der Fits, I. B. M., Otten, E., Klip, A. W. J., Eykern, L. A. V., Hadders-Algra, M.: The development of postural adjustments during reaching in 6- to 18-month-old infants Evidence for two transitions. Exp. Brain Res. **126**(4), 517–528 (1999) http://doi.org/10.1007/s002210050760
23. Assaiante, C., Woollacott, M., Amblard, B.: Development of postural adjustment during gait initiation: kinematic and EMG analysis. J. Motor Behav. **32**(3), 211–226 (2000) http://doi.org/10.1080/00222890009601373
24. Hugon, M., Massion, J., Wiesendanger, M.: Anticipatory postural changes induced by active unloading and comparison with passive unloading in man. Pflugers Archiv Eur. J. Physiol. **393**(4), 292–296 (1982) http://doi.org/10.1007/BF00581412
25. Von Hofsten, C., Vishton, P., Spelke, E. S., Feng, Q., Rosander, K.: Predictive action in infancy: tracking and reaching for moving objects. Cognition **67**(3), 255–285 (1998) http://doi.org/10.1016/S0010-0277(98)00029-8
26. Schum, N., Jovanovic, B., Schwarzer, G.: Ten- and twelve-month-olds' visual anticipation of orientation and size during grasping. J. Exp. Child Psychol. **109**(2), 218–231 (2011) http://doi.org/10.1016/j.jecp.2011.01.007
27. Fagard, J., Jacquet, A. Y.: Changes in reaching and grasping objects of different sizes between 7 and 13 months of age. Br. J. Dev. Psychol. **14**(1), 65–78 (1996) http://doi.org/10.1111/j.2044-835X.1996.tb00694.x
28. Knudsen, B., Henning, A., Wunsch, K., Weigelt, M., Aschersleben, G.: The end-state comfort effect in 3- to 8-year-old children in two object manipulation tasks. Cognition 3, **445** (2012) http://doi.org/10.3389/fpsyg.2012.00445
29. Elsner, C., Bakker, M., Rohlfing, K., Gredebäck, G.: Infants' online perception of give-and-take interactions. J. Exp. Child Psychol. **126**, 280–294 (2014) http://doi.org/10.1016/j.jecp.2014.05.007
30. Baldwin, D. A.: Interpersonal understanding fuels knowledge acquisition. Current Dir. Psychol. Sci. **9**(2), 40–45 (2000) http://doi.org/10.1111/1467-8721.00057
31. Falck-Ytter, T., Gredebäck, G., von Hofsten, C.: Infants predict other people's action goals. Nat. Neurosci. **9**(7), 878–879 (2006) http://doi.org/10.1038/nn1729
32. Kanakogi, Y., Itakura, S.: Developmental correspondence between action prediction and motor ability in early infancy. Nat. Commun., **2**, 341 (2011) http://doi.org/10.1038/ncomms1342
33. Gredebäck, G., Melinder, A.: Infants' understanding of everyday social interactions: a dual process account. Cognition **114**(2), 197–206 (2010) http://doi.org/10.1016/j.cognition.2009.09.004
34. Ambrosini, E., Reddy, V., de Looper, A., Costantini, M., Lopez, B., Sinigaglia, C.: Looking Ahead: Anticipatory Gaze and Motor Ability in Infancy. PLoS ONE **8**(7), e67916 (2013) http://doi.org/10.1371/journal.pone.0067916
35. Cannon, E. N., Woodward, A. L., Gredebäck, G., von Hofsten, C., Turek, C.: Action production influences 12-month-old infants' attention to others' actions. Dev. Sci. **15**(1), 35–42 (2012) http://doi.org/10.1111/j.1467-7687.2011.01095.x
36. Gredebäck, G., Stasiewicz, D., Falck-Ytter, T., Rosander, K., von Hofsten, C.: Action type and goal type modulate goal-directed gaze shifts in 14-month-old infants. Dev. Psychol. **45**(4), 1190–1194 (2009) http://doi.org/10.1037/a0015667
37. American Psychiatric Association: Diagnostic and Statistical Manual of Mental Disorders, 5th edn. American Psychiatric Association, Arlington (2013)

38. Forti, S., Valli, A., Perego, P., Nobile, M., Crippa, A., Molteni, M.: Motor planning and control in autism. A kinematic analysis of preschool children. Res. Autism Spectr. Disord. **5** (2), 834–842 (2011) http://doi.org/10.1016/j.rasd.2010.09.013

39. Martineau, J., Schmitz, C., Assaiante, C., Blanc, R., Barthélémy, C.: Impairment of a cortical event-related desynchronisation during a bimanual load-lifting task in children with autistic disorder. Neurosci. Lett. **367**, 298–303 (2004)

40. Gliga, T., Senju, A., Pettinato, M., Charman, T., Johnson, M. H.: The BASIS team: spontaneous belief attribution in younger siblings of children on the autism spectrum. Dev. Psychol. **50**(3), 903–913 (2014) http://doi.org/10.1037/a0034146

41. Lelord, G., Sauvage, G.: L'autisme de l'enfant. Masson, Paris (1991)

42. Ornitz, E. M., Guthrie, D., Farley, A. H.: The early development of autistic children. J. Autism Child. Schizophr. **7**(3), 207–229 (1977) http://doi.org/10.1007/BF01538999

43. Perrot-Beaugerie, A., Hameury, L., Adrien, J.L., Garreau, B., Pepe, M., Sauvage, D.: Autisme du nourrisson et du jeune enfant: intérêt du diagnostic précoce. Annales de pédiatrie **5**, 287–293 (1990)

44. Sauvage, D.: Autisme du nourrisson et du jeune enfant. Masson, Paris (1988)

45. Sauvage, D., Faure, M., Adrien, J.L., Hameury, L., Barthélémy, C., Perrot, A.: Autisme et films familiaux. Ann. de Psychiatr. **3**(4), 418–424 (1988)

46. Trevarthen, C., Daniel, S.: Disorganized rhythm and synchrony: early signs of autism and Rett syndrome. Brain Dev. **27**(1), S25–S34 (2005) http://doi.org/10.1016/j.braindev.2005.03.016

47. Gangi, D. N., Ibañez, L. V., Messinger, D. S.: Joint attention initiation with and without positive affect: risk group differences and associations with ASD symptoms. J. Autism Dev. Disord. **44**(6), 1414–1424 (2014) http://doi.org/10.1007/s10803-013-2002-9

48. Parlade, M. V., Messinger, D. S., Delgado, C. E. F., Kaiser, M. Y., Van Hecke, A. V., Mundy, P. C.: Anticipatory smiling: linking early affective communication and social outcome. Infant Behav. Dev. **32**(1), 33–43 (2009) http://doi.org/10.1016/j.infbeh.2008.09.007

49. Venezia, M., Messinger, D. S., Thorp, D., Mundy, P.: The development of anticipatory smiling. Infancy **6**(3), 397–406 (2004) http://doi.org/10.1207/s15327078in0603_5

50. Frith, U., Happé, F.: Autism: beyond "theory of mind". Cognition **50**(1–3), 115–132 (1994)

51. Mottron, L.: L'autisme: une autre intelligence. Mardaga, Liège (2004)

52. Gepner, B.: «Malvoyance» du mouvement dans l'autisme infantile? La psychiatrie de l'enfant **44**(1), 77–126 (2001)

53. Lainé, F., Rauzy, S., Tardif, C., Gepner, B.: Slowing down the presentation of facial and body movements enhances imitation performance in children with severe autism. J. Autism Dev. Disord. **41**(8), 983–996 (2011) http://doi.org/10.1007/s10803-010-1123-7

54. Gepner, B., Féron, F.: Autism: a world changing too fast for a mis-wired brain? Neurosci. Biobehav. Rev. **33**(8), 1227–1242 (2009) http://doi.org/10.1016/j.neubiorev.2009.06.006

55. Tardif, C., Lainé, F., Rodriguez, M., Gepner, B.: Slowing down presentation of facial movements and vocal sounds enhances facial expression recognition and induces facial-vocal imitation in children with autism. J. Autism Dev. Disord. **37**(8), 1469–1484 (2007) http://doi.org/10.1007/s10803-006-0223-x

56. Lyons, V., Fitzgerald, M.: A typical sense of self in autism spectrum disorders: a neuro-cognitive perspective. In: Fitzgerald, M. (ed.), Recent Advances in Autism Spectrum Disorders, vol. I, InTech (2013) http://doi.org/10.5772/53680

57. Dapretto, M., Davies, M.S., Pfeifer, J.H., Scott, A.A., Sigman, M., Bookheimer, S.Y., Iacoboni, M.: Understanding emotions in others: mirror neuron dysfunction in children with autism spectrum disorders. Nat. Neurosci. **9**, 28–30 (2005)

58. Chin, K.: Evaluation of early social contingency behavior in children with autism. Dissertation Abstracts International: Section B: The Science and Engineering, vol. 70 (1-B), p. 682 (2009)

59. Pierce, K., Conant, D., Hazin, R., Stoner, R., Desmond, J.: Preference for geometric patterns early in life as a risk factor for autism. Arch. Gen. Psychiatry **68**(1), 101–109 (2011). doi:10.1001/archgenpsychiatry.2010.113

60. Rogers S., Benetto L.: Le fonctionnement moteur dans le cas d'autisme. Enfance **1**, 63–73 (2001)
61. Damasio, A.R., Maurer, R.G.: A neurological model for childhood autism. Arch. Neurol. **35** (12), 777–786 (1978)
62. Hughes, C.: Planning problems in autism at the level of motor control. J. Autism Dev. Disord. **26**, 99–109 (1996)
63. Rogers, S.J., Bennetto, L., McEvoy, R., Pennington, B.F.: Imitation and pantomime in high functioning adolescents with autism spectrum disorders. Child Dev. **67**, 2060–2073 (1996)
64. Bibby, P., Eikeseth, S., Martin, N.T., Mudford, O.C., Reeves, D.: Progress and outcomes for children with autism receiving parent-managed intensive interventions. Res. Dev. Disabil. **22**, 425–447 (2001)
65. Boyd, B.A., Odom, S.L., Humphreys, B.P., Sam, A.M.: Infants and toddlers with autism spectrum disorder: early identification and early intervention. J. Early Interv. **32**(2), 75–98 (2010)
66. Corsello, C.M.: Early intervention in autism. Infant Young Child. **18**(2), 74–85 (2005)
67. Fenske, E.C., Zalenski, S., Krantz, P.J., McClannahan, L.E.: Age at intervention and treatment outcome for autistic children in a comprehensive intervention program. Anal. Interv. Dev. Disabil. **5**, 49–58 (1985)
68. Harris, S.L., Handleman, J.S.: Age and IQ at intake as predictors of placement for young children with autism: a four- to six-year follow-up. J. Autism Dev. Disord. **30**(2), 137–142 (2000)
69. Makrygianni, M.K., Reed, P.: A meta-analytic review of the effectiveness of behavioural early intervention programs for children with autistic spectrum disorders. Res. Autism Spectr. Disord. **4**(4), 577–593 (2010)
70. Woods, J.J., Whetherby, A.M.: Early identification of and intervention for infants and toddlers who are at risk for autism spectrum disorder. Lang. Speech Hear. Serv. Sch. **34**(3), 180–193 (2003)
71. Järbrink, K.: Knapp. M.: the economic impact of autism in Britain. Autism **5**(1), 7–22 (2001)
72. Ornitz, E.M., Guthrie, D., Farley, A.H.: The early development of autistic children. J. Autism Schizophr. **7**, 207–229 (1977)
73. Rogers, S.J., DiLalla, D.L.: Age of symptom onset in young children with pervasive developmental disorders. J. Am. Acad. Child Adolesc. Psychiatry **29**(6), 863–872 (1990)

Anticipation and the Neural Response to Threat

Nathaniel G. Harnett, Kimberly H. Wood, Muriah D. Wheelock, Amy J. Knight and David C. Knight

Abstract An important function of emotion is that it allows one to respond more effectively to threats in our environment. The response to threat is an important aspect of emotional behavior given the direct biological impact it has on survival. More specifically, survival is dependent upon the ability to avoid, escape, or defend against a threat once it is encountered. Anticipatory processes supported by neural circuitry that includes the prefrontal cortex and amygdala are critical for the expression and regulation of the emotional response. Further, these anticipatory processes appear to regulate the response to the threat itself. Healthy emotional function is characterized by anticipatory processes that diminish the emotional response to threat. In contrast, emotional dysfunction is characterized by anticipatory processes that lead to an exaggerated threat response. Thus, anticipatory mechanisms play an important role in both healthy and dysfunctional emotional behavior.

Keywords Anticipation · Conditioning · Fear · Emotion · Threat · Regulation

A valuable function of emotion is that it motivates effective responses to important events in our environment. For example, fear motivates defensive responses

N.G. Harnett · K.H. Wood · M.D. Wheelock · D.C. Knight (✉)
Department of Psychology, University of Alabama at Birmingham,
Birmingham, AL, USA
e-mail: knightdc@uab.edu

N.G. Harnett
e-mail: nharnett@uab.edu

K.H. Wood
e-mail: khwood@uab.edu

M.D. Wheelock
e-mail: mwheel@uab.edu

A.J. Knight
Department of Physical Medicine and Rehabilitation,
University of Alabama at Birmingham, Birmingham, AL, USA
e-mail: ajk@uab.edu

© Springer International Publishing Switzerland 2017   219
M. Nadin (ed.), *Anticipation and Medicine*, DOI 10.1007/978-3-319-45142-8_13

(e.g., fight or flight) and promotes rapid associative learning of warning cues and the threats they predict, which is critical for survival. More specifically, the knowledge that a threat is imminent allows one to execute preparatory behaviors in anticipation of the impending threat. Thus, survival is promoted when threats can be anticipated and effectively managed. However, maladaptive anticipatory and threat-elicited responses appear to be linked to anxiety and stress-related disorders. A key goal, then, in the study of emotion is to understand the relationship between the anticipatory response and the response to the threat itself. Specifically, understanding how anticipatory functions influence threat-evoked behavior is important for understanding emotion learning, expression, and regulation processes that mediate anxiety and stress-related disorders.

1 Anticipatory Response

Pavlovian fear conditioning is a procedure that is frequently used to investigate emotion learning, memory, expression, and regulation processes [1–5]. During Pavlovian fear conditioning, an originally innocuous warning cue (i.e., a conditioned stimulus; CS) is typically paired with an innately aversive threat (i.e., an unconditioned stimulus; UCS) that produces a reflexive unconditioned response (UCR). Repeated pairing of the warning cue (CS) and threat (UCS) then elicits an anticipatory response (i.e., a conditioned response; CR) in anticipation of the threat. Thus, CR expression during Pavlovian conditioning reflects anticipation of the forthcoming threat. Skin conductance response (SCR), a measure of sweat gland activity that reflects sympathetic activation of the autonomic nervous system (ANS), is often used as an index of the peripheral emotional response in human fear conditioning research [6–9]. An anticipatory SCR (i.e., the CR) to the CS (i.e., the warning cue) occurs when one has learned the CS-UCS (i.e., cue-threat) contingency and serves as an objective measure of associative learning. Previous Pavlovian fear conditioning research has demonstrated that anticipation of threat initiates preparatory responses that promote behavioral and physiological reactions that minimize harm [10–14]. For example, conditioned hypoalgesia (decreased sensitivity to painful stimuli) develops during fear conditioning, reducing the pain produced by noxious stimuli [12, 15]. A similar process appears to diminish the ANS response to the threat itself during fear conditioning [16–20]

2 Threat-Elicited Response

In contrast to the anticipatory response (CR), the response (UCR) elicited by an aversive threat (UCS) is typically considered an innate and automatic reaction that does not require learning. However, learning-related changes in the response to threat itself frequently develop during conditioning [10, 17, 20–25]. Specifically, the predictability of threat (UCS) modulates the magnitude of the threat-elicited response such that a diminished response is produced by predictable threat (UCS that follows a CS) compared to unpredictable threat (UCS presented alone). Thus, there is a conditioned reduction in the response to predictable versus unpredictable threat (conditioned UCR diminution) [16, 19, 23, 25, 26]. These findings are consistent with learning theory, which states:

1. Learning occurs when there is a discrepancy between expectations and outcomes.
2. The CS gains discriminatory control over the UCR to the UCS.
3. The UCR is diminished by predictable compared to unpredictable threat [5, 27, 28].

Thus, an enhanced anticipatory response to the warning versus safety cue demonstrates that the cue-threat association has been learned.

The anticipatory response (the CR to the CS+) appears to be essential for the diminution of the response to the threat itself (UCS). Specifically, conditioned diminution of the response to threat develops when the threat follows the warning cue (CS+), but not when the threat follows the safety cue (CS−) or when the threat is presented alone [16, 18, 19, 23, 25, 26]. Further, as the magnitude of the anticipatory response increases, the magnitude of the response to predictable threat decreases [23, 25]. However, a similar relationship is not observed when the threat is unpredictable. That is, the anticipatory response does not vary with the response to threat itself when threat unexpectedly follows a learned safety cue [23, 25]. These findings suggest that an anticipatory response specific to the warning cue is necessary for conditioned diminution of the response to threat.

3 Neural Substrates of Anticipatory and Threat-Elicited Responses

A neural network that includes the amygdala and prefrontal cortex (PFC) supports anticipatory processes and appears to regulate the emotional response to threat. The amygdala is a critical component of the neural circuit that mediates fear learning and expression of conditioned fear [2, 7, 15, 17, 29–34]. The amygdala receives information about the warning cue and the threat (CS and UCS), forms the cue-threat association, and projects to other brain regions (such as the periaqueductal gray, hypothalamus, and ventral tegmental area) to control the peripheral expression of emotion [15, 32, 35]. The conditioned emotional response mediated by the amygdala appears to be regulated by projections from the PFC [1, 36]. This PFC-amygdala circuitry is critical for the expression and regulation of conditioned changes in the peripheral emotional response. Further, the PFC and amygdala

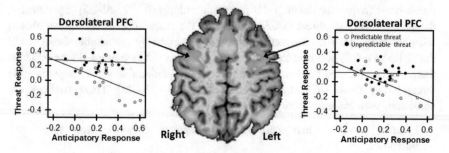

Fig. 1 Relationship between anticipatory and threat-related brain activity. Anticipatory activity was inversely related to threat-related activity on predictable trials within dorsolateral prefrontal cortex (PFC). In contrast, anticipatory activity did not modulate the response to unpredictable threat. These findings suggest that anticipatory activity inhibits threat-related activity within the dorsolateral PFC

appear to support processes that mediate the modulatory effect the anticipatory response (CR) has on the response to threat itself (UCR).

The amygdala and dorsal regions of the PFC (dorsomedial and dorsolateral PFC) show increased anticipatory activity to the warning cue (CR to the CS+) during fear conditioning [7, 30, 31, 37–40]. These same brain regions show diminished responses to predictable compared to unpredictable threat during fear conditioning [17, 24, 25, 41], which closely mirrors the behavioral response to threat itself [17, 21–23, 41]. Further, the dorsolateral PFC demonstrates an inverse relationship between anticipatory activity to the warning cue and the response to the threat (Fig. 1), such that as anticipatory activity increases, threat-evoked activity decreases [24, 25]. The anticipatory activity within the dorsolateral PFC appears to be particularly important for regulating threat-related activity within other brain regions such as the ventromedial PFC and the amygdala [25]. Further, our prior work demonstrates greater dorsolateral PFC connectivity to the dorsomedial PFC, ventromedial PFC, and amygdala during predictable compared to unpredictable threat [42]. Thus, anticipatory dorsolateral PFC activity appears to regulate threat-related responses, and may support healthy emotion regulation (Fig. 2).

Threat predictability and controllability appear to interact to influence the neural response to threat. Specifically, ventromedial PFC and hippocampal activity varies with the predictability and controllability of threats [43]. Activity within these brain regions is diminished when threats are both predictable and controllable. In contrast, ventromedial PFC and hippocampal activity is enhanced when threats are unpredictable and/or uncontrollable. Further, the stress-ameliorating effects observed when one has control over a threat appear to be mediated by the ventromedial PFC [44] and hippocampus [45]. For example, a prior encounter with a controllable threat modifies the ventromedial PFC function that regulates amygdala activity and controls the emotional response elicited by future threats [46–49]. These findings suggest that the ability to predict and control threat has an important impact on ventromedial PFC and hippocampal function. Given that the

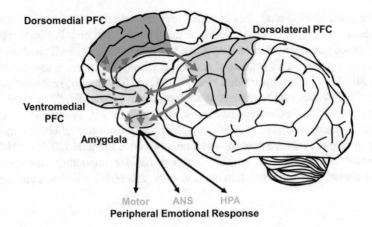

Fig. 2 Neural circuit hypothesized to regulate the emotional response to threat. Anticipatory dorsolateral PFC activity regulates threat-elicited activity within the PFC and amygdala (*solid lines*). Connectivity between other regions of the PFC and amygdala (*dotted lines*) also appears to influence emotion expression. In turn, the amygdala controls learning-related changes in the peripheral emotional response. *PFC* prefrontal cortex; *ANS* autonomic nervous system; *HPA* hypothalamic-pituitary-adrenal axis

ventromedial PFC and hippocampus are important components of the neural circuit that regulates emotion [45–47, 50–54], these brain structures may support processes that mediate the stress resilience that develops when one can control an imminent threat.

4 Understanding Internalizing Disorders

Neural circuitry that supports the cue-threat association may be important for understanding internalizing disorders. For example, individuals with anxiety and stress-related disorders display greater amygdala activity to warning cues compared to healthy controls [55–57]. Further, anxious individuals tend to show greater anticipatory activity to both warning and safety cues compared to healthy individuals [58]. This hyperarousal to warning cues persists even once the cue-threat contingency changes [54, 59, 60]. Thus, enhanced anticipatory activity observed in patient populations may fail to regulate the response to actual threats within the environment. Specifically, individuals with anxiety and stress-related disorders may show enhanced brain (amygdala) activity to predictable threat instead of the diminished response typically observed in healthy individuals. The enhanced threat-elicited response may be mediated, in part, by disruption of functional connectivity of the PFC-amygdala network. In turn, hyperactivity within this neural circuitry may mediate the hyperarousal associated with anxiety and stress-related disorders.

Disruption of PFC-amygdala circuitry may mediate the emotional disinhibition that characterizes many anxiety and stress-related disorders. For example, the hyperarousal associated with anxiety and stress may be due to insufficient top-down regulatory control. In fact, anxiety and stress disorders are often associated with hypoactivation of the ventromedial PFC [50, 54, 61, 62]. In turn, ventromedial PFC hypoactivation may lead to hyperactivation of the amygdala [50, 54, 63–65]. Thus, dysfunction of the PFC-amygdala circuit may mediate key symptoms of anxiety and stress-related disorders [52, 66–68]. Prior work from our laboratory indicates anticipatory processes regulate the emotional response to threat [20, 23, 25]. Thus, anticipatory processes that typically support healthy regulatory functions may instead disrupt emotion regulation and increase susceptibility to stress and anxiety.

5 Conclusion

Anticipation of threat is an important process that facilitates the healthy regulation of the emotional response to threat. Associative learning of the cue-threat relationship supports anticipatory processes that mediate the conditioned diminution of the response to threat. Conditioned diminution of the emotional response to threat appears to be mediated by a PFC-amygdala network. Anticipatory processes supported by the dorsolateral PFC regulate the dorsomedial PFC, ventromedial PFC, and amygdala response to threat. The ability to predict and control threats is a critical aspect of emotional resilience [43, 69–73]. Evidence suggests that the ventromedial PFC and hippocampus play an important role in the emotion regulation process [51, 52, 68, 74–76]. Thus, these brain regions appear to mediate functions that are important for stress resilience. Therefore, dysfunction of the PFC-amygdala network may result in maladaptive anticipatory processes that disrupt emotion regulation in the face of threat, and may be responsible for the emotional dysfunction associated with anxiety and stress-related disorders.

References

1. Delgado, M.R., Nearing, K.I., LeDoux, J.E., Phelps, E.A.: Neural circuitry underlying the regulation of conditioned fear and its relation to extinction. Neuron 59, 829–838 (2008)
2. Knight, D.C., Cheng, D.T., Smith, C.N., Stein, E.A., Helmstetter, F.J.: Neural substrates mediating human delay and trace fear conditioning. J. Neurosci. 24, 218–228 (2004)
3. Knight, D.C., Smith, C.N., Stein, E.A., Helmstetter, F.J.: Functional MRI of human Pavlovian fear conditioning: patterns of activation as a function of learning. NeuroReport 10, 3665–3670 (1999)
4. LeDoux, J.E.: Emotion circuits in the brain. Annu. Rev. Neurosci. 23, 155–184 (2000)
5. Wagner, A. R., Brandon, S. E., Klein, S., Mowrer, R.: Evolution of a structured connectionist model of Pavlovian conditioning (AESOP). Contemporary learning theories: Pavlovian conditioning and the status of traditional learning theory 149–189 (1989)

6. Critchley, H.D., Elliott, R., Mathias, C.J., Dolan, R.J.: Neural activity relating to generation and representation of galvanic skin conductance responses: a functional magnetic resonance imaging study. J. Neurosci. **20**, 3033–3040 (2000)
7. Knight, D.C., Nguyen, H.T., Bandettini, P.A.: The role of the human amygdala in the production of conditioned fear responses. Neuroimage **26**, 1193–1200 (2005)
8. Lykken, D.T., Venables, P.H.: Direct measurement of skin conductance: a proposal for standardization. Psychophysiology **8**, 656–672 (1971)
9. Mangina, C.A., Beuzeron-Mangina, J.H.: Direct electrical stimulation of specific human brain structures and bilateral electrodermal activity. Int. J. Psychophysiol. **22**, 1–8 (1996)
10. Domjan, M.: Pavlovian conditioning: a functional perspective. Annu. Rev. Psychol. **56**, 179–206 (2005)
11. Franchina, J.J.: Escape behavior and shock intensity: within-subject versus between-groups comparisons. J. Comp. Physiol. Psychol. **69**, 241–245 (1969)
12. Helmstetter, F.J., Bellgowan, P.S.: Lesions of the amygdala block conditional hypoalgesia on the tail flick test. Brain Res. **612**, 253–257 (1993)
13. Kamin, L.J.: Traumatic avoidance learning: the effects of CS-US interval with a trace-conditioning procedure. J. Comp. Physiol. Psychol. **47**, 65–72 (1954)
14. Kim, J.J., Jung, M.W.: Neural circuits and mechanisms involved in Pavlovian fear conditioning: a critical review. Neurosci. Biobehav. Rev. **30**, 188–202 (2006)
15. Helmstetter, F.J.: The amygdala is essential for the expression of conditional hypoalgesia. Behav. Neurosci. **106**, 518–528 (1992)
16. Baxter, R.: Diminution and recovery of the UCR in delayed and trace classical GSR conditioning. J. Exp. Psychol. **71**, 447–451 (1966)
17. Dunsmoor, J.E., Bandettini, P.A., Knight, D.C.: Neural correlates of unconditioned response diminution during Pavlovian conditioning. Neuroimage **40**, 811–817 (2008)
18. Marcos, J.L., Redondo, J.: Effects of conditioned stimulus presentation on diminution of the unconditioned response in aversive classical conditioning. Biol. Psychol. **50**, 89–102 (1999)
19. Rust, J.: Unconditioned response diminution in the skin resistance response. J. Gen. Psychol. **95**, 77–84 (1976)
20. Knight, D.C., Waters, N.S., King, M.K., Bandettini, P.A.: Learning-related diminution of unconditioned SCR and fMRI signal responses. NeuroImage **49**, 843–848 (2010)
21. Canli, T., Detmer, W.M., Donegan, N.H.: Potentiation or diminution of discrete motor unconditioned responses (rabbit eyeblink) to an aversive pavlovian unconditioned stimulus by two associative processes: conditioned fear and a conditioned diminution of unconditioned stimulus processing. Behav. Neurosci. **106**, 498–508 (1992)
22. Canli, T., Donegan, N.H.: Conditioned diminution of the unconditioned response in rabbit eyeblink conditioning: Identifying neural substrates in the cerebellum and brainstem. Behav. Neurosci. **109**, 874–892 (1995)
23. Knight, D.C., Lewis, E.P., Wood, K.H.: Conditioned diminution of the unconditioned skin conductance response. Behav. Neurosci. **125**, 626–631 (2011)
24. Wood, K. H., Kuykendall, D., Ver Hoef, L. W., Knight, D. C.: Neural substrates underlying learning-related changes of the unconditioned fear response. Open Neuroimaging J. **7**, 41–52 (2013)
25. Wood, K. H., Ver Hoef, L. W., Knight, D. C.: Neural mechanisms underlying the conditioned diminution of the unconditioned fear response. Neuroimage **60**, 787–799 (2012)
26. Kimmel, E.: Judgments of UCS intensity and diminution of the UCR in classical GSR conditioning. J. Exp. Psychol. **73**, 532–543 (1967)
27. Rescorla, R.A.: Pavlovian conditioning: It's not what you think it is. Am. Psychol. **43**, 151–160 (1988)
28. Rescorla, R. A., Wagner, A. R.: A theory of Pavlovian conditioning: variations in the effectiveness of reinforcement and nonreinforcement. Class. Conditioning: Current Res. Theory (1972)

29. Amaral, D. G., Price, J. L., Pitkanen, A., Carmichael, S. T.: Anatomical organization of the primate amygdaloid complex. In: The Amygdala: Neurobiological Aspects of Emotion, Memory, and Mental Dysfunction 1–66 (1992)

30. Cheng, D.T., Knight, D.C., Smith, C.N., Helmstetter, F.J.: Human amygdala activity during the expression of fear responses. Behav. Neurosci. **120**, 1187–1195 (2006)

31. Cheng, D.T., Knight, D.C., Smith, C.N., Stein, E.A., Helmstetter, F.J.: Functional MRI of human amygdala activity during Pavlovian fear conditioning: stimulus processing versus response expression. Behav. Neurosci. **117**, 3–10 (2003)

32. Davis, M., Walker, D.L., Miles, L., Grillon, C.: Phasic vs sustained fear in rats and humans: role of the extended amygdala in fear vs anxiety. Neuropsychopharmacology **35**, 105–135 (2010)

33. Knight, D.C., Nguyen, H.T., Bandettini, P.A.: The role of awareness in delay and trace fear conditioning in humans. Cogn. Affect. Behav. Neurosci. **6**, 157–162 (2006)

34. LeDoux, J.E., Cicchetti, P., Xagoraris, A., Romanski, L.M.: The lateral amygdaloid nucleus: sensory interface of the amygdala in fear conditioning. J. Neurosci. **10**, 1062–1069 (1990)

35. Fendt, M., Fanselow, M.: The neuroanatomical and neurochemical basis of conditioned fear. Neurosci. Biobehav. Rev. **23**, 743–760 (1999)

36. Urry, H.L., Van Reekum, C.M., Johnstone, T., Kalin, N.H., Thurow, M.E., Schaefer, H.S., Jackson, C.A., Frye, C.J., Greischar, L.L., Alexander, A.L.: Amygdala and ventromedial prefrontal cortex are inversely coupled during regulation of negative affect and predict the diurnal pattern of cortisol secretion among older adults. J. Neurosci. **26**, 4415–4425 (2006)

37. Büchel, C., Morris, J., Dolan, R.J., Friston, K.J.: Brain systems mediating aversive conditioning: an event-related fMRI study. Neuron **20**, 947–957 (1998)

38. Dunsmoor, J.E., Bandettini, P.A., Knight, D.C.: Impact of continuous versus intermittent CS-UCS pairing on human brain activation during Pavlovian fear conditioning. Behav. Neurosci. **121**, 635–642 (2007)

39. Hugdahl, K., Berardi, A., Thompson, W.L., Kosslyn, S.M., Macy, R., Baker, D.P., Alpert, N. M., LeDoux, J.E.: Brain mechanisms in human classical conditioning: a PET blood flow study. NeuroReport **6**, 1723–1728 (1995)

40. Knight, D.C., Smith, C.N., Cheng, D.T., Stein, E.A., Helmstetter, F.J.: Amygdala and hippocampal activity during acquisition and extinction of human fear conditioning. Cogn. Affect. Behav. Neurosci. **4**, 317–325 (2004)

41. Knight, D.C., Waters, N.S., Bandettini, P.A.: Neural substrates of explicit and implicit fear memory. Neuroimage **45**, 208–214 (2009)

42. Wheelock, M., Sreenivasan, K., Wood, K., Ver Hoef, L., Deshpande, G., Knight, D.: Threat-related learning relies on distinct dorsal prefrontal cortex network connectivity. NeuroImage **102**, 904–912 (2014)

43. Wood, K. H., Wheelock, M. D., Shumen, J. R., Bowen, K. H., VerHoef, L. W., Knight, D. C.: Controllability modulates the neural response to predictable but not unpredictable threat in humans. NeuroImage **119**, 371–381 (2015)

44. Baratta, M.V., Zarza, C.M., Gomez, D.M., Campeau, S., Watkins, L.R., Maier, S.F.: Selective activation of dorsal raphe nucleus-projecting neurons in the ventral medial prefrontal cortex by controllable stress. Eur. J. Neurosci. **30**, 1111–1116 (2009)

45. Amat, J., Matus-Amat, P., Watkins, L.R., Maier, S.F.: Escapable and inescapable stress differentially and selectively alter extracellular levels of 5-HT in the ventral hippocampus and dorsal periaqueductal gray of the rat. Brain Res. **797**, 12–22 (1998)

46. Baratta, M.V., Christianson, J.P., Gomez, D.M., Zarza, C.M., Amat, J., Masini, C.V., Watkins, L.R., Maier, S.F.: Controllable versus uncontrollable stressors bi-directionally modulate conditioned but not innate fear. Neuroscience **146**, 1495–1503 (2007)

47. Baratta, M.V., Lucero, T.R., Amat, J., Watkins, L.R., Maier, S.F.: Role of the ventral medial prefrontal cortex in mediating behavioral control-induced reduction of later conditioned fear. Learn. Mem. **15**, 84–87 (2008)

48. Amat, J., Paul, E., Watkins, L.R., Maier, S.F.: Activation of the ventral medial prefrontal cortex during an uncontrollable stressor reproduces both the immediate and long-term protective effects of behavioral control. Neuroscience **154**, 1178–1186 (2008)
49. Maier, S.F., Amat, J., Baratta, M.V., Paul, E., Watkins, L.R.: Behavioral control, the medial prefrontal cortex, and resilience. Dialogues Clin. Neurosci. **8**, 397–407 (2006)
50. Rauch, S.L., Shin, L.M., Phelps, E.A.: Neurocircuitry models of posttraumatic stress disorder and extinction: human neuroimaging research–past, present, and future. Biol. Psychiatry **60**, 376–382 (2006)
51. Milad, M.R., Wright, C.I., Orr, S.P., Pitman, R.K., Quirk, G.J., Rauch, S.L.: Recall of fear extinction in humans activates the ventromedial prefrontal cortex and hippocampus in concert. Biol. Psychiatry **62**, 446–454 (2007)
52. Hartley, C.A., Phelps, E.A.: Changing fear: the neurocircuitry of emotion regulation. Neuropsychopharmacology **35**, 136–146 (2010)
53. Schiller, D., Kanen, J.W., LeDoux, J.E., Monfils, M.-H., Phelps, E.A.: Extinction during reconsolidation of threat memory diminishes prefrontal cortex involvement. Proc. Natl. Acad. Sci. USA **110**, 20040–20045 (2013)
54. Milad, M.R., Pitman, R.K., Ellis, C.B., Gold, A.L., Shin, L.M., Lasko, N.B., Zeidan, M.A., Handwerger, K., Orr, S.P., Rauch, S.L.: Neurobiological basis of failure to recall extinction memory in posttraumatic stress disorder. Biol. Psychiatry **66**, 1075–1082 (2009)
55. Etkin, A., Wager, T. D.: Functional neuroimaging of anxiety: a meta-analysis of emotional processing in PTSD, social anxiety disorder, and specific phobia. Am. J. Psychiatry **164**, 1476–1488 (2007)
56. Indovina, I., Robbins, T.W., Núñez-Elizalde, A.O., Dunn, B.D., Bishop, S.J.: Fear-conditioning mechanisms associated with trait vulnerability to anxiety in humans. Neuron **69**, 563–571 (2011)
57. Phan, K.L., Fitzgerald, D.A., Nathan, P.J., Tancer, M.E.: Association between amygdala hyperactivity to harsh faces and severity of social anxiety in generalized social phobia. Biol. Psychiatry **59**, 424–429 (2006)
58. Lissek, S., Powers, A.S., McClure, E.B., Phelps, E.A., Woldehawariat, G., Grillon, C., Pine, D.S.: Classical fear conditioning in the anxiety disorders: a meta-analysis. Behav. Res. Ther. **43**, 1391–1424 (2005)
59. Milad, M.R., Orr, S.P., Lasko, N.B., Chang, Y., Rauch, S.L., Pitman, R.K.: Presence and acquired origin of reduced recall for fear extinction in PTSD: results of a twin study. J. Psychiatr. Res. **42**, 515–520 (2008)
60. Milad, M.R., Rauch, S.L., Pitman, R.K., Quirk, G.J.: Fear extinction in rats: implications for human brain imaging and anxiety disorders. Biol. Psychol. **73**, 61–71 (2006)
61. Bremner, J.D., Vermetten, E., Vythilingam, M., Afzal, N., Schmahl, C., Elzinga, B., Charney, D.S.: Neural correlates of the classic color and emotional stroop in women with abuse-related posttraumatic stress disorder. Biol. Psychiatry **55**, 612–620 (2004)
62. Haas, B.W., Garrett, A., Song, S., Reiss, A.L., Carrio, V.G.: Reduced hippocampal activity in youth with posttraumatic stress symptoms: an fMRI study. J. Pediatr. Psychol. **35**, 559–569 (2010)
63. Jovanovic, T., Ressler, K.J.: How the neurocircuitry and genetics of fear inhibition may inform our understanding of PTSD. Am. J. Psychiatry **167**, 648–662 (2010)
64. Briscione, M.A., Jovanovic, T., Norrholm, S.D.: Conditioned fear associated phenotypes as robust, translational indices of trauma-, stressor-, and anxiety-related behaviors. Front. Psychiatry **5**, 1–9 (2014)
65. Robison-Andrew, E.J., Duval, E.R., Nelson, C.B., Echiverri-Cohen, A., Giardino, N., Defever, A., Norrholm, S.D., Jovanovic, T., Rothbaum, B.O., Liberzon, I., Rauch, S.A.M.: Changes in trauma-potentiated startle with treatment of posttraumatic stress disorder in combat Veterans. J. Anxiety Disord. **28**, 358–362 (2014)
66. Sripada, R.K., King, A.P., Welsh, R.C., Garfinkel, S.N., Wang, X., Sripada, C.S., Liberzon, I.: Neural dysregulation in posttraumatic stress disorder: evidence for disrupted equilibrium between salience and default mode brain networks. Psychosom. Med. **74**, 904–911 (2012)

67. Birn, R.M., Patriat, R., Phillips, M.L., Germain, A., Herringa, R.J.: Childhood maltreatment and combat posttraumatic stress differentially predict fear-related fronto-subcortical connectivity. Depress. Anxiety **31**, 880–892 (2014)
68. Gilmartin, M. R., Balderston, N. L., Helmstetter, F. J.: Prefrontal cortical regulation of fear learning. Trends Neurosci. **37**, 1–10 (2014)
69. Maier, S.F., Seligman, M.E.: Learned helplessness: theory and evidence. J. Exp. Psychol. Gen. **105**, 3–46 (1976)
70. Foa, E.B., Zinbarg, R., Rothbaum, B.O.: Uncontrollability and unpredictability in post-traumatic stress disorder: an animal model. Psychol. Bull. **112**, 218–238 (1992)
71. Chorpita, B.F., Barlow, D.H.: The development of anxiety: the role of control in the early environment. Psychol. Bull. **124**, 3–21 (1998)
72. McNally, G.P., Johansen, J.P., Blair, H.T.: Placing prediction into the fear circuit. Trends Neurosci. **34**, 283–292 (2011)
73. Kerr, D.L., McLaren, D.G., Mathy, R.M., Nitschke, J.B.: Controllability modulates the anticipatory response in the human ventromedial prefrontal cortex. Front. Psychol. **3**, 1–11 (2012)
74. Franklin, T.B., Saab, B.J., Mansuy, I.M.: Neural mechanisms of stress resilience and vulnerability. Neuron **75**, 747–761 (2012)
75. Russo, S.J., Murrough, J.W., Han, M.-H., Charney, D.S., Nestler, E.J.: Neurobiology of resilience. Nat. Neurosci. **15**, 1475–1484 (2012)
76. Maren, S.: Fear of the unexpected: hippocampus mediates novelty-induced return of extinguished fear in rats. Neurobiol. Learn. Mem. **108**, 88–95 (2014)

Part VI
Anticipation and Ubiquitous Computing

Smart Watches for Physiological Monitoring: A Case Study on Blood Pressure Measurement

Viswam Nathan, Simi Susan Thomas and Roozbeh Jafari

Abstract Given the close coupling between wearable devices and the human body, a natural application for these devices is to inform the user of their physiological status. An important point is the potential for wearables to empower the user to monitor his/her own health rather than rely solely on medical professionals or sophisticated medical equipment. This makes it more convenient for the user to monitor certain vital signs more often and this continuous monitoring can prove crucial in diagnosing certain conditions. Pervasive monitoring allows for collection of a large amount of data, which in turn can allow treatments in medicine that are anticipatory rather than reactionary. We emphasize the importance of the design of convenient, wearable physiological sensors by looking in-depth at one specific wearable device called BioWatch, which can be used to non-invasively and continuously measure blood pressure from the wrist.

Keywords Anticipatory · Wearable sensors · Non-invasive blood pressure measurement · Pulse transit time

1 Introduction

Medicine has made progress in considering health, and the associated healthcare, from an anticipatory perspective (see Nadin [1]). Progress will continue if the expectation of providing significant data about the state of an individual is met.

V. Nathan · R. Jafari (✉)
Center for Remote Health Technologies and Systems College Station, Texas A&M
University, College Station, TX, USA
e-mail: rjafari@tamu.edu

V. Nathan
e-mail: viswamnathan@tamu.edu

S.S. Thomas
Intel Corporation, Hillsboro, OR, USA
e-mail: simi.susan.thomas@intel.com

© Springer International Publishing Switzerland 2017
M. Nadin (ed.), *Anticipation and Medicine*, DOI 10.1007/978-3-319-45142-8_14

Although we do not discuss the broader subject of anticipation in this text, we focus on a particular application: a wristwatch-based platform for continuous blood pressure monitoring. Our contribution takes place at the point where science (physiology in this case) and technology (what it takes to provide a small device that does not interfere with an individual's activities) meet. Once the BioWatch is deployed, knowledge concerning predictive performance will be gained.

With recent advancements in sensor design, miniaturization of hardware and improvements in battery life, the adoption of wearable sensors among the general populace has increased significantly. This provides a great opportunity to open a new dimension for healthcare and diagnostics. Too often, patients go to a doctor and get treatment only after a serious medical problem has manifested. An approach that anticipates these conditions before they happen would prove invaluable. We have already seen commercial solutions that can measure the wearer's heart rate or step count, providing useful feedback. However, we need to go beyond this and fully exploit the continuous, close coupling between the sensor and the user in order to anticipate imminent illness and radically change our perception of medicine.

Diagnosis through wearable sensors has several advantages over the traditional model of diagnosis via clinical visits. Firstly, wearable sensors allow for pervasive, round-the-clock monitoring as opposed to sporadic visits to a doctor or medical facility. For example, heart rate variability (HRV) is a condition that requires repeated monitoring throughout the day to detect trends, some of which might indicate myocardial ischemia. Thus, continuous monitoring could potentially allow early identification of problems and more timely delivery of medicines. Secondly, well-designed wearable sensors can be very convenient to use compared to current medical grade equipment; they are likely to be relatively noninvasive and easier to use correctly without expert knowledge. We will see in this case study itself how blood pressure can be conveniently measured from the wrist rather than through an invasive cuff-based one-time measurement. Another important advantage of wearable sensors is that they can monitor the users in their natural environment as they go about their daily lives. This is in contrast to clinical settings, in which patients may not always be in their natural state; a commonly cited example is that of the "white coat syndrome," wherein patients experience stress in the presence of the doctor, which leads to a blood pressure reading that is higher than usual. Finally, one of the most important potential advantages of wearable sensors is that they could empower the users with more information about their own physiological status, which in turn could lead to proactive healthcare and treatment that is more preventative than reactive.

In this article, we provide a detailed look at BioWatch, a device developed by our research group that attempts to meet the previously mentioned criteria for a wearable physiological monitor. The device, in the form of a watch, is capable of measuring the electrocardiogram (ECG) and photoplethysmogram (PPG) from the wrist. Using these two modalities in conjunction, we can measure the pulse transit time (PTT), which is inversely proportional to blood pressure. Using appropriate calibration for the specific subject and posture, blood pressure can be continuously measured from the wrist. We believe that this eases self-monitoring of blood pressure and can potentially aid both users and their respective clinicians by providing a wealth of data that can be used to build a patient history, observe trends and potentially anticipate life threatening conditions.

2   Background

According to statistics from the Centers for Disease Control and Prevention (CDC), heart disease is the leading cause of death for both men and women. According to the CDC, about 600,000 people die of heart disease in the United States every year, which is 1 in every 4 deaths [2]. One of the most important and easiest ways to keep track of the heart's health is by monitoring blood pressure (BP). With the increased interest in personal health monitoring products and the development of new wearable sensors, a continuous noninvasive wearable BP device would be a great asset for health-conscious persons or someone who is diagnosed with heart related ailments. Even though different studies have proposed noninvasive solutions, such as measuring BP from the pulse transit time (PTT) [3] or from the radial artery [4, 5], an easily wearable product for this purpose is still not available in the market. Such a device should not only provide accurate and reliable readings of BP, but also be easy to use in a convenient form that does not unduly cause the user discomfort for daily use. Gearing towards this goal, we dedicated our effort to develop a noninvasive wearable BP monitoring device using PTT.

Over the years researchers have observed that PTT correlates well with systolic BP (SBP) [6–9]. Since PTT can be calculated using electrocardiogram (ECG) and photoplethysmogram (PPG) [10], it can be leveraged to obtain a continuous and non-invasive measurement of SBP. Diastolic BP (DBP) on the other hand has been shown in previous studies, such as [6], to not correlate as well with PTT. This was also confirmed by our own experiments (described later in this text). However, we can get around this restriction by using pulse pressure (PP) instead. PP is simply the difference between SBP and DBP and it has been shown to correlate better with PTT [9, 11]. Once we estimate SBP and PP through regression techniques, we can obtain DBP using simple subtraction.

2.1   The BioWatch

In order to measure PTT we need the ECG and PPG; in many of the previous studies, the modules, which measured these two signals were separate. This arrangement results in the challenge of proper synchronization and these systems were not independently wearable [12, 13]. Among the studies attempting to introduce wearable BP measurement devices, the wearable solution in one of the studies [14] is not convenient enough because of its wired connection extension to the other arm, which led to discomfort when worn all day. The study by Kim et al. [15] did not

Fig. 1 Biowatch with two electrodes underneath the watch making contact with the left arm and the top electrode making contact with the right arm via a finger touch

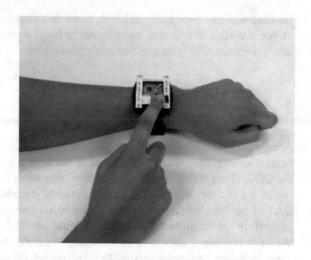

provide a clear validation on the measured BP. The BioWatch, a wearable device in the form of a wrist watch, can monitor the user's BP noninvasively. BioWatch can also monitor heart rate using either ECG or PPG and measure blood oxygenation by easily replacing the PPG sensors with a different set. The system uses Bluetooth for wireless transmission, which makes BioWatch an easily wearable device (Fig. 1). The user wears it on the left hand and touches the ECG electrode with the right hand to complete the electrical connection for the ECG.

While this study focuses on the relationship between PTT and BP, it must be noted that there are studies showing that there are other PPG characteristics that also correlate with BP. In [11] it is shown that the RSD (the time ratio of systole to diastole), RtArea (area ratio of systole to diastole), TmBB (time span of PPG cycle), and TmCA (diastolic duration) are also correlated with BP. According to our study, using more than one of these indices will improve the BP estimation. However, the PPG data in the above mentioned work is collected at the fingertip while the PPG collected in our experiments is from the wrist. PPG waveforms acquired from different positions of body can be different in appearance, which in turn can affect the recognition and validity of certain features. Therefore, more extensive analysis is required in order to fully explore the possibilities of using alternate features with a wrist-based PPG. This text focuses on wearable system design and development of techniques to translate PTT to BP.

The human body is complex and there are many physiological factors, such as physical characteristics of blood vessels, which influence BP and PTT. Some studies show that factors like mood or mental stress [16], race [17], and posture can influence BP or PTT [18–20]. The effect of these varies from individual to individual and can change with age or health condition in each individual. Because of these different and complex factors, coming up with a single solution for accurately calculating BP from PTT is extremely difficult without controlling some of these factors. Calibration is one of the viable methods addressing this issue.

A simple one-point calibration that uses a fixed BP shift to compensate for the bias of an empirical formula is proposed in [21]. Some other empirical formula-based works have investigated an adaptive Kalman filter (AKF) for continuous calibration [22], and Hilbert-Huang Transformation (HHT) for better PTT generation [23]. Apart from the empirical formula-based method, linear regression is also applied to derive the relationship between BP and PTT [24, 25]. Nevertheless, first, there is no clear justification behind the assumption of a linear relationship for PTT-based BP estimation. Second, most PTT-BP relationship studies are done while assuming only one posture. Third, one of the most important reasons for the demand of continuous calibration in empirical formula-based methods is the lack of the ability to track the height difference between heart and measurement position.

2.2 Innovative Aspects

The main contributions of our work are the following:

1. Multiple closed-form regression relationships between BP and PTT are investigated.
2. Analysis of the effects of inter-subject and posture variations on the BP estimation validates the need for training for each individual and each posture.
3. The method to detect height difference between the heart and the point of measurement is designed to provide feedback to the user in order to correct the arm position in real time.

Every time the heart beats, there is a rush of blood from the heart to all parts of the body. The speed of this movement is directly proportional to the BP. So the time taken for the blood to travel from the heart to any specific location in the human body is inversely proportional to BP, and this corresponds to the PTT. The speed of this travel corresponds to the pulse wave velocity (PWV). By measuring PTT noninvasively and continuously, we can then estimate SBP and DBP.

2.2.1 Definition of PTT

PTT can be defined as the time between the ECG "R peak" and the corresponding maximum inclination in the PPG [10]. This is illustrated in Fig. 2, which shows filtered ECG and PPG waveforms from the BioWatch. PTT divided by the distance of travel, which is approximately the arm length in our case, gives us the PWV. The PTT measured here also includes the pre-ejection period (PEP), which is the interval between the onset of the QRS complex in ECG and actual cardiac ejection of the blood from the heart. Ideally, we would have to remove PEP from PTT to get the true PWV.

According to one study [26], PEP varies with different postures. Since our method calibrates and trains the PTT-BP equation to each individual and posture, this should eliminate the effect of PEP on the resulting BP. Keeping this in mind, we decided not to assume the extra burden to measure PEP separately.

Fig. 2 PTT obtained from filtered ECG & PPG collected through BioWatch

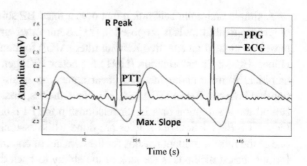

3 Hardware Description

Our first goal was to integrate both ECG and PPG measuring devices into one unit in a wristwatch form factor. The platform used for BioWatch is designed by our lab, the Embedded Signal Processing Lab (ESP), in partnership with Texas Instruments (TI). In an earlier project for TI (called Health Hub), we developed a wristwatch-based Heart Rate and SPO2 monitor. In that system, we used a flat PPG sensor with both infra-red (IR) and Red LEDs. Taking this design as the base, we added the ECG circuitry. To get the ECG signal across the heart, we need two differential electrodes in order to make contact with the body on opposite sides of the heart. A third bias electrode is also required for this system in order to bias the signals to within proper operating range for the amplifiers. The positive differential electrode and the bias electrode are placed underneath the watch, where they will make contact with the left arm; and the negative differential electrode is placed on top so that the user can touch it with the right hand.

BioWatch comes with two analog front ends (AFE): the TI ADS1292 for acquiring ECG signal; and the TI AFE4400 for reading PPG. The board also contains a nine-axis (Accelerometer + Gyro + Magnetometer) MEMS inertial sensor (MPU-9150 from InvenSense, Inc.), which allows to sense body movements and detect posture. More details about the hardware and its operation are available in our earlier papers [27, 28].

4 Methods

As a first step in validating our hardware, we verified the correlation between the measured PTT and SBP, as well as between PTT and PP. We then looked into how SBP and PTT changes for different postures on different individuals. Conclusions from these steps led to the training of PTT to BP equations for each individual and for each posture. This training was done to generate fitted equations for SBP and DBP. The results were analyzed to quantify how much the specific posture and subject training affected the resulting BP calculation. Finally, the accelerometer data available from BioWatch was used to develop arm position detection techniques.

Fig. 3 ABP collected from
radial artery

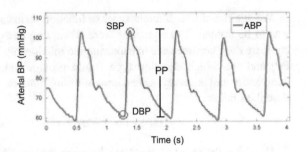

Time (s)

4.1 Measuring Reference BP

Through utilization of a reference device—the Colin Continuous Blood Pressure
Monitor CBM-7000—arterial blood pressure (ABP) is measured continuously from
the radial artery. (More details about this reference device are provided in
Sect. 5.1.) Figure 3 shows the beat-to-beat ABP. In a window of one heartbeat, the
peak of the measured signal corresponds to the SBP, whereas the valley corresponds to the DBP. The difference between these corresponds to the pulse pressure
(PP). We use this to get beat-to-beat reference measurements for BP.

4.2 Translating PTT to BP

We know PTT is correlated with SBP and PP, and the exact relationship is a
function of various factors such as age, height, thickness of blood vessels, and so
on. Instead of identifying all the factors and coming up with a fixed relationship, we
train equations using curve fitting and regression techniques to capture this relationship in a training data set. This method also allows us the flexibility to easily
generate different equations for different postures, which we will show to be
important.

The PTT from the BioWatch is measured along with the ABP from a reference
device, which gives us the ground truth. PTT can change depending on individual
arm lengths, but PWV can give a better stable value across different individuals
having the same BP, and will also help in comparing results across different
individuals. So before beginning the curve fitting process, the PTT is first transformed to the pulse wave velocity (PWV) as defined below:

$$PWV = \frac{d}{PTT} \tag{1}$$

Here d is the distance from heart to the wrist and is calculated as 50 % of the
height of the individual [21]. This measure is also supported by studies such as [29],
where a high correlation between arm-span and body height is shown.

We attempted five different types of function formats for the fitting function that was to be trained. The equations were of varying complexity and the idea was to compare their performance in capturing the relationship between PWV and SBP (or PP) and see which equation type would provide the optimum balance between complexity and accuracy. One common fitting function is polynomial regression as described below:

$$y = a_0 + a_1 \times x + a_2 \times x^2 + \cdots + a_k \times x^k \tag{2}$$

Here, k is the order of the fitting function, x is the PWV, and y is either the SBP or PP, depending on which reference was used. In this work, we attempted only polynomial equations of orders 1 and 2.

In [10], BP is calculated as 70 % of the total pressure drop ΔBP in the body, where ΔBP is calculated as shown below.

$$\Delta BP = \frac{1}{2} \rho \frac{d^2}{PTT^2} + \rho g h \tag{3}$$

Here ρ is the density of the blood, d is the distance from heart to the wrist, g is Earth's gravity, and h is the height difference between the two sites: the heart and the wrist. We trained a generalized equation based on this BP model as defined below:

$$y = ax^2 + b \tag{4}$$

where x is the PWV, y is the SBP (or PP) measure, and a, b are constants that are determined by the curve fitting process. In addition, we also tested two exponential equation formats. Table 1 summarizes all the equations used to train the fitting function. Parameters a, b, and c are determined after the training. Evidently, more complex equations would be able to fit a given training data set better. But there are two factors that would favor opting for a simpler equation: First, the system is expected to be a real-time implementation on a wearable device with a micro-controller, thus restricting its ability to perform complex computations; second, simpler equations would be less susceptible to over-fitting to a particular training data set and prove less effective for the remaining testing data. We also primarily considered polynomial and exponential equations, since many previous works, such as [10, 21], used similar models.

In order to test and compare the performance of the trained equations, we divided the data set into 10 segments, of which 9 were used for training. The effectiveness of each equation is tested on the untrained segment. A 10-fold cross validation is done to compute the root mean square error (RMSE) between the calculated SBP and the actual measured SBP as measured by the reference device. The process is also repeated for DBP: the trained estimate of PP is subtracted from the corresponding trained estimate of SBP to get the calculated DBP, which is then compared to the reference DBP. This procedure is repeated for all 3 postures on all subjects.

Table 1 Different fitting functions used for training

Fitting eq. no:	Equations
1	$y = ax + b$
2	$y = ax^2 + bx + c$
3	$y = ax^2 + b$
4	$y = ax^b + c$
5	$y = ae^{bx}$

4.3 Position Detection

While measuring the PTT, users are encouraged to keep their arm with the BioWatch across the chest at position C, as shown in Fig. 4. This is the position used for training the equations. Changing this position every time the user measures the BP can lead to less accurate results. Studies show that differences in the height of the sensor position relative to the heart can change the PTT values [30]. We verified this phenomenon using a simple test: we asked one subject to hold the arm in 5 different positions, as shown in Fig. 4, and measured the PTT and resulting BP measure for each position. This test was conducted over a span of 2–3 min. The results detailing the BP variation are shown in Table 9 in Sect. 6.5. After this experiment, we heuristically defined a range of ±2 cm as the acceptable arm position deviation from the training position in order that results remain consistent. A wider range means the BP measurement will be less reliable and a shorter range can make proper positioning of the arm difficult for the user.

We wanted to correlate this range of acceptable arm position with a range of values on the accelerometer. The 3-axis acceleration is captured by the accelerometer on the BioWatch and is sent to the PC during the data collection. Since the BP is measured when the subject is stationary (e.g., standing, sitting, supine), the accelerometer measures only the effect of gravity on the three axes. In

Fig. 4 Different arm positions with BioWatch. In this case position C is the position during the training

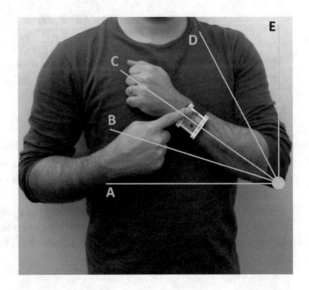

Fig. 5 Measuring g_y range

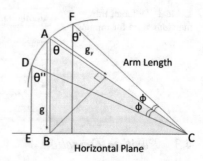

this work, the accelerometer orientation and position are assumed to be fixed, since the user is expected to wear the watch in a relatively similar fashion each time. This means that arm position changes along the plane, as shown in Fig. 4, were best captured by the y-axis readings of the accelerometer, denoted as g_y, and this is effectively the projection of gravity along the y-axis. To define the acceptable g_y range, we used the lower arm length (defined as the length from the elbow to the wrist where the subject wore the BioWatch), g_y value during training, Earth gravity g, and the previously defined range of acceptable sensor height change (± 2 cm).

In Fig. 5, CA is the arm position during training, CF is the upper limit of the arm position, and CD is the lower limit. In other words, arcs FA and AD are 2 cm long in our case.

We can then see that:

$$\theta = \arccos\left(\frac{g_y}{g}\right) \tag{5}$$

where θ is the angle made by the arm to the axis of gravity. The angle ϕ corresponding to each arc of 2 cm is computed as follows:

$$\phi = \frac{AF \times 360}{2\pi \times AC} \tag{6}$$

It is then trivial to obtain all the other relevant angles.

Finally the upper g_y limit can be computed as follows:

$$g_y' = g \times \cos\theta' \tag{7}$$

Here g_y' is the projection of gravity on the y-axis when the arm is placed at the upper limit of the sensor height change, and θ' is the corresponding angle made by the arm to the axis of gravity. Similarly the lower limit of g_y is also computed. Here we assume that the position of the elbow is relatively fixed.

Using this method, we find the acceptable range of g_y for each individual during the training phase. This range can then be used to indicate to the users whether they are placing their arm properly during the BP measurement. The application can indicate to the user whether he/she should move the arm up or down in order to bring it to the proper position. We should clarify here that the current system is limited to detecting this particular movement of the arm up and down across the chest; we assume that the arm is still close to the chest and the elbow remains in the same position.

We also further looked into our device to see how accurately it can classify the arm position using y-axis accelerometer data. In a separate, more comprehensive test, readings for g_y were taken at several different arm positions, and a k-nearest neighbor (k-NN) classifier was used to classify each g_y value to one of about 16 different arm positions. It should be noted here that the k-NN classifier is used merely to validate our accelerometer's accuracy in the absence of a gold standard. The classifier is not used as part of the angle or gravity measurements and is not part of the usual signal processing flow of the BioWatch. The experimental protocol is explained in detail in Sect. 5.3, while the accuracy of the classifier is reported in Sect. 6.5.

5 Experimental Setup

The experiments for gathering PTT and BP data were conducted at the University of North Texas Health Science Center. A total of 11 healthy subjects participated in the experiment, with written consent obtained from all of them. The protocol and consent were approved by the IRB at the UNT Health Science Center. The tests for position detection were conducted at The University of Texas at Dallas with a total of 4 subjects participating. The protocol and consent were approved by the IRB at The University of Texas at Dallas.

5.1 Measurement Procedure for PTT-BP

For measuring the PTT and the reference BP, we needed to use both hands of the subject. The BioWatch was worn on the left wrist to collect both ECG and PPG. With the help of a trained technician, we used the Colin Continuous Blood Pressure Monitor (CBM-7000) as the reference device for ABP. It collects beat-to-beat ABP from the radial artery using the right arm and the right wrist. This device itself is calibrated to a standard BP cuff at the beginning of each measurement cycle, and then continues to collect indirect ABP from the wrist using its wrist-worn module. The ABP recorded is sent to the PC through BIOPAC MP150, which simultaneously also records another ECG signal from the subject. This ECG is taken across the heart and is also used to align the data from both the BioWatch and the reference device. To control the height of the sensor and keep it consistent, we decided to keep all the measurements at chest level. Since in this experiment the right arm was used to collect the ABP, we used a wire connecting from the top ECG electrode to the right hand to complete the electrical connection for ECG on the BioWatch.

5.2   Protocol for PTT-BP Study

Once the subject is ready for data collection, he/she is asked to assume the standing position for the first set of data collection. Our aim was to make sure that the BP is accurately measured during normal conditions as well as during periods of BP fluctuation. Keeping this in mind, we added the Valsalva maneuver to the experimental protocol [31]. This helps in bringing sharp changes in BP for a brief amount of time. The Valsalva maneuver required the subject to exhale against the closed airway after the maximal inhalation (at the maximal lung volume) to induce the required change. The subject is at rest during the first 4 min of the data collection. This gives us enough time to get the PTT and BP measurements using normal conditions. After that, the Valsalva maneuver was performed 5 times. A 45-s to 1-min gap is maintained between each one. The total time taken for data collection for one posture for each subject is around 10–12 min. This same procedure is repeated in sitting position and supine position. The collected data was imported to MATLAB for further processing. The reference SBP and DBP are also obtained from the ABP device. With the help of the two ECG signals from both the BioWatch and the reference device, the data can be aligned. The data is visually checked and any data segments with motion artifacts are removed for accurate PTT calculation. Using the calculated PTT and the reference SBP and PP, we trained PTT-SBP and PTT-PP equations. For validation we found the RMSE between the calculated BP and the real measured BP on an untrained data segment. Figure 7 shows the PTT and SBP measured simultaneously during the Valsalva maneuver and the results are reported in Sect. 6.

5.3   Protocol Used for Development of Arm Position Classifier

For this test we used 3 test subjects. We noted arm length, initial arm position, and position A in Fig. 6, which is used for training and the corresponding y-axis measurement g_y. Based on these measurements, the allowable g_y range is calculated using the methods described in Sect. 4.3. The subject is asked to keep his/her arm in different positions 2 cm apart. Measurements in each position were repeated 5 times and the corresponding g_y values were noted. A total of 15–17 positions are used for testing, as indicated by the circles in Fig. 6, where every two circles are 2 cm apart. Using the k-NN algorithm and leave-one-out cross validation (LOOCV), we further use this data to see how well the device can classify among these positions. We used 4 out of the 5 values in each position as the training set and the left-out value from each position is tested and classified using the k-NN algorithm. This is repeated for each of the 5 values as per LOOCV.

Fig. 6 Different positions used for arm position detection test

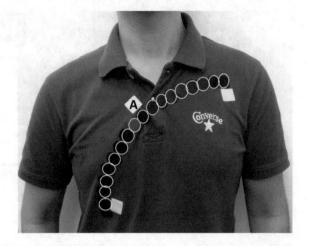

6 Discussion and Results

6.1 Correlation Between PTT and BP

As we expected, we observed a negative correlation between PTT and SBP. The highest negative Pearson correlation coefficient we observed was 0.88, and the lowest negative correlation coefficient was 0.25. The average across the 11 subjects yielded the observed negative correlation coefficients of 0.64, 0.55 and 0.74 for supine, sitting, and standing postures respectively. The fact that the correlations were different for each subject and posture shows that any calibration or algorithm developed to calculate BP from PTT needs to take into consideration both the posture and the individual. Figure 7 shows how PTT varies as SBP varies for one of the individuals during the Valsalva maneuver. We also found that the average correlation coefficient between PTT and reference DBP was as low as 0.38 ± 0.21, confirming the findings from previous studies. The correlation coefficient between PTT and PP on the other hand was 0.65 ± 0.17, thus making it a good proxy to train and use in order to estimate DBP.

6.2 Change in PTT and SBP During Posture Changes

Table 2 shows the mean systolic BP from the reference device and the mean PTT collected during each posture for each individual. Each value is the average of the continuous SBP and PTT collected during the entire time segment for each posture. Table 2 shows how both SBP and PTT change differently across different individuals. For example, in the sitting posture, Subject 1 and Subject 3 have the same SBP, but their PTTs differ by 35 ms. Moreover, when the posture changes from

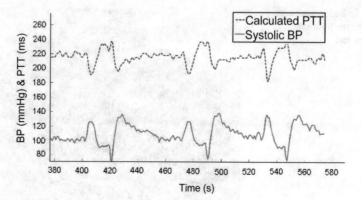

Fig. 7 PTT and BP collected simultaneously

Table 2 Average systolic BP collected from individuals and their corresponding measured average pulse transit time

Subject no.	Systolic BP (mmHg) and pulse transit time (ms)					
	Supine		Sitting		Standing	
	SBP	PTT	SBP	PTT	SBP	PTT
1	105	256	118	293	133	271
2	96	270	107	262	112	256
3	105	221	118	224	81	222
4	93	292	104	291	103	300
5	129	264	113	253	110	276
6	111	267	95	277	107	259
7	108	277	97	289	110	281
8	112	236	113	245	128	245
9	150	248	113	270	104	279
10	102	272	110	249	98	255
11	124	279	118	277	115	274

sitting to standing, SBP increases in the case of Subject 1, but decreases in the case of Subject 3. The PTT, however, decreases for both subjects. Finally, Fig. 8 shows the scatter plot of BP and PTT for a single individual for 3 different postures; and we can see that there are significant differences. These observations indicate that a PTT-SBP equation trained for one posture/subject will not hold true for another. This notion will be further analyzed and quantified in Sect. 6.4.

6.3 Root Mean Square Error

For each subject, we trained the PTT-SBP equation on data from each posture and calculated the SBP using this equation. Similarly, we trained PTT-PP equations and computed DBP using the calculated SBP and PP. The RMSE between the

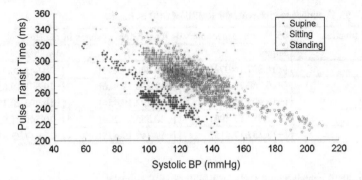

Fig. 8 Scatter plot of PTT & BP collected from the same subject during supine, sitting and standing postures

Table 3 RMSE between calculated SBP and measured SBP

Fitting function	RMSE for different postures in mmHg (mean ± *std*)		
	Supine	Sitting	Standing
1	7.90 ± 1.60	8.88 ± 2.30	9.36 ± 2.06
2	7.86 ± 1.64	8.86 ± 2.39	9.14 ± 2.19
3	7.92 ± 1.64	8.90 ± 2.30	9.35 ± 2.03
4	7.83 ± 1.64	8.80 ± 2.42	9.06 ± 2.21
5	7.95 ± 1.65	8.90 ± 2.30	9.37 ± 2.07

Table 4 RMSE between calculated DBP and measured DBP

Fitting function	RMSE for different postures in mmHg (mean ± *std*)		
	Supine	Sitting	Standing
1	5.88 ± 1.62	5.97 ± 1.15	6.90 ± 2.02
2	5.77 ± 1.71	5.97 ± 1.28	6.68 ± 1.95
3	5.86 ± 1.64	5.96 ± 1.17	6.87 ± 1.99
4	5.78 ± 1.66	6.09 ± 1.22	6.72 ± 1.95
5	5.86 ± 1.63	5.97 ± 1.18	6.89 ± 2.02

calculated BP from the BioWatch and the actual measured BP from the reference device was computed on the untrained segments for each of the postures. These results were then averaged across all subjects for each of the fitting functions, as shown in Tables 3 and 4 for SBP and DBP, respectively. We observed low average RMSEs across all postures ranging between 7.83 and 9.37 mmHg for SBP, and between 5.77 and 6.90 mmHg for DBP. The maximal RMSE over all subjects and postures is 14.27 and 9.68 mmHg for SBP and DBP, respectively. The 95 % confidence intervals for the RMSE for both SBP and DBP are shown in Tables 5 and 6. These results validate the feasibility of PTT-based BP measurement on a wrist-based device. Figure 9 shows the fitted systolic BP and the measured systolic

Table 5 95 % confidence intervals for RMSE of SBP estimates

Fitting function	95 % confidence intervals for different postures in mmHg		
	Supine	Sitting	Standing
1	[−16.837,16.971]	[−19.240,19.410]	[−20.045,20.214]
2	[−16.760,16.749]	[−19.284,19.416]	[−19.691,19.836]
3	[−16.908,16.933]	[−19.292,19.454]	[−19.995,20.154]
4	[−17.232,17.496]	[−20.555,21.208]	[−20.546,21.020]
5	[−18.137,17.815]	[−19.749,19.675]	[−22.917,22.410]

Table 6 95 % confidence intervals for RMSE of DBP estimates

Fitting function	95 % confidence intervals for different postures in mmHg		
	Supine	Sitting	Standing
1	[−13.120,13.133]	[−13.003,13.142]	[−15.340,15.408]
2	[−12.938,12.936]	[−13.112,13.169]	[−15.148,15.199]
3	[−13.102,13.112]	[−12.997,13.133]	[−15.303,15.371]
4	[−13.626,13.939]	[−14.705,15.355]	[−17.248,18.330]
5	[−13.312,13.234]	[−13.376,13.517]	[−15.415,15.551]

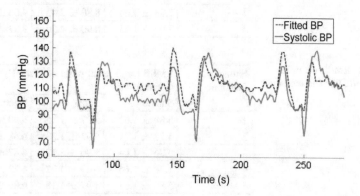

Fig. 9 Fitted BP & corresponding systolic BP

BP during the Valsalva maneuver for one of the subjects. Similarly, Fig. 10 shows the fitted and measured DBP during the Valsalva maneuver for a subject.

In general, the DBP estimates were better, both in terms of average RMSE as well as the tightness of the confidence interval. Moreover, the "standing" position resulted in relatively poorer performance compared to the other postures.

If we consider the results across different fitting equations, we do not find any large variations in performance between the equations. This means that in eventual deployment and online processing on a microcontroller, a simple equation might suffice. Figure 11 shows the plot of PWV against the actual systolic BP, and the

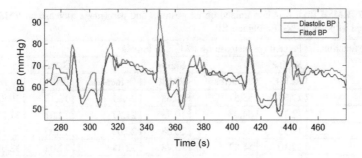

Fig. 10 Fitted BP and corresponding diastolic BP

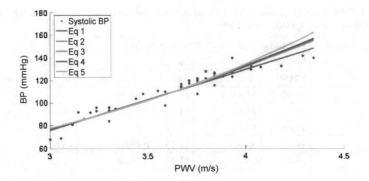

Fig. 11 PWV versus actual systolic BP and fitted BP

fitted BP from the 5 fitting equations on one subject. The results are largely similar for DBP as well.

In [32], the authors conclude that the calibration that accounts for a variety of possible BP points is important for the overall accuracy of the device. Accordingly, the PTT equations developed in this work are calibrated across a wide range of BP variations. The average SBP change during the Valsalva maneuver across all subjects is 90 mmHg. Consequently, the methods described here can be expected to provide more accurate results under varying conditions.

6.4 Individual-Specific and Posture-Specific Training

Apart from the method, shown in Sect. 6.3, for generating Tables 3 and 4, we also generated Tables 7 and 8 to test the importance of training the equation to specific postures and individuals. The equations were first trained using data from all the postures of the same subject. This equation was tested on the untrained segment of the data on each posture of the same individual. The RMSEs calculated using this

Table 7 RMSE from equation trained on all postures and percentage increase on RMSE when compared to posture specific trained BP

Fitting function	Percentage increase on RMSE in mmHg					
	Supine		Sitting		Standing	
	RMSE	% increase	RMSE	% increase	RMSE	% increase
1	12.30	55.51	10.68	20.17	12.13	29.58
2	12.11	54.03	10.78	21.57	12.01	31.37
3	12.33	55.63	10.68	20.06	12.09	29.38
4	12.10	54.57	10.78	22.48	12.02	32.58
5	12.38	55.76	10.65	19.63	12.07	28.75

Table 8 RMSE from equation trained on all subjects and percentage increase on RMSE when compared to subject specific trained BP

Fitting function	Percentage increase on RMSE in mmHg					
	Supine		Sitting		Standing	
	RMSE	% increase	RMSE	% increase	RMSE	% increase
1	14.58	84.44	11.14	25.37	16.81	79.56
2	14.56	85.18	11.15	25.77	16.69	82.60
3	14.64	84.71	11.14	25.14	16.86	80.37
4	14.51	85.22	11.15	26.64	16.62	83.35
5	14.65	84.35	11.13	25.04	16.84	79.61

method and their percentage increase from the Table 3 SBP results are shown in Table 7. The increase in RMSE shows that posture-specific training always yields better results than training without regard to posture.

We also wanted to evaluate the results when training the equation on the same posture across all subjects. This equation is then tested on the untrained data from the same posture of each individual. The results in Table 8 again show a significant increase in the RMSE. This led to the conclusion that not only posture-specific, but also individual-specific training significantly improves the performance of the system.

6.5 Position Detection Accuracy

Table 9 shows the value of PTT, SBP, and g_y measured on one individual for the different arm positions defined in Fig. 4. This test was done to confirm the effects of arm position changes on the PTT. Here the SBP was calculated from the measured PTT. The table shows that the PTT and BP changes significantly with different arm positions. About 15 % change in SBP can be noted from position E to position A. A simple position detection of the arm can ensure that the accuracy is maximized.

Arm positions	g_y, PTT and SBP values for different arm positions		
Table 9 g_y, PTT and BP values at different arm positions	y-axis (g_y)	PTT (mS)	SBP (mmHg)
A	0	266	115
B	0.25	273	110
C	0.58	279	105
D	0.7	282	103
E	0.955	286	100

The g_y value shown in position C in Fig. 4 is considered to be the proper position of the arm. This value will differ slightly between individuals since the arm length and other body characteristics of each individual can be different. During the training phase for each individual, this g_y value was recorded and the acceptable range of g_y was calculated (as explained in Sect. 4.3) to indicate proper arm position for any future readings. The GUI displaying the measured BP can then indicate whether the user placed the arm in the proper position that was used for training by simply monitoring the g_y value of the accelerometer and ensuring it is in the correct range.

To test the effectiveness of this arm position detection based on g_y measurements, we tested the accuracy of the k-NN classifier defined previously in Sect. 5.3. The value of k is set to be the total number of positions tested for each subject, which varied from 15 to 17. Any time there is a tie between two or more sets of neighbors belonging to different classes, we chose the class with the set of neighbors that has the lowest cumulative distance to the tested input. When evaluating the ability to classify within ± 2 cm of the actual position, we found that the accuracy of the classifier was 98 %.

The arm position detection described in Sect. 4.3 can be used for both sitting and standing postures. Currently, this technique cannot be used for the supine posture since the effect of gravity on all three axes during arm movement remains the same, and more sophisticated signal processing would be required. For arm position detection, the current implementation is just an initial foray into dealing with certain arm position changes. Further improvement is also desirable through developing algorithms such that the PTT/PWV- BP equation adjusts itself with the change in the arm position. Such extensive techniques were not in the current scope of the research presented here, but will be investigated in the future.

7 Conclusion

In this article, we presented a wrist-based platform for continuous BP monitoring. The device measures PTT, which is then converted to BP using appropriate fitting functions. We observed that PTT and BP values change significantly across different postures and different individuals. We proposed a unified method to train, develop, and calibrate the PTT-BP equation to each individual and posture. We are yet to evaluate the long-term effectiveness of this approach and the necessity and frequency of future training or calibrations due to advancement in age and progress of ailments. However, in the short term, this method gives us a way to measure individual- and posture-calibrated BP without the need for considering many other characteristics, such as PEP, age, etc. RMSEs in the range of 5.77–6.90 mmHg for DBP, and 7.8–9.37 mmHg for SBP are obtained using the described techniques. This method also relies on correct arm position when measuring BP, and we provided a built-in solution based on an accelerometer to automatically detect the arm position with an accuracy of 98 %.

The proposed noninvasive wireless solution of the BioWatch could potentially provide a convenient, wearable solution for continuous BP monitoring, which in turn could lead to effective anticipatory measures for patients at risk of cardiac disease.

References

1. Nadin, M.: Medicine: the decisive test of anticipation. In: Nadin, M. (ed.) Anticipation and Medicine, pp. 1–25. Springer, Cham (2016)
2. Murphy, S.L., Xu, J., Kochanek, K.D.: Deaths: final data for 2010. Nat. Vital Statist. Rep. **61** (4), 1–118 (2013)
3. Poon, C., Zhang, Y.: Cuff-less and noninvasive measurements of arterial blood pressure by pulse transit time> In: 27th Annual International Conference of the, IEEE-EMBS Engineering in Medicine and Biology Society. , pp. 5877–5880, Wiley-IEEE Press, Hoboken (2006)
4. Ciaccio, E.J., Drzewiecki, G.M.: Tonometric arterial pulse sensor with noise cancellation. IEEE Trans. Biomed. Eng. **55**(10), 2388–2396 (2008). (Wiley-IEEE Press, Hoboken)
5. Park, M., Kang, H. J., Huh, Y., Kim, K-C.: Cuffless and noninvasive measurement of systolic blood pressure, diastolic blood pressure, mean arterial pressure and pulse pressure using radial artery tonometry pressure sensor with concept of korean traditional medicine. In: 29th Annual International Conference of the IEEE-EMBS 2007 Engineering in Medicine and Biology Society, pp. 3597–3600. Wiley-IEEE Press, Hoboken (2007)
6. Hennig, A., Patzak, A.: Continuous blood pressure measurement using pulse transit time. Somnologie-Schlafforschung und Schlafmedizin **17**(22), 104–110 (2013)
7. Cattivelli, F. S., Garudadri, H.: Noninvasive cuffless estimation of blood pressure from pulse arrival time and heart rate with adaptive calibration. In: Sixth International Workshop on Body Sensor Networks Wearable and Implantable Body Sensor Networks, pp. 114–119 Wiley-IEEE Press, Hoboken (2009)
8. Ahlstrom, C., Johansson, A., Uhlin, F., Länne, T., Ask, P.: Noninvasive investigation of blood pressure changes using the pulse wave transit time: a novel approach in the monitoring of hemodialysis patients". J. Artif. Organs **8**(3), 192–197 (2005)
9. Chandrasekaran, V., Dantu, R., Jonnada, S., Thiyagaraja, S., Subbu, K.: Cuffless differential blood pressure estimation using smart phones. IEEE Trans. Biomed. Eng. **60**, 1080–1089 (2013). (Wiley-IEEE Press, Hoboken)
10. Fung, P., Dumont, G., Ries, C., Mott, C., Ansermino, M.: Continuous noninvasive blood pressure measurement by pulse transit time. In: 26th Annual International Conference of the IEEE-EMBS Engineering in Medicine and Biology Society. IEMBS '04, vol. 1, pp. 738–741. Wiley-IEEE Press, Hoboken (2004)

11. Li, Y., Wang, Z., Zhang, L., Yang, X., Song, J.: Characters available in photoplethysmogram for blood pressure estimation: beyond the pulse transit time Australas. Phys. Eng. Sci. Med. **37**, 367–376 (2014)
12. Lass, J., Meigas, K., Karai, D., Kattai, R., Kaik, J., Rossmann, M.: Continuous blood pressure monitoring during exercise using pulse wave transit time measurement. In: 26th Annual International Conference of the IEEE Engineering in Medicine and Biology Society. IEMBS'04, vol. 1, pp. 2239–2242, Wiley-IEEE Press, Hoboken (2004)
13. Ma, T., Zhang, Y.: A correlation study on the variabilities in pulse transit time, blood pressure, and heart rate recorded simultaneously from healthy subjects. In: 27th Annual International Conference of the IEEE-EMBS Engineering in Medicine and Biology Society, pp. 996–999. Wiley-IEEE Press, Hoboken (2005)
14. Zheng, Y., Yan, B. P., Zhang, Y., Yu, C., Poon, C. C.: Wearable cuff-less ptt-based system for overnight blood pressure monitoring. In: 35th Annual International Conference of the IEEE- EMBC Engineering in Medicine and Biology Society (EMBC) pp. 6103–6106. Wiley-IEEE Press, Hoboken (2013)
15. Kim, Y., Lee, J.: Cuffless and non-invasive estimation of a continuous blood pressure based on ptt. In: 2nd International Conference on ITCS Information Technology Convergence and Services (ITCS), pp. 1–4, Wiley-IEEE Press, Hoboken (2010)
16. Furedy, J.J., Szabo, A., Péronnet, F.: Effects of psychological and physiological challenges on heart rate, t-wave amplitude, and pulse-transit time. Int. J. Psychophysiol. **22**(3), 173–183 (1996)
17. Gellman, M., Spitzer, S., Ironson, G., Llabre, M., Saab, P., Pasin, R.D., Weidler, D.J., Schneiderman, N.: Posture, place, and mood effects on ambulatory blood pressure. Psychophysiology **27**(55), 544–551 (1990)
18. Caird, F., Andrews, G., Kennedy, R.: Effect of posture on blood pressure in the elderly. Br. Heart J. **35**(5), 527 (1973)
19. Nardo, C. J., Chambless, L. E., Light, K. C., Rosamond, W. D., Sharrett, A. R., Tell, G. S, Heiss, G.: Descriptive epidemiology of blood pressure response to change in body position the aric study. Hypertension **33**(5), 1123–1129 (1999)
20. Muehlsteff, J., Aubert, X., Morren, G.: Continuous cuff-less blood pressure monitoring based on the pulse arrival time approach: The impact of posture. In: 30th Annual International Conference of the IEEE-EMBS Engineering in Medicine and Biology Society, EMBS, pp. 1691–1694. Wiley-IEEE Press, Hoboken (2008)
21. Gesche, H., Grosskurth, D., Küchler, G., Patzak, A.: Continuous blood pressure measurement by using the pulse transit time: comparison to a cuff-based method. Eur. J. Appl. Physiol. **112** (1), 309–315 (2012)
22. Jadooei, A., Zaderykhin, O., Shulgin, V.: Adaptive algorithm for continuous monitoring of blood pressure using a pulse transit time. In: XXXIII International Scientific Conference Electronics and Nanotechnology (ELNANO), IEEE, pp. 297–301. Wiley-IEEE Press, Hoboken (2013)
23. Zhang, Q., Shi, Y., Teng, D., Dinh, A., Ko, S-B., Chen, L., Basran, J., Bello-Haas, D., Choi, Y., et al.: Pulse transit time-based blood pressure estimation using hilbert-huang transform. In: Annual International Conference of the IEEE- EMBC Engineering in Medicine and Biology Society, pp. 1785–1788 Wiley-IEEE Press, Hoboken (2009)
24. Yoon, Y., Cho, J.H., Yoon, G.: Non-constrained blood pressure monitoring using ecg and ppg for personal healthcare. J. Med. Syst. **33**(4), 261–266 (2009)
25. Chan, G.S., Middleton, P.M., Celler, B.G., Wang, L., Lovell, N.H.: Change in pulse transit time and pre-ejection period during head-up tilt-induced progressive central hypovolaemia. J. Clin. Monit. Comput. **21**(5), 283–293 (2007)
26. Houtveen, J.H., Groot, P.F., Geus, E.J.: Effects of variation in posture and respiration on rsa and pre-ejection period. Psychophysiology **42**(6), 713–719 (2005)
27. Thomas, S. S., Nathan, V., Zong, C., Aroul, P., Philipose, L., Soundarapandian, K., Shi, X., Jafari, R.: Demonstration abstract: biowatch: a wrist watch based physiological signal

acquisition system. In: Proceedings of the 13th international symposium on Information processing in sensor networks, pp. 349–350. Wiley-IEEE Press, Hoboken (2014)

28. Thomas, S. S., Nathan, V., Zong, C., Akinbola, E., Aroul, A. L. P., Philipose, L., Soundarapandian, K., Shi, X., Jafari, R.: Biowatch—a wrist watch based signal acquisition system for physiological signals including blood pressure. In: 36th Annual International Conference of the IEEE-EMBC Engineering in Medicine and Biology Society (EMBC), pp. 2286–2289. Wiley-IEEE Press, Hoboken (2014)

29. Nygaard, H.A.: Measuring body mass index (bmi) in nursing home residents: the usefulness of measurement of arm span. Scand. J. Prim. Health Care **26**(1), 46–49 (2008)

30. Poon, C. C., Zhang, Y-T.: Using the changes in hydrostatic pressure and pulse transit time to measure arterial blood pressure. In: 29th Annual International Conference of the IEEE-EMBS Engineering in Medicine and Biology Society, EMBS, pp. 2336–2337. Wiley-IEEE Press, Hoboken (2007)

31. Gorlin, R., Knowles, J.H., Storey, C.F.: The valsalva maneuver as a test of cardiac function: pathologic physiology and clinical significance. Am. J. Med. **22**(2), 197–212 (1957)

32. Yan, I. R., Poon, C. C., Zhang, Y.: A protocol design for evaluation of wearable cuff-less blood pressure measuring devices. In: Annual International Conference of the IEEE-EMBC Engineering in Medicine and Biology Society, EMBC, pp. 7045–7047. Wiley-IEEE Press, Hoboken (2009)

Anticipatory Mobile Digital Health: Towards Personalized Proactive Therapies and Prevention Strategies

Veljko Pejovic, Abhinav Mehrotra and Mirco Musolesi

Abstract Recent advances in healthcare illuminated the role that individual traits and behaviors play in a person's health. Consequently, a need has arisen for, currently expensive and non-scalable, continuous long-term patient monitoring and individually tailored therapies. Equipped with an array of sensors, high-performance computing power, and carried by their owners at all times, mobile computing devices promise to enable continuous patient monitoring, and, with the help of machine learning, build predictive models of patient's health and behavior. Moreover, through their close integration with a user's lifestyle, mobiles can be used to deliver personalized proactive therapies. In our work we develop the concept of anticipatory mobile-based healthcare (anticipatory mobile digital health) and examine the opportunities and challenges associated with its practical realization.

Keywords Mobile sensing · Anticipatory mobile digital health · Anticipatory mobile computing · Ubiquitous computing · Machine learning

The original version of the chapter was revised: The updated corrections have been incorporated. The erratum to the chapter is available at 10.1007/978-3-319-45142-8_23

V. Pejovic (✉)
University of Ljubljana, Ljubljana, Slovenia
e-mail: Veljko.Pejovic@fri.uni-lj.si

A. Mehrotra
University of Birmingham, Birmingham, UK
e-mail: AXM514@bham.ac.uk

A. Mehrotra · M. Musolesi
University College London, London, UK
e-mail: m.musolesi@ucl.ac.uk

© Springer International Publishing Switzerland 2017
M. Nadin (ed.), *Anticipation and Medicine*, DOI 10.1007/978-3-319-45142-8_15

253

1   Introduction

Mobile computing devices, such as smartphones and wearables[1] represent more than occasionally used tools, and nowadays coexist with their users throughout the day. In addition, these devices host an array of sensors, such as GPS receivers, accelerometers, heart rate sensors, microphones, and cameras, to name a few [1]. When data from these sensors are processed through machine learning algorithms, they can reveal the context in which a device finds itself. The context can include anything from a device's location to a user's physical activity, even stress levels and emotions [2, 3]. Thus, the personalization and the sensing capabilities of today's mobiles can provide a close view of a user's behavior and well-being.

Above all, mobile devices are always connected. They represent the most direct point of contact for the majority of the world's population. Mobile phones, for example, provide an opportunity for intimate, timely communication, unimaginable just 20 years ago. One of the positive results is that mobile devices are becoming a new channel for the delivery of health and well-being therapies. For instance, digital behavior change interventions (dBCIs) harness smartphones to deliver personally tailored coaching to participants seeking behavioral change pertaining to smoking cessation, depression or weight loss [4]. Communication through a widely used, yet highly personal device ensures that a person can be contacted at all times, which might be crucial in case of suicide prevention interventions. In addition, the smartphone is used for numerous purposes, which protects a dBCI participant from stigmatization that may arise if the device is used exclusively for therapeutic purposes.

In addition to inferring the current state of the sensed context, an ever-increasing amount of sensor data, advances in machine learning algorithms, and powerful computing hardware packed into mobile devices allow for predictions[2] of the future state of the context. *Context predictions* have already been shown in the domains of human mobility [5–7], but also population health state [8]. Every next generation of mobile devices comes equipped with new sensors, and soon we may expect galvanic skin response (GSR), heart rate, and body temperature oxymetry sensors as standard features.[3] This would open up the ability to accurately predict an individual's health state.

---

[1]In this paper, *wearables* refer to smartwatches, smartglasses, e-garments, and similar clothing and accessory items equipped with computing and sensing capabilities.

[2]Editor's note: It is questionable whether the current state of algorithmic computing can actually "predict." For a detailed explanation, see [5, 6], which also informs why "project" is the better word to apply to the current ability of mobile sensing devices.

[3]For example, Intel's proposal: http://iq.intel.co.uk/glimpse-of-the-future-the-healthcare-smartwatch/.

1.1 Anticipatory Mobile Computing

Anticipatory mobile computing is a novel concept that (similar to prediction) relies on mobile sensors to provide information upon which the models of context evolution are built. It extends the idea with reasoning and actioning upon such predictions. The concept is inspired by biological systems that often use the past, present and the projected future state of itself and its environment in order to change the state at an instant, so as to steer the future state in a preferred direction [9]. Anticipatory mobile computing has a potential to revolutionize proactive healthcare. Health and well-being problems could be anticipated from personalized sensor readings, and preventive actions could be taken even before the onset of a problem. We term this new paradigm *Anticipatory Mobile Digital Health* (AMDH); and in this paper we discuss the challenges and opportunities related to its practical realization.

First, we examine the key enablers (i.e., mobile and wearable sensors) that provide the contextual data, which can be leveraged to infer a user's health state. Then, we discuss machine learning techniques utilized for building predictive models of the user's (health) context. We are particularly interested in the models that describe how the context might change after an intervention or a therapy. We investigate the challenges related to unobtrusive learning of the impact of an intervention to a person, and the opportunities for highly personalized healthcare. We also take into account individual differences among users, and the potential for capturing and including genetic pre-determinants into the system. We continue with the examination of human-computer interaction issues related to the therapy delivery, and conclude with a consideration of ethical issues in AMDH. Finally, although we earlier examined the potential for inducing change in a person's behavior through anticipatory mobile computing [10], this paper extends the idea to the much larger domain of digital healthcare and elaborates on particular challenges and opportunities in the area.

2 Mobile Sensing for Healthcare

The use of wireless and wearable sensors represents a novel and a rapidly evolving paradigm in healthcare. These sensors have the potential to revolutionize the way of assessing a person's health. Sensor-embedded devices are given to the patients in order to obtain their health related data remotely. These devices do not only help a patient in reducing the number of visits to the clinic, but also offer unprecedented opportunities to practitioners for diagnosing diseases and tailoring treatments through continuous real-time sampling of the patient's heath data. Furthermore, some of these devices empower the users with the ability to self-monitor and curb certain well-being issues on their own.

2.1   Advantages of Mobile Sensing Devices

Today's mobile phones are laden with sensors that are able to monitor various contextual modalities such as physical movement, sound intensity, environment temperature and humidity, to name a few. Some previous studies have shown the potential of mobile phones for providing data that can be used to infer the health state of a user [11–14]. Houston [11] and UbiFit [12] are the early examples of mobile sensing systems designed to encourage users to increase their physical activity. Houston monitors a user's physical movement by counting the number of steps taken via an accelerometer that serves as a pedometer; whereas, UbiFit relies on the Mobile Sensing Platform (MSP) [15] to monitor a user's varied physical activities. MSP is capable of inferring physical activities that include walking, running, cycling, cardio and strength exercise, and other non-exercise physical activities, such as housework. BeWell is a mobile application that continuously monitors a user's physical activity, social interaction, and sleeping patterns and helps users manage their well-being [13]. Bewell relies on sensors such as accelerometer, microphone, and GPS, which are embedded in mobile phones. In [14] the authors show that the depressive states of users can be inferred purely from location and mobility data collected via mobile phones. The above examples demonstrate the close bond between smartphone sensed data and different aspects of human health and well-being.

2.2   The User's Role

A particularly interesting example of mobile healthcare monitoring is given by LifeWatch (www.lifewatch.com), a smartphone equipped with health sensors that constantly monitor the user's vital parameters, including ECG, body temperature, blood pressure, blood glucose, heart rate, oxygen saturation, body fat percentage, and stress levels. A user has to perform a specific action in order to take health measurements. For example, a user should hold the phone's thermometer against the forehead in order to measure the body temperature; to take ECG readings, the user should clutch the phone horizontally with thumb and forefinger placed directly on top of a set of sensors placed on the sides of the phone. The sensor data are sent to the cloud for the analysis and the results are delivered back to the user within a short time interval. Such a phone can prove to be extremely useful in the healthcare domain. (*Caveat*: there is still no proof of the accuracy of its results.) Although within the owner's reach for most of the time, smartphones do not stay in a constant physical contact with the user, and consequently are limited with respect to personal data they can provide.

More recently, mobile phone companies have introduced smart-watches that link with mobile phones and enable the users to perform actions on the mobiles without actually interacting with them. These devices open up new possibilities for health data sensing. First, they maintain continuous physical contact with their users, and second, they host a new set of sensors, usually unavailable on traditional smartphones. In general, these devices come with accelerometer, heart rate and body temperature sensors. Smart-watches are inspired by the concept of a smart wristband, a device that monitors the health state of a user and presents it in a visual form on the linked mobile phone. Smart-wristbands enable real-time health state monitoring, and have achieved considerable commercial success among the health-aware population (e.g., the UP line of wristbands by Jawbone[4]). Initially these bands were able to report only a user's physical activity. However, new sensors, such as body temperature and heart rate, have been introduced, together with more sophisticated data analytics and presentation to the user.

2.3 A Plethora of Smart Devices

Mobile sensing on the phone is for the majority of readings limited by the amount of physical contact the user makes with the phone. Smart-watches and smart-wristbands ensure that the contact is there, yet are limited to a particular part of the users body (the wrist). Sensor embedded smart-wearables designed to dedicatedly monitor specific health related parameter from a specific part of a user's body, have appeared recently and promise more reliable sensing. Such smart-wearables could enable healthcare practitioners to obtain the patient's health data continuously and in the patient's natural environment. These devices come with a variety of health sensors. Pulse and oxygen in blood sensor, airflow sensor, body temperature sensor, electrocardiogram sensor, glucometer sensor, galvanic skin response sensor (GSR), blood pressure sensor (sphygmomanometer), and electromyography sensor (EMG), are some examples of the health sensors embedded in the smart-wearables. Other examples of smart-wearables include Epoc Emotiv [16], an EEG headset capable of capturing brain signals that can be analyzed to infer a user's thoughts, feelings, and emotions. MyoLink (https://somaxisdotcom.wordpress.com) is another wearable that can continuously monitor the user's muscles and heart. It can capture muscle energy output, which in turn can be used to quantify the user's fatigue, endurance, and recovery level. Also, it can be placed on the chest to continuously track the user's heart rate. ViSi mobile (www.soterawireless.com), worn on a wrist, measures blood pressure, hemoglobin level, heart rate, respiration rate, and skin temperature. The device is highly portable and enables users to monitor their health at any time and anywhere.

2.4 Future Steps

The next step in wearable computing is the one in which devices become completely stealth, and as in the Weiser's vision of pervasive computing, completely

---

[4]https://jawbone.com/up.

integrated with people's lives [17]. Shrinking the size of smart wearables—for example reducing the size of a device from something obtrusive to a small adaptive device that users can wear on their bodies and forget about—is a step in that direction. BioStamp [18] is a device composed of small and flexible electronic circuits that stick directly to the skin like a temporary tattoo and monitors the user's health. It is a stretchable sensor capable of measuring body temperature, monitoring exposure to ultraviolet light, and checking pulse and blood oxygen levels. The company envisions future versions of BioStamp able to monitor changes in blood pressure, analyze sweat, and obtain signals from the user's brain and heart in order to use them in electroencephalograms and electrocardiograms [20].

These wearable sensors enable the continuous measurement of health metrics *and* deliver treatment to the patients on time. Yet, the difficulty of continuous monitoring is not the only problem in modern healthcare. Recent studies have shown that around 50 % of prescribed drugs are never taken [19, 20], and thus prescribed therapies fail to improve patient health [21]. In order to address this problem, Hafezi et al. [22] proposed Helius, a novel sensor for detecting the ingestion of a pharmaceutical tablet or a capsule. The system is basically an integrated-circuit micro-sensor developed for daily ingestion by patients, and as such allows real-time measurement of medication ingestion and adherence patterns. Moreover, Helius enables practitioners to measure the correlation between drug ingestion and patients' health parameters, e.g. physical activity, heart rate, sleep quality, and blood pressure—all of which can be sensed by mobile sensors.

The ecosystem of devices supporting health sensing is already substantial and constantly increasing. Soon, healthcare practitioners will have a remote multi-faceted view of a patient's health in real time. The key enabler is the unob-trusiveness of these smart sensing devices. Furthermore, issues such as the accuracy of measurements, accountability for mistakes and the security of a user's privacy, need to be thoroughly addressed before these devices can be accepted in official medical practice.

In the next section we discuss the novel concept of anticipatory mobile digital health, outlining the challenges and opportunities in this promising field. Although smart health sensing devices are still in their infancy, we believe that we shall witness a rapid evolution of this research area in the coming years.

3 Anticipatory Mobile Computing

Anticipation is the ability to reason upon past and present information together with possible future states (See Nadin[5] and [23]). To date, anticipation exists only in living systems. Rosen described an anticipatory system as one "containing a

---

[5]Nadin, M.: Medicine: The Decisive Test of Anticipation. In: Nadin, M.: (ed.) Anticipation and Medicine, pp. 1–25. Springer, Cham (2016).

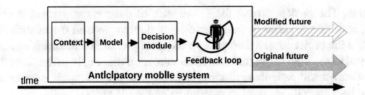

Fig. 1 Anticipatory mobile systems predict context evolution and the impact that current actions can have on the predicted context. The feedback loop consisting of a mobile and a human enables the system to affect the future

predictive model of itself and/or its environment, which allows it to change state at an instant in accord with the model's predictions pertaining to a later instant" [9]. He thus points out that there is an internal predictive model that an anticipatory system builds and maintains. The concept of an anticipatory computing system envisions digital implementation of such a model and automated action based on the model's predictions. Yet, an anticipatory computing system is of interest only if the anticipation carries a value, or meaning, for the end-user [24, 25].

We argue that modern mobile computing devices fulfill the necessary prerequisites for anticipatory computing. First, thanks to built-in sensors and personalized usage, these devices can gather data about their own state—and indirectly the user's state—and the state of the environment. Second, their computing capabilities allow devices to build predictive models of the evolution of the state. Third, the bond between a device and its end-user is so tight that the automated suggestions (based on the anticipation) that a device might convey to a user are *likely* to influence the user's acting upon them. (After all, if people already look into their smartphones when they need to navigate in a new environment or choose a restaurant, they are more likely to turn to smartphones when it comes to health issue consultation.)

To clarify the concept of anticipation on mobile devices (termed Anticipatory Mobile Computing—AMC), in Fig. 1 we sketch a system that senses the context and builds a model of the environment evolution, which gives it the original predicted future. The system then evaluates the possible outcome of its actions on the future. An action that leads to the preferred modified future is realized through the feedback loop that involves interaction between the system and the user.

4 Anticipatory Healthcare System Architecture Design

The opportunity to infer an individual's health and well-being with the help of mobile sensing, together with the perspective of AMC, paves the way for preventive healthcare through anticipatory mobile healthcare systems. We sketch the main ideas behind such a system in Fig. 2. Physiological (e.g., heart rate, GSR) and conventional mobile sensors (e.g., GPS, accelerometer) provide training data for machine learning models of the context (e.g., a user's depression level) and its evolution. The models project the future state of the context, termed the original future, and the state after an intervention or a therapy, termed the modified future. Based on the projections, a therapy with the most preferred outcome is selected and conveyed to the user. Since different users may react differently to the same therapy, close sensor-based patient monitoring, together with a priori inputs, such as a user's genetic background, are used to custom tailor the therapies.

A practical realization of an anticipatory mobile digital health system requires that the following building blocks be present:

- *Mobile sensing.* The role of this block is to manage which of a number of available mobile sensors are sampled, and how often. The mobile device's sensors were originally envisioned as occasionally used features, and their frequent sampling can quickly deplete a device's battery. At the same time, important events may be missed if sampling is too coarse.
- *Therapy and prevention toolbox.* This block contains definitions of possible therapies and prevention strategies that can be delivered to the user. Although we envision further automatization of this module, for now, we believe that a professional therapist's expertise should be harnessed to limit the number of possible therapies and to oversee their deployment.
- *Machine learning core.* Anticipatory mobile digital health employs machine learning for two separate aspects of health state evolution modeling: *context evolution model* and *therapy/prevention-effect model*. The former connects sensor data with higher-level context, and provides a predictive model of how the context might evolve. The latter provides a picture of how different therapies might affect a user's health state. (These models will be discussed in detail in the next section.)
- *User interaction interface.* The success of an anticipatory mobile digital healthcare system is limited by the user's compliance with the provided therapy and prevention strategy. The look, feel, and behavior of the mobile application that delivers the therapy or prevention strategy to a user is crucial in this step.

In the following section, we also discuss the challenges in designing a successful user interaction interface.

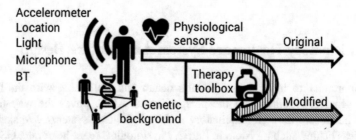

Fig. 2 Anticipatory mobile systems predict context evolution and the impact their actions can have on the predicted context. The feedback loop consisting of a mobile and a human enables the system to affect the future

5 Challenges and Opportunities

Numerous challenges obstruct the path towards implementing anticipatory mobile digital healthcare systems. Rooted in mobile sensing, anticipatory mobile digital health faces challenges such as resource, primarily energy, inefficiency of continuous sensing, and the difficulty of reliable context modeling. Yet, these challenges are common for a larger field of mobile sensing, and a thorough discussion of these issues is available elsewhere [1, 26, 27]. Instead, here we focus on aspects that are unique to anticipatory mobile healthcare.

The use of machine learning algorithms to model and project user behavior and the effect of a therapy or a prevention strategy on the future health state of a specific user is the main challenge. The value of machine learning models, for instance, increases with the amount of available training data for the user. Second, the mobile monitors the user, and may suggest courses of action; yet, it is the user alone who decides whether to follow the therapy or to take certain preventive measures—or not.

In addition to machine learning, future anticipatory mobile digital health developers should pay special attention to human-computer interaction issues in this field. They must try to resolve what is the best way to convey advice/therapy to a user, so that compliance with the proposed therapy or prevention strategy is the highest. Last but not least, issues of ethics, responsibility, and entity roles in anticipatory mobile digital health remain uncharted territory. Further on, we discuss each of the challenges individually and provide positional guidelines for overcoming the challenges.

5.1 Machine Learning in Anticipatory Mobile Digital Health

As stated above, anticipatory mobile digital health employs machine learning for two separate aspects of health state evolution modeling: *context evolution model* and *therapy/prevention-effect model*.

5.1.1 First, a Model of a User Current and Projected Future Health State is Needed

In this model, a relationship is built between mobile sensor data and high-level health state. The model can be direct, if certain values of physiological sensor readings indicate a certain health state, or indirect, if sensor readings reveal contextual aspects that can be connected to an individual's health state. For instance, GPS readings can reveal user mobility, which in turn can hint a user's depression state [14]. In the next step, inference models are extended to provide possible future states of the health state directly, or indirectly through predictions regarding the future context. Anticipating a user's next location is an active area of research, with substantial achievements [5–7]. For many other aspects of the user's behavior and health state, reliable predictive models still do not exist, and even the possibility of their being built remains an open question.

5.1.2 Second Major Machine-Learning Model

The second major machine-learning model in anticipatory mobile digital health is the model of the impact of a possible therapy or prevention strategy on the predicted future health state of a user. There are two non-exclusive ways to construct such a model: one is to harness the existing expertise in healthcare in order to map available therapies to health state transitions. For example, we could map antidepressants to a transition from depressive states to a healthy state.

However, these rules are not suitable for preventive healthcare. Anticipatory mobile digital health operates on projections (predictions, in Rosen's sense), and consequently, therapies should aim for prevention. In addition, although mobile devices remain highly personal, and the sensor data uncovers fine-grained individual health state information, these general rules limit the system's ability to deliver personalized healthcare. An alternative approach is to build a therapy/prevention-effect model by monitoring the evolution of a user state after a proactive therapy or prevention strategy is delivered. By comparing the original predicted state with the actual state recorded some time after the therapy (or prevention strategy), we can identify the relationship between the therapy (or prevention strategy) and the future health state change. Built this way, a model reveals successful proactive therapies, which is difficult to achieve in traditional practice.

It must be kept in mind that what works for one patient may not work for another. These models are highly personalized and can reveal therapies that are useful for a particular kind of a person only. Still, we argue that these models should not be built from scratch. The available therapies that could be automatically suggested to a particular patient in a particular situation should be determined by the rules stemming from existing medical expertise.

5.1.3 Learning with a User

Automated tool-effect modeling in anticipatory mobile digital health requires that a therapy (or prevention strategy) be induced in a user so that its effects can be observed. This outcome is then used to train and refine the model. *Reinforcement learning*, where an agent uses a tool in the intervention environment (which can be represented through a Markov decision process, for example), is a natural way to model the problem [28]. In every step, a certain tool is selected and applied; the observed change in the health state elicits a reward that reflects how positive the change is.

5.1.4 Measuring Health State

Thus, there is a need for a suitable metric for *measuring the health state change*. Here we need to evaluate the effect of a proactive therapy or prevention strategy, that is, basically *compare the original predicted health state and the modified predicted health state*. We argue that the comparison metric has to be domain dependent. For example, if an anti-stress therapy is evaluated, the difference between the predicted skin conductivity and heart rate values without intervention, and the actual values after the intervention, is a reasonable measure of stress level change [29]. However, system designers should keep in mind that the metric has to be both suitable for machine learning algorithms and relevant from the healthcare point of view.

5.1.5 Learning Without Interfering

Reinforcement learning uncovers the mapping between therapies and health state changes. Delivering a previously unused therapy or prevention strategy refines the model as we learn more about how the user reacts to this tool. From the practical point of view, however, we face a dilemma: use a tool that is known to result in a positive health change outcome, or experiment with an unused tool that *might* yield an even better outcome. In reinforcement learning this dilemma is known as the *exploration versus exploitation trade-off*.

Providers of strategies for solving the dilemma in an anticipatory mobile digital health setting must be aware of the possible irreversible negative consequences of a wrong therapy or prevention strategy. Preferably, the system should learn as much as possible without explicit delivery of therapies to a user. Such a learning concept is called *latent learning*. It is a form of learning where a subject is immersed into an unknown environment or a situation without any reward or punishment associated with them [30]. Latent learning has been demonstrated in living beings, who form a cognitive map of the environment solely because they are immersed in the environment, and later use the same map in decision making.

We argue that mobile computing devices, through multimodal sensing, can harness latent learning to build a model of the user reaction with respect to certain actions or environmental changes that correspond to those targeted by the therapies. This is particularly relevant for therapies that are not based on medications, such as behavioral change interventions [10]. For example, suppose a depression prevention system can provide the user with the suggestion to go out for dinner with friends. We can get a priori knowledge of how this suggestion would affect the user, if, for example, on a separate occasion we detected that the participant went out for a dinner with friends, and we gauged the depression levels, estimated through mobility and physical activity metrics, before and after the dinner. Defining how the expected action—going out with friends—should manifest from the point of view of sensors—e.g., a number of Bluetooth contacts detected, location, time of the day—is one of the prerequisites for practical latent learning. Again, interdisciplinary efforts are crucial to ensure that the detected reaction corresponds to the one that should be elicited by the tool.

5.2 Personalized Healthcare

Current therapies are often created as "one size fits all," even though the healthcare industry is well aware that in many cases individuals react differently. For example, antidepressants are ineffective in 38 % of the population, while cancer drugs work for only 25 % of the patient population [31]. Personalized therapies promise to revolutionize healthcare, by avoiding the traditional trial-and-error therapy prescription, minimizing adverse drug reactions, revealing additional or alternative uses for medicines and drug candidates [32] and curbing the overall cost of healthcare [31].

Anticipatory mobile digital health is poised to bring personalized healthcare closer to mainstream practice. Mobile sensing can provide a glimpse into individual behavioral patterns, identifying risky lifestyles. Therapy/prevention-effect machine learning models can also take into account a patient's genetics in order to individualize the therapy or prevention strategy. Investigation of which genes impact the occurrence and reaction to a treatment of a certain disease is a very active area of research. The potential for healthcare improvement is immense, having in mind that with some conditions, such as melanoma tumors, the majority of cases are driven by certain person-specific genetic mutations, and could be targeted by specific drugs [31].

The relationship is not one way, and anticipatory mobile digital health could also help with pharmacogenomics—the study of how genes affect a person's response to drugs. Identifying common traits of genetic background in populations who reacted to an anticipatory therapy or prevention strategy in the same way would help in discovering the relationship between genes and health treatments. Finally, the inclusion of the genetic background in common medical practice is not far from reality. For example, in 2014 a human genome sequencing for less than USD1000 became available.

5.3 HCI Issues in Anticipatory Mobile Digital Health

Despite the automation that anticipatory mobile digital health brings, in the end, it is up to a user to comply with the given therapy or prevention strategy. This is particularly important for behavioral change intervention therapies, i.e., those that are delivered in cases where the health state is directly influenced by the patient's behavior. Consequently, communication between the system and the patient has to be seamless. Users are an important part of the system, and their inclusion requires an appropriate interface between the participant and the system. As noted by Russell et al., [33] a system that autonomously brings decisions and evolves over the course of its lifetime needs to be transparent to the user. The user interface must make such a system understandable to the user and capable of review, revision, and alteration. In addition, the content should be framed to emphasize that the tool can help; it is of fundamental importance to avoid any hint of harassing and patronizing the participant.

The timing of a therapy or a prevention strategy is also important for its successful delivery. This is particularly true for automated therapies delivered via a mobile device.

An inappropriately timed intervention that comes, for instance, when a patient is in a meeting, or riding a bicycle, may lead to annoyance, or may be completely overlooked by the patient. Mobile sensing helps with identifying opportune moments for delivering therapies. The context in which a user is, such as location, physical activity, or engagement in a task, to an extent determines whether the patient is able or willing to interrupt [34, 35]. Machine learning and mobile sensing are harnessed for monitoring a user's reaction to an interruption arriving when the user is in a certain context, and from there on a model of personal interruptibility is built.

Querying the model with a momentarily value of a user's context returns the estimated *interruptibility* at the moment. While practical implementations of the above models already exist [34], we envision predictive models of user interruptibility. Finally, we highlight that opportune moments denote those times at which a patient is likely to quickly acknowledge/read the content of a delivered message. Identifying moments when the delivered information will have the highest medical impact is even more important; yet due to the difficulty of obtaining the training data—we would need to deliver the same therapy or prevention strategy at different times to the same user—identifying such moments remains very challenging.

5.4 Ethics and Accountability

Privacy issues in mobile sensing emerged soon after the proliferation of smart-phones about a decade ago. Misuse and leaking of information that can be collected by a mobile device—such as a user's location, collocation with other people, physical activity of a user—may deter people from trusting mobile application, especially in a domain as sensitive as health. Trust is a key component for the success of anticipatory mobile digital health applications, and every care should be taken that personal information not leak. Ensuring that sensor data do not leave the device at which they are collected is one way to minimize the risk. However, this complicates the construction of the joint machine learning models discussed earlier.

The responsibility chain in the domain of anticipatory mobile digital health is yet to be defined. Unsuccessful therapies can have serious consequences. It is unclear who would be to blame if a delivered therapy or prevention strategy does not improve the patient's health state, or even worse, endangers the person's life. A therapist who designed the therapy, a software architect who devised the underlying machine learning components, together with the patient, play a role in the process.

6 Conclusions

Personalized and proactive healthcare brings undisputed benefits in terms of therapy and/or prevention strategy, efficiency, and cost effectiveness of the healthcare system. Mobile devices have a potential to become both our most vigilant observers and closest advisors. Anticipatory mobile digital health harnesses the sensing capabilities of mobiles in order to learn about the user's health state and to anticipate its evolution so that proactive therapies tackling predicted health issues can be delivered to the user in advance. With the help of machine learning that takes into account rich sensor data and a user's genetic background, anticipatory mobile digital health applications can tailor personalized therapies. In addition, the concept can be used to learn more about how therapies affect different demographics, users who behave in a certain way, or have a particular genetic background. Generalizing from a larger pool of users and therapies can identify groups for which a therapy or prevention strategy is successful, basically discovering new facts about drugs. Last but not least, we believe anticipatory mobile health applications warrant a discussion on their inclusion into the health insurance frameworks.

Acknowledgments This work was supported by the EPSRC grants: "UBhave: ubiquitous and social computing for positive behaviour change" (EP/I032673/1) and "Trajectories of Depression: Investigating the Correlation between Human Mobility Patterns and Mental Health Problems by means of Smartphones" (EP/L006340/1). The authors would like to thank the participants of the Anticipation and Medicine Workshop (Hanse-Wissenschaftskolleg, Delmenhorst, Germany, September 2015) for a fruitful discussion that has shaped our views of the topic, and to Professor Mihai Nadin for his suggestions regarding the final manuscript.

References

1. N. D. Lane, E. Miluzzo, H. Lu, D. Peebles, T. Choudhury, and A. T. Campbell. A Survey of Mobile Phone Sensing. IEEE Communications Magazine, 48(9):140–150, 2010.
2. K. K. Rachuri, M. Musolesi, C. Mascolo, P. J. Rentfrow, C. Longworth, and A. Aucinas. EmotionSense: A Mobile Phones based Adaptive Platform for Experimental Social Psychology Research. In UbiComp'10, Copenhagen, Denmark, September 2010.
3. H. Lu, G. T. C. Mash_qui Rabbi, D. Frauendorfer, M. S. Mast, A. T. Campbell, D. Gatica-Perez, and T. Choudhury. StressSense: Detecting Stress in Unconstrained Acoustic Environments using Smartphones. In UbiComp'12, Pittsburgh, PA, USA, September 2012.
4. N. Lathia, V. Pejovic, K. Rachuri, C. Mascolo, M. Musolesi, and P. J. Rentfrow. Smartphones for Large-Scale Behaviour Change Intervention. IEEE Pervasive Computing, 12(3), 2013.
5. D. Ashbrook and T. Starner. Using GPS to Learn Signficant Locations and Predict Movement Across Multiple Users. Journal of Personal and Ubiquitous Computing, 7(5):275-286, October 2003.
6. S. Scellato, M. Musolesi, C. Mascolo, V. Latora, and A. T. Campbell. Nextplace: A spatio-temporal prediction framework for pervasive systems. In Pervasive'11, San Francisco, CA, USA, June 2011.
7. Y. Chon, E. Talipov, H. Shin, and H. Cha. SmartDC: Mobility Prediction-based Adaptive Duty Cycling for Everyday Location Monitoring. IEEE Transactions on Mobile Computing, 13:1, 2013.
8. A. Madan, M. Cebrian, D. Lazer, and A. Pentland. Social Sensing for Epidemiological Behavior Change. In UbiComp'10, Copenhagen, Denmark, September 2010.
9. R. Rosen. Anticipatory Systems. Pergamon Press, Oxford, UK, 1985.
10. V. Pejovic and M. Musolesi. Anticipatory Mobile Computing for Behaviour Change Interventions. In Workshop on Mobile Systems for Computational Social Science (MCSS'14), Seattle, WA, USA, September 2014.
11. S. Consolvo, K. Everitt, I. Smith, and J. A. Landay. Design requirements for technologies that encourage physical activity. In CHI'06, Quebec, Canada, April 2006.
12. S. Consolvo, D. W. Mcdonald, T. Toscos, M. Y. Chen, J. Froehlich, B. Harrison, P. Klasnja, A. Lamarca, L. Legr, R. Libby, I. Smith, and J. A. Landay. Activity Sensing in the Wild: a Field Trial of UbiFit Garden. In CHI'08, Florence, Italy, April 2008.

13. N. D. Lane, T. Choudhury, A. Campbell, M. Mohammod, M. Lin, X. Yang, A. Doryab, H. Lu, S. Ali, and E. Berke. BeWell: A Smartphone Application to Monitor, Model and Promote Wellbeing. In Pervasive Health'11, Dublin, Ireland, May 2011.
14. L. Canzian and M. Musolesi. Trajectories of Depression: Unobtrusive Monitoring of Depressive States by means of Smartphone Mobility Traces Analysis. In UbiComp'15, Osaka, Japan, September 2015.
15. J. Lester, T. Choudhury, and G. Borriello. A practical approach to recognizing physical activities. In Pervasive Computing, pages 1–16. Springer, 2006.
16. T. Fraga, M. Pichiliani, and D. Louro. Experimental art with brain controlled interface. In Universal Access in Human Computer Interaction. Design Methods, Tools, and Interaction Techniques for eInclusion, pages 642–651. Springer, 2013.
17. M. Weiser. The Computer for the 21st Century. Scientfic American, 265(3):94–104, 1991.
18. T. S. Perry. Giving your body a "check engine" light. Spectrum, IEEE, 52(6):34–84, 2015.
19. K. Nasseh, S. G. Frazee, J. Visaria, A. Vlahiotis, and Y. Tian. Cost of medication nonadherence associated with diabetes, hypertension, and dyslipidemia. American Journal of Pharmacy Bene_ts, 4(2):e41–e47, 2012.
20. L. Osterberg and T. Blaschke. Adherence to medication. New England Journal of Medicine, 353(5):487–497, 2005.
21. J. N. Rasmussen, A. Chong, and D. A. Alter. Relationship between adherence to evidence-based pharmacotherapy and long-term mortality after acute myocardial infarction. Jama, 297(2):177–186, 2007.
22. H. Hafezi, T. L. Robertson, G. D. Moon, K. AuYeung, M. J. Zdeblick, and G. M.Savage. An Ingestible Sensor for Measuring Medication Adherence. IEEE Transactions on Biomedical Engineering, 62(1):99–109, 2015.
23. Nadin, M.: The anticipatory profile. An attempt to describe anticipation as process. In: Nadin, M. (ed.) International Journal of General Systems, vol. 41, no. 1. pp. 43–75. Taylor & Francis, London, (2012).
24. Nadin, M.: Anticipation and Computation. Is Anticipatory Computing Possible? In: Nadin, M. (ed.) Anticipation Across Disciplines, Cognitive Science Monographs, vol. 26, pp. 163–257. Springer, Cham (2015).
25. Nadin, M.: (2014) Can Predictive Computation Reach the Level of Anticipatory Computing? In: International Journal of Applied Research on Information Technology and Computing, vol. 5, pp. 171–200 (2014).
26. V. Pejovic and M. Musolesi. Anticipatory Mobile Computing: A Survey of the State of the Art and Research Challenges. ACM Computing Surveys (CSUR), 47(3), 2015.
27. P. Klasnja and W. Pratt. Healthcare in the Pocket: Mapping the Space of Mobile- Phone Health Interventions. Journal of Biomedical Informatics, 45(1):184–198, 2012.
28. R. S. Sutton and A. G. Barto. Reinforcement Learning: An Introduction. MIT Press, Cambridge, MA, USA, 1998.
29. J. Healey and R. W. Picard. Detecting stress during real-world driving tasks using physiological sensors. IEEE Transactions on Intelligent Transportation Systems, 6(2):156–166, 2005.
30. C. Tavris and C. Wade. Psychology in Perspective. Longman, New York, NY, USA, 1997.
31. Personalised Medicine Coalition. The Case for Personalized Medicine, 2014. http://www.personalizedmedicinecoalition.org/Userfiles/PMC-Corporate/file/pmc_case_for_personalized_medicine.pdf.
32. L. Mancinelli, M. Cronin, and W. Sade. Pharmacogenomics: the promise of personalized medicine. AAPS Pharmsci, 2(1):29–41, 2000. Former 16.
33. D. M. Russell, P. P. Maglio, R. Dordick, and C. Neti. Dealing with Ghosts: Managing the User Experience of Autonomic Computing. IBM Systems Journal, 42:177–188, 2003.
34. V. Pejovic and M. Musolesi. InterruptMe: Designing Intelligent Prompting Mechanisms for Pervasive Applications. In UbiComp'14, Seattle, WA, USA, September 2014.
35. A. Mehrotra, M. Musolesi, R. Hendley, and V. Pejovic. Designing Content-driven Intelligent Noti_cation Mechanisms for Mobile Applications. In UbiComp'15, Osaka, Japan, September 2015.

Intelligent Support for Surgeons in the Operating Room

Rainer Malaka, Frank Dylla, Christian Freksa, Thomas Barkowsky, Marc Herrlich and Ron Kikinis

Abstract Modern technology gives surgeons the possibility to plan operations using complex 3D information tools providing data integration, analysis and visualization. However, in the operating room, most of the tools are not at hand. It might be useful to access the data and to visualize certain surgical procedures during the actual surgery. We investigate such situations and look for novel solutions for intra-operative support for surgeons to access 3D information: what they need, when they need it. We integrate medical image processing, cognitive modeling and human- computer interaction in order to anticipate the surgeons' needs.

R. Malaka (✉) · M. Herrlich
TZI, Digital Media Lab, University of Bremen, Bremen, Germany
e-mail: malaka@tzi.de

M. Herrlich
e-mail: mh@tzi.de

F. Dylla · C. Freksa · T. Barkowsky
Cognitive Systems, University of Bremen, Bremen, Germany
e-mail: dylla@sfbtr8.uni-bremen.de

C. Freksa
e-mail: freksa@sfbtr8.uni-bremen.de

T. Barkowsky
e-mail: barkowsky@sfbtr8.uni-bremen.de

R. Kikinis
Medical Image Computing, University of Bremen, Bremen, Germany
e-mail: kikinis@uni-bremen.de

R. Malaka · F. Dylla · C. Freksa · T. Barkowsky · M. Herrlich · R. Kikinis
Creative Unit Intra-Operative Information, University of Bremen, Bremen, Germany

R. Kikinis
Surgical Planning Laboratory, Brigham and Women's Hospital and Harvard
Medical School, Boston, MA 02138, USA

R. Kikinis
Fraunhofer MEVIS, Bremen, Germany

© Springer International Publishing Switzerland 2017   269
M. Nadin (ed.), *Anticipation and Medicine*, DOI 10.1007/978-3-319-45142-8_16

We address three issues for developing such systems: how to identify what information the surgeon needs; how to adapt pre- and intra-procedure information to the surgical situation; how to present the relevant information to the surgeon. This paper presents the vision and preliminary results of a collaborative research project.

Keywords Human-computer interaction · Surgery · Cognitive systems · 3D medical data

1   Introduction

Computer assistance for surgical procedures and interventions is a rapidly growing field. Typically, pre-operative image information is used for planning surgical interventions. Software tools support this planning by registering data from different sources and segmenting structures of interest, e.g., for building 3D model representations of tumors, surrounding tissue, and organs [1]. The surgical plan is then used during the procedure to guide the surgeon's determination of what to remove and from where [2]. However, it is a huge challenge to realign pre-operative data (e.g., volume data, 3D models) with the patient during the procedure. Furthermore, during the intervention, tissue shift, resections and cutting, intentional displacement and deformation of organs, and other actions will distort the information obtained in the pre-operative phase [3].

The goal of an intelligent intra-operative support would thus be to present appropriate information to a surgeon during a procedure. Specifically, we aim to present the proper guidance information at the correct time in an optimal manner. To succeed in this endeavor it is necessary to anticipate the surgeon's information needs and to allow for quick, easy and robust interaction. To anticipate desired information is not trivial. An anticipatory system depends not only on the current and previous states, but also on possible future states [4]. Therefore, it requires to observe the situation, model the context and to predict potential next states. This way, some of the information can be anticipated; that is, inferred based upon experience [5]. A surgeon who needs additional information from 3D data in a concrete situation cannot start searching through a number of menus of a computer program, interact intensively with 3D interfaces and finally select the information. Ideally, computer systems are aware of the situation in the surgery room and, depending on the progress of the operation anticipate what is needed and suggest entries to specific visualizations of data, e.g. with respect to content, granularity, or perspective. The surgeon then selects the desired information with interfaces that need only a low level of attention. The selected data subset would be displayed for immediate use, e.g., directly connected to some surgical device in order to minimize focus shifts. Such an approach will also include mixed reality technologies that have not yet been studied deeply in the international research community.

2 State of the Art

Space and conditions in the operating room (OR) limit the ways in which information can be presented and what kinds of standard input devices can be used (Fig. 1). Recently, touchscreens and mobile devices covered with sterile sleeves have enabled the surgeon to directly steer and interact with the information provided [2, 6]. Scientists have investigated the use of projectors, in combination with hand/finger tracking, to turn parts of a room into interactive surfaces, and have looked at touchless interaction for circumventing the problem of unsterile input devices [7].

In current systems, neither is the information presented automatically adapted to a phase or specific situation in the OR, nor do current systems take into account the cognitive load[1] [8]. In addition, each input device processes input independently of other devices (cf. e.g. [2, 6]); there is no shared contextual information between different systems or interfaces, and synchronization is left to the users.

Despite the fact that research in Medical Image Computing and Computer Assisted Interventions is a well-established topic, there has been little focus on the integration of mixed reality technologies, novel interaction paradigms, and 3D simulation algorithms in the OR. Information presentation and interaction based on human-computer interaction (HCI) underwent a number of paradigm shifts in the last decades. In the early days of computing, users were expected to adapt to machines; the user perspective was widely ignored. From the mid-1980s on, usability methods focused on the ergonomics of the users' work environment [9]. More recent trends look into the "hedonistic" aspects of HCI and the user experience (UX) [10].

Considering the cognitive aspects of HCI, studies show that there are severe limitations to the amount of information that a person can simultaneously perceive and process, how the data are distributed over various channels of perception and processed, and how different modalities (e.g., visual vs. haptic) interfere with each other [11, 12]. There are also individual differences concerning whether auditory or visual information is more easily and quickly processed, and how easily condensed abstract visuo-spatial information can be processed [13, 14]. Moreover, since the setup in the OR can vary substantially depending on the kind of intervention, it is important to provide flexible multi-modal interaction possibility. Abstract models of multi-modalities and intelligent presentation planning can bring in flexibility [15].

For adapting not only the dialog modality but the content itself, the presented information has to be adapted to the user, the dialog situation, and the context. It has been shown that entrainment can enhance dialog effectiveness [16]. Models of pragmatic knowledge, the domain and contexts can be used for building context-dependant systems [17–19]. Such models are built on various AI techniques (both symbolic and sub-symbolic) and as with many intelligent interactive

---

[1]Cognitive load refers to the mental effort in human working memory to solve a specific problem.

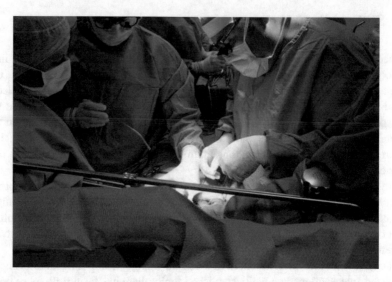

Fig. 1 Situation in the OR: very limited space for HCI that, moreover, has to be sterile. The image shows an abdominal surgical treatment

systems there is also a tradeoff when integrating too much intelligence into the system [20].

3   Towards Intelligent Support in the Operating Room

Previous research has paid relatively little attention to the integration of novel interaction paradigms for effortless interaction with 3D data in the OR and 3D simulation algorithms necessary to present accurate information after "opening" the patient's body and making cuts. We are proposing to investigate these important scientific and technical challenges. We believe that these are topics with a rich potential for producing significant scientific progress and will lead to improved intra-operative support for surgeons. This technological research requires knowledge about and understanding of the information needs of a surgeon during surgery. In order to advance the research agenda, we propose to focus on three central questions:

(a) What information does the surgeon need?
(b) How can we adapt pre- and intra-procedure information to the surgical situation?
(c) How should we present the relevant information to the surgeon?

The answers to these questions are key to novel systems that help the surgeon with decision-making during surgery (e.g., by improving access to preoperatively acquired information). Help with the actual manual tasks (e.g., by adapting the information to the current situation) is also important. OR environments are challenging to all participants in a procedure: the surgeon and other personnel in the room must integrate multiple information streams that aid in performing the surgery. Information overload is a real issue with potentially detrimental consequences for the outcome of the procedure. In this scenario, it is critical to present the right information at the right time.

3.1 How to Identify What Information the Surgeon Needs

We need to find out about objects, concepts, processes, and their interrelation when applied in a surgical context and how they can be derived and represented in an assistance system. This requires applying and further developing novel HCI methods based on natural user interaction (NUI) and contextual computing in the domain of the OR. Even though recent HCI methods are promising for complex and professional environments, which are characterized by many constraints, there is a substantial need for advancing existing methods beyond the state of the art. The current decision workflow and procedures that lead to a certain course of action must be observed and discussed with scrutiny. An analysis and models of the context and workflow, in which an assistive intra-operative system is to be deployed, are necessary. This comprises, for instance, analysis of clinical requirements, analysis of human input and feedback requirements, as well as proposals for information architecture (user interfaces) and integration into the workflow.

3.2 Adapting Pre- and Intra-procedure Information to the Surgical Situation

For modeling the results, we propose a qualitative approach to the conceptualization of knowledge, as this is considered to be cognitively more adequate than quantitative approaches (e.g., [21]). Qualitative Spatio-Temporal Reasoning (QSTR) provides well-defined representations and reasoning techniques to deal with imprecise and incomplete knowledge on a symbolic level, i.e., without numerical values [22, 23]. Qualitative representations have been successfully applied to disciplines such as intelligent service robotics, architecture, nautical and pedestrian navigation, airport apron observation, and biomedicine, by linking representations to the specific concepts of each domain [24, 25]. For these reasons, we use this approach to build the representational basis of information architecture. We thus aim for:

(1) investigating the conceptual structure of entities, relations, and spatial reference frames used in surgery;
(2) deriving suitable conceptualizations including (potentially new) reasoning techniques to support planning and simulation of surgical procedures; and
(3) extracting qualitative context and process formalizations of selected processes.

3.3   Presenting Relevant Information to the Surgeon

In order to advance current techniques for context-based aggregation, clustering, and presentation of available information based on the conceptual models, it is necessary to automatically adapt the type and amount of information that is presented to immediate situational needs. Presenting the appropriate information at the correct time on the right device or display is the goal.

Thus, the first step for intelligent information presentation is the detection and analysis of the context. Therefore, approaches to monitor a surgeon's cognitive capabilities with respect to various sensory modalities (visual, auditory, and haptic perception) are necessary. The results of this monitoring can be used to steer the way that additional pieces of information are conveyed to the surgeon. The OR is a very special environment that poses a number of specific challenges for user interaction: surgeons need to concentrate on the procedure; everything must be sterile, and OR personnel cannot touch conventional IT systems. This poses a number of challenging research questions for user interaction metaphors and technologies. For example, intuitive metaphors for controlling the display of medical 2D and 3D content without third-party help are needed. In contrast to existing work, an ideal system would integrate different input devices and modalities within one interaction context.

In addition, interactions with IT systems should be touchless, if possible; otherwise maintaining sterility is an issue that has to be addressed [7]. Interaction design in the OR can be guided by the concept of embodied interaction in context [20]. The combination of NUIs for 3D interaction [26, 27] and user-centric methods brings together new technological means for interaction that would feel natural from a user perspective.

4   Building Intelligent Interactive Information Systems for the OR

An ideal system would be design according to how information is mentally acquired, processed, organized, and stored, and which representation structures and cognitive processes are involved. In mental information processing, there is an intense interplay between the perception and action capabilities of a cognitive agent

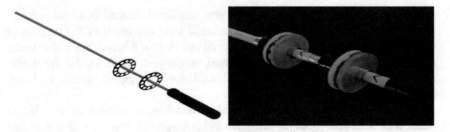

Fig. 2 A prototype of an ablation probe with two LED rings

and the physical environment in which cognition is embedded [28]. So it is essentially due to the situational context in the given environment in which a task is performed that influences the cognitive effort induced by a task or the corresponding information processing.

Depending on context or information needs, a variety of devices and interaction methods must be available in flexible combinations. This allows for optimization to the actual setting in the context of a particular operation. Therefore, a practical approach towards our research goals it not to follow the idea of just one very mighty and highly automated system, but rather to build smaller and feasible systems for particular contexts. Such systems could be particular devices dedicated to singular treatments or operation types, but optimized to present the right information at the right time. Building a number of such solutions can, in turn, be extended and integrated into unified approaches with more complex models and a whole spectrum of input and output devices.

An example of such a small step towards larger systems is a device for radio frequency ablation to present navigational information regarding where to position the instrument's needle tip and how to align the device exactly in 3D, for instance, in order to ablate a tumor [29].

The solution in this case proposes an LED ring mounted on the device (Fig. 2). Thus the surgeons do not need to look at a monitor, but rather get all required information when and where it is needed.

We currently work on a number of projects addressing visualization and interaction possibilities that respect the particular settings of the OR, e.g. through foot interaction devices or even brain-computer interfaces [30, 31].

5   Discussion and Conclusions

This paper presents an approach to building systems for intelligent support for surgeons. In contrast to existing projects and interfaces in this area, our attempt is unique in that we want to integrate different devices into an interaction context in order to enable a fluent switch between different input devices and interaction techniques and metaphors. The goal will be to base the information/visualization with which the surgeon interacts upon the situational context (e.g., the specific phase of the surgery), current task, or personal style and needs of the surgeon or other team members. The input devices used can include a broad range of devices, such as portable interactive surfaces, fixed touchscreen monitors, 3D full body, hand and finger tracking, as well as combinations of projectors with body, hand, and finger tracking.

In the future, various interfaces and/or information streams of established medical devices present in the OR have to be integrated, thus providing a master interface that can be intuitively and dynamically adjusted both manually and automatically to the surgeon's and other team members' needs, personal operation style, and technique. For example, the surgeon could point at an important device or organ area and directly drag the most important information or interface to a suitable location across the available displays or projections.

In order to achieve unobtrusive interaction with the IT systems in the OR, one additional goal is to develop and investigate a range of novel methods and algorithms to achieve precise and robust full degree-of-freedom hand tracking.

Acknowledgements The research reported on in this contribution is funded within the Creative Unit "Intra-Operative Information: What Surgeons Need, When They Need It" in the context of the Excellence Initiative at the University of Bremen.

References

1. Schenk, A., Haemmerich, D., Preusser, T.: Planning of image-guided interventions in the Liver. IEEE Pulse **2**(5), 48–55 (2011)
2. Peterhans, M., Oliveira, T., Banz, V., Candinas, D., Weber, S.: Computer-assisted liver surgery: clinical applications and technological trends. Crit. Rev. Biomed. Eng. **40**(3), 199–220 (2012)
3. Heizmann, O., Zidowitz, S., Bourquain, H., Potthast, S., Peitgen, H.-O., Oertli, D., Kettelhack, C.: Assessment of intraoperative liver deformation: prospective clinical Study. World J. Surg. **34**(8), 1887–1893 (2010)
4. Nadin, M.: Anticipation and dynamics: Rosen's anticipation in the perspective of time. Int. J. Gen. Syst. **39**(1), 3–33 (2010)
5. Nadin, M.: Not everything we know we learned. In: Butz, M. (ed.) Adaptive behavior in anticipatory learning systems. LNAI 2684, pp. 23–43. BertelsmanSpringer, Heidelberg (2003)
6. Schenk, A., Köhn, A., Matsuyama, R., Endo, I.: Transfer of liver surgery planning into the operating room: Initial experience with the iPad. 10th Congress European-African Hepato Pancreato Biliary Association (E-AHPBA), Belgrade, Serbia (2013)
7. O'Hara, K., Gonzalez, G., Sellen, A., Penney, G., Varnavas, A., Mentis, H., Criminisi, A., Corish, R., Rouncefield, M., Dastur, N., Carrel, T.: Touchless interaction in surgery. Commun. ACM **57**(1), 70–77 (2014)
8. Sweller, J.: Instructional design in technical areas. Camberwell, Australian Council for Educational Research (1999)
9. Dix, A., Finlay, J., Abowd, G., Beale, R.: Human Computer Interaction, 3rd edn. Pearson (2003)
10. Hassenzahl, M., Tractinsky, N.: User experience-a research agenda. Behav. Inf. Technol. **25**(2), 91–97 (2006)

11. Anderson, J.R.: Cognitive Psychology and its Implications. Worth Publishers (2009)
12. Kahneman, D.: Attention and Effort. Prentice-Hall, New Jersey (1973)
13. Richardson, A.: Verbalizer-visualizer: a cognitive style dimension. J. Mental Imagery **1**, 109–126 (1977)
14. Kozhevnikov, M., Kosslyn, S.M., Shephard, J.: Spatial versus object visualizers: A new characterization of visual cognitive style. Mem. Cogn. **44**(3), 710–726 (2005)
15. Elting, C., Zwickel, J., Malaka, R.: What are multimodalities made of? Modeling output in a multimodal dialogue system. In: International Conference on Intelligent User Interfaces IUI, pp. 13–16 (2002)
16. Porzel, R., Scheffler, A., Malaka, R.: How entrainment increases dialogical efficiency. In Proceedings of Workshop on Effective Multimodal Dialogue Interfaces at IUI 2006, Sydney (2006)
17. Gurevych, I., Porzel, R., Malaka, R.: Modeling domain knowledge: Know-how and know-what. In: SmartKom: Foundations of Multimodal Dialogue Systems, pp. 71–84. Springer, Berlin (2006)
18. Porzel, R., Gurevych, I., Malaka, R.: In context: Integrating domain-and situation-specific knowledge. In: SmartKom: Foundations of Multimodal Dialogue Systems, pp. 269–84. Springer, Berlin (2006)
19. Porzel, R., Zorn, H.-P., Loos, B., Malaka, R.: Towards a separation of pragmatic knowledge and contextual information. In: Proceedings of ECAI-06 Workshop on Contexts and Ontologies (2006)
20. Malaka, R.: Intelligent User Interfaces for Ubiquitous Computing. In: Handbook of Research on Ubiquitous Computing Technology for Real Time Enterprises, pp. 470–486 (2008)
21. Renz, J., Rauh, R., Knauff, M.: Towards Cognitive Adequacy of Topological Spatial Relations. In: Freksa, C., Brauer, W., Habel, C., Wender, K.F. (eds.) Spatial Cognition II, Integrating Abstract Theories, Empirical Studies, Formal Methods, and Practical Applications, pp. 184–197. Springer, London, UK, (2000)
22. Freksa, C.: Qualitative spatial reasoning. In: Mark, D.M., Frank, A. (eds.) Cognitive and Linguistic Aspects of Geographic Space, pp. 361–372. Kluwer, Dordrecht (1991)
23. Dylla, F., Mossakowski, T., Schneider, T., Wolter, D.: Algebraic properties of qualitative spatio-temporal calculi. In: Conference On Spatial Information Theory (COSIT- 13). North Yorkshire, UK, Springer (2013)
24. Cohn, A. G., Renz, J.: Qualitative spatial reasoning. In: Harmelen, F., Lifschitz, V., Porter B. (eds.) Handbook of Knowledge Representation. Elsevier (2007)
25. Ligozat, G.: Qualitative Spatial and Temporal Reasoning. Wiley-ISTE (2011)
26. Walther-Franks, B., Herrlich, M., Malaka, R.: A multi-touch system for 3D modelling and animation. In: Smart Graphics. vol. 6815 of Lecture Notes in Computer Science, Chap. 5, pp. 48–59. Springer, Berlin (2011)
27. Walther-Franks, B., Malaka, R.: An interaction approach to computer animation. Entertainment Comput. **5**(4), 271–283 (2014)
28. Freksa, C.: Understanding cognition through synthesis and analysis. Int. J. Softw. Inform. **7**(1), 3–18 (2013)
29. Herrlich, M., Benker, J., Black, D., Dylla, F., Malaka, R.: Tool-mounted ring displays for intraoperative navigation. In: Proceedings of CURAC 2015, Bremen, Germany (2015)
30. Fitzke, T., Krail, N., Kroll, F., Ohlrogge, L., Schröder, F., Spillner, L. Voll, A., Dylla, F., Herrlich, M., Malaka, R.: Fußbasierte interaktionen mit computersystemen im operationssaal. In: Proceedings of CURAC 2015, Bremen, Germany (2015)
31. Ritter, F., Porzel, R., Mostajeran Gourtani, F., Malaka, R.: Brain-computer-interface for software control in medical interventions. In: Proceedings of CURAC 2015, Bremen, Germany (2015)

Part VII
Anticipation and Alternative Medicine

Unorthodox Forms of Anticipation

Dean Radin

Abstract Prediction involves the act of mentally projecting into possible futures based on knowledge of the past and influenced by present wants and needs. Most scientists assume that prediction is sufficient to account for forms of behavior in which the future is represented by wants and needs. Experiences that are labeled intuitive hunches, gut feelings, premonitions, or presentiments are suggestive of time-reversed forms of anticipation. Despite the seeming impossibility of genuine time-reversed effects, a growing body of empirical data in psychology, psychophysiology, and physics suggests that despite the disquiet associated with the concept of retrocausality, such influences may nevertheless exist.

Keywords Anticipation · Prediction · Presentiment · Retrocausation · Teleology

1 Introduction

Preparing for the future is a central preoccupation of human beings. Adults plan for retirement; children plan for Halloween. Physicians plan a patient's course of healing; patients would like to know if treatments will be effective. Epidemiologists expect epidemics; geologists would like to be able to predict earthquakes, meteorologists forecast weather, and so on.

To anticipate entails capabilities different from those involved in planning, expectation, forecasting, prediction, intuition [1]. A human being's ability to anticipate allows him/her to successfully hit a baseball with a bat without actually watching the ball. Through anticipation, a catcher runs to where the ball will be. A tennis pro returns the serve in anticipation of where the tennis ball is heading [2]. The reactive mode—waiting to see where the ball lands and then acting on that information—spells failure in sports. It prevents us from passing out when we stand up from a sitting position [3]. It determines what we see or fail to see [4], and it

D. Radin (✉)
Institute of Noetic Sciences, Petaluma, CA, USA
e-mail: dean@noetic.org

© Springer International Publishing Switzerland 2017
M. Nadin (ed.), *Anticipation and Medicine*, DOI 10.1007/978-3-319-45142-8_17

forms the basis for an entire class of humor [5]. In sum, anticipation is a key feature that distinguishes living from non-living systems, as well as much of human activity from that of other living organisms (e.g., animals, insects). As such, it is important to understand the full range of anticipatory behavior.

Conventional models of anticipation assume that this ubiquitous behavior can be fully understood by common sense notions of causality. But some forms of anticipatory experience, variously called intuitive hunches, flashes of insight, gut feelings, precognition, premonitions, or presentiments, appear to violate ordinary causality and suggest teleological pulls from the future. Are such appearances of retrocausation merely telic veneers, or is it possible that some experiences do involve genuine influence from the future?

The orthodox answer is that experiences of retrocausation are necessarily illusory because reversed causation is impossible. Indeed, scientific explanations are predicated on the assumption of unidirectional and inviolate causality, so claims of precognition must be mistaken because they would presumably violate one or more natural laws [6]. However, despite such common sense assumptions, both empirical and theoretical reasons can be brought to bear to challenge this orthodox stance.

2   Methods

One of the hallmarks of science is that it has repeatedly revealed that many of our intuitions about the nature of reality—including such foundational concepts as space, time, matter and energy—are wrong. For example, everyday experience tells us that Nature is based on three core principles: locality, causality, and reality [7, 8]. *Locality* refers to the idea that all interactions between physical systems occur through physical contact. This disallows any form of "spooky action at a distance," to use a phrase made famous by Einstein. *Causality* tells us that the cause-effect order is sacrosanct, i.e., that time moves strictly forward. *Reality* means that the moon (or any object) is still there even when you are not looking at it, i.e., that the world consists of objects with real properties that are completely independent of observers.

From an everyday perspective, all of these principles are self-evident. The problem is that developments in physics over the course of the 20th century (primarily relativity and quantum theory) have established to very high degrees of confidence that one or more of these three principles are simply wrong [7–10]. To date there is no widespread consensus about whether we need to relinquish locality, causality, or reality, or all three; but it is abundantly clear that *something* about our understanding of the deep nature of reality is radically at odds with common sense. This opens the door to thinking about new, previously unthinkable, possibilities, including retrocausal experience. We will refer to such experiences as forms of "unorthodox anticipation."

Fortunately we are not limited to discussion of anecdotes. These experiences are perfectly amenable to scientific study in a variety of rigorous ways, including

(a) consciously predicting future events that cannot be inferred via ordinary means, and where the probability of a chance outcome is known; (b) similar studies conducted while the participant is dreaming, in which unconscious responses are measured by implicit behavior and physiological manifestations.

2.1 Forced-Choice Tasks

The protocol in these experiments involves asking participants to guess the outcome of a future random decision, like the tossing of a pair of dice or its modern equivalent, generation of a random number by a truly random process instantiated within a hardware-based electronic circuit. The source of randomness in these modern random number generators (RNGs) includes radioactive decay times, electron tunneling, and other quantum-randomness events.

A meta-analysis of these forced-choice experiments conducted from 1935 to 1987 [11], based on 309 publications, found a small overall average effect size (Rosenthal effect size $r = 0.02$ [12]), but due to the large statistical power, the deviation from chance was highly significant (associated with a standard normal deviate of $z = 6.02$, or $p < 1.1 \times 10^{-9}$). The possibility that this outcome was inflated due to selective reporting practices was addressed by calculating how many unreported or unretrieved studies averaging a null effect would be necessary to reduce the effect to a non-significant level. That number turned out to be 14,268 studies, which was deemed implausible given the number of researchers known to have conducted these studies. It was further found that while experimental methods had significantly improved from 1935 and 1987, the effect size remained constant, which argues against the potential that the results were biased by differences in study quality. Also, studies with participants selected for better performance produced significantly larger effects as compared to unselected participants, which is consistent with the observation that human performance displays wide variations in natural talent.

While this literature provides evidence for a form of unorthodox anticipation, forced-choice experiments eventually declined in popularity for two main reasons. First, repeated-guessing tasks are boring, and as such they encourage participants to guess the next target based on the gambler's fallacy rather than on intuitive impressions. Second, this type of test constrains the impressionist and spontaneous way that these abilities manifest in everyday life [13, 14]. These limitations led to the development of new experimental designs.

2.2 Free-Response Tasks

In a free-response task, participants are asked to describe a photo, video clip, or a geographic location that will be randomly selected and displayed or visited in the future. Independent judges then compare the participant's impressions against a pool of five targets, one of which was the actual (randomly selected) target and four were decoys. The five-item target pool is devised in advance so the possible targets are as different from one another as possible. The judge's task is to rank-order the participant's impressions to the best match of the five, the next best, and so on. In the simplest form of analysis, if the actual target is ranked first, then that trial would be classified a "hit;" otherwise it would be a "miss." Many other, more sophisticated methods of analysis have also been applied to this type of data. Most of the free-response trials based on this general protocol were performed by two groups. The first was (at the time) a classified project housed first at the Stanford Research Institute (SRI) from 1973 to 1988, and then later continued at Science Applications International Corporation (SAIC) from 1988 to 1995 [15]. The second was conducted by the Princeton (University) Engineering Anomalies Research Laboratory (PEAR Lab) from 1978 through the late 1990s [16].

Analysis of the trials conducted at SRI, consisting of 770 individual sessions, resulted in a mean effect size of (Rosenthal's) $r = 0.21$, associated with $z = 5.8$, $p < 3.3 \times 10^{-9}$. Some 445 tests conducted later at SAIC resulted in a mean effect size of $e = 0.23$, $z = 4.85$, $p < 6.1 \times 10^{-7}$ [15]. A total of 653 sessions conducted at about the same time at the PEAR Lab resulted in a mean effect size of $r = 0.21$, $z = 5.42$, $p < 3.0 \times 10^{-8}$) [16]. The similar effect sizes observed in these three sets of data suggest the presence of similar underlying phenomena, and the magnitude of these effects as compared to that observed in the forced-choice tasks confirms the suspicion that experimental designs based more closely on how this information spontaneously arises in everyday life might produce stronger effects.

2.3 Dream Experiments

Precognitive dreams, to which Burk made reference during the conference Anticipation and Medicine[1] (see Burk[2]) are one of the more frequent spontaneous forms of unorthodox anticipation [17, 18]. To explore these experiences under controlled conditions, experiments have been conducted while participants were in the dream state. A participant would go to bed in a sleep lab and periodically be awakened to report his/her dream when exhibiting REM (rapid eye movements). If the dreamer was at home, he/she would simply be asked to write down the dreams upon spontaneously awaking. In the morning, a target would be randomly selected from a pool of prepared targets, and the selected target would be shown to the dreamer. As with the free-response technique, independent judges would blindly

---

[1]*Anticipation and Medicine*. Third International Conference: Anticipation Across Disciplines. Hanse Institute for Advanced Study/Hanse Wissenschaftskolleg, September 28–30, 2015. http://www.h-w-k.de/index.php?id=2181.

[2]Burk, L.: Anticipating the Diagnosis of Breast Cancer: Screening Controversies and Warning Dreams. In: Nadin, M.: (ed.) Anticipation and Medicine, pp. 285–297. Springer, Cham (2016).

compare the dream content against the actual target and the decoy targets; and if the actual target was assigned a rank of 1, then that would be considered a "hit."

Three of four published experiments using this technique reported significant results based on simple counting statistics ($p < 0.05$, two-tailed tests [19–21]). The fourth study did not achieve a statistically significant outcome, but the result was in the predicted direction.

In two other dream precognition experiments [22, 23], rather than showing the dreamer just the actual target, all of the target images were shown and the dreamer had to rank the similarity of his/her dreams to each of the items in the target pool. This design may have introduced some confusion because the dreamer's future experience included both the actual target and the decoys. This may be why both of these studies produced non-significant results. With each of three of four well-designed studies producing significant outcomes, this limited empirical data-base suggests that information from the future may be present below the level of awareness.

2.4 Implicit Behavioral Responses

Implicit anticipation experiments investigate whether present-time behavior is unconsciously influenced by events in the future. For example, the phenomenon of "mere exposure" indicates that people who are exposed to one of two equally preferably items (e.g., photographs of similar-looking people) will tend to prefer the one they have already seen, even when that exposure is subliminal [24]. An unorthodox anticipation version of the mere exposure experiment first asks a participant to select one of two images, and then a computer randomly selects one of the images and presents it subliminally. If mere exposure in the future influences present-time behavior, then the participant's freely selected present choice should be biased to match the randomly selected future image.

This paradigm was popularized by Bem, who in 2011 reported a series of nine such experiments with overall highly significant results [25]. Two years later a meta-analysis collected 90 studies using similar implicit designs, as well as replications of Bem's method, conducted between 2000 and 2013 by laboratories around the world. The results showed that the effect was independently repeatable and highly significant overall (*Hedges' g* = 0.09; $p < 1.2 \times 10^{-10}$) [26].

After categorizing the 90 studies according to the cognitive style required by the task (known as "fast-thinking" versus "slow-thinking" [27]), 61 of the experiments were determined to be fast-thinking and 29 were slow-thinking. The former refers to snap judgments performed without conscious effort, whereas the latter refers to conscious deliberation. Over all, the fast-thinking implicit anticipation tasks were highly significant ($z = 7.11$, $p < 6 \times 10^{-13}$), but the slow-thinking tasks were not ($z = 1.38$, $p > 0.15$). This difference was consistent with the observation that unorthodox anticipatory phenomena appear to arise first in the unconscious mind and only rarely bubble up to the level of conscious awareness [28–30].

2.5 Physiological Responses

If unorthodox anticipation does indeed reside in the unconscious, then the phenomenon should also be detectable by monitoring unconscious bodily changes in the nervous and circulatory systems. Psychophysiological tasks examining these purported effects have been dubbed "presentiment effects," i.e., pre-feeling as opposed to pre-cognitive responses [28].

Unlike the implicit anticipation tasks, these studies do not require behavioral responses or decisions. Instead, the participant is simply exposed to random dichotomous stimuli, e.g., a series of unpredictable weak vs. strong electrical shocks, or calm vs. emotional photographs, while an aspect of their physiology is monitored. The hypothesis is that the physiological measure will begin to react in a manner consistent with the future stimulus. Thus, seconds before an emotional photo is randomly selected and displayed, the participant's sympathetic nervous system (SNS) activity is expected to increase, reflected by say, a rise in skin conductance level, whereas before a calm picture the SNS will remain calm and skin conductance level will show no unexpected deviations from the baseline.

One of the first presentiment experiments used a reaction time task to test if contingent negative variation (CNV), an unconscious brainwave indicator of anticipation, would detect a randomly timed stimulus in the immediate future [31]. The experiment showed a small but statistically significant difference. Shortly thereafter, two independent replication attempts obtained outcomes in the expected direction, but not to statistically significant degrees [32, 33]. At about the same time, an experiment was reported that included presentiment as a possible factor with electric shock as the stimulus [34]. Based on skin conductance measures, 6 of 10 experimental sessions individually showed significant results, each at $p < 0.01$.

Two decades later a presentiment experiment was conducted using skin conductance as the main measure and calm versus emotional photographs as stimuli [28]. That study resulted in a statistically significant outcome, which was soon independently and successfully replicated [35]. That sparked many new replications using physiological measures, including skin conductance, heart rate, peripheral blood flow, pupil dilation, brain electrical activity, and brain blood oxygenation [36–57]. The basic protocol in these studies was conceptually similar, but the stimuli ranged from photographs to cartoons, audio tones, light flashes, and electrical shock. As in the other experiments mentioned herein, the future stimuli in most of the presentiment studies were selected by hardware-based random number generators (RNG).

By 2011, over three dozen presentiment replications had been reported. The first meta-analysis retrieved 37 experiments involving a total of 1064 participants [58]. The overall average effect size was a Cohen's d of 0.26 ($CI^{95\%} = 0.19$ to 0.37), and the combined statistical result was $p < 1.6 \times 10^{-18}$. A Bayes factor was also calculated, providing a Bayesian interpretation of the strength of evidence for or against the hypothesis. According to Jeffreys [59], for a Bayes factor less than 3 to 1 the hypothesis under test may be interpreted as "barely worth mentioning." If it reaches 10 to 1, the evidence may be considered "substantial;" above 30 to 1, it may be considered "strong;" above 100 to 1, it is "very strong;" and above 100 to 1, the evidence can be regarded as "decisive." For the presentiment studies, the Bayes factor was an unambiguous 28 trillion to 1. The worst case file-drawer was estimated to be 954, a ratio of 26 hidden, unpublished, or non-retrievable studies for each of the known 37 experiments. Such a degree of selective reporting was judged implausible.

A second meta-analysis found 49 published and unpublished presentiment experiments through 2010 [60]. To help narrow the scope of that analysis, each study included was required to have specified a preplanned analysis, use human physiological measures, and contain clear expectations (or *desiderata*) for the physiological outcomes both before and after the stimuli. Of the 49 studies, 26 reported by seven laboratories fit these criteria. The result was an effect size similar to that observed in the first meta-analysis (Cohen's $d = 0.21$), and the overall probability was again highly significant with $p < 2.7 \times 10^{-12}$. Higher quality studies were associated with larger effect sizes, and the file-drawer estimates ranged from a conservative 87 studies to a more liberal 256 studies, with both estimates judged as implausible. Finally, among those studies that had explicitly investigated the possibility that mundane anticipatory strategies may have been responsible for the significant outcomes, no evidence was found.

3 Discussion

The orthodox response to the experiments reviewed here is that retrocausality—a reversal of the ordinary cause-and-effect relationship—violates common sense, and thus apparently positive evidence can be understood only as flaws or flukes. This reaction is not unreasonable because retrocausation strongly challenges the everyday sense of the unidirectional flow of time. But the history of science has amply demonstrated that "naïve reality" is often revealed as a special case of a more comprehensive reality the moment we glimpse beyond the ordinary senses. For example, Einstein showed that matter, energy, space, and time are not the absolutes suggested by common sense, but rather they are intimate relationships [61]. Likewise, quantum theory informs us that quanta (i.e., elementary particles) do not have definite properties when no one is looking—at least not in the way we understand either "properties" or "looking" in common sense terms [7].

But perhaps the oddest challenge to what we take as self-evident is the nature of causality. This topic has generated more restlessness among scientists and philosophers than is commonly appreciated. As Bertrand Russell put it,

> All philosophers imagine that causation is one of the fundamental axioms of science, yet oddly enough, in advanced sciences, the word 'cause' never occurs .... The law of causality, I believe, is a relic of bygone age, surviving, like the monarchy, only because it is erroneously supposed to do no harm [62, p. 337].

Or as mathematician John von Neumann wrote in 1955,

> We may say that there is at present no occasion and no reason to speak of causality in nature—because no [macroscopic] experiment indicates its presence ... and [because] quantum mechanics contradicts it [63, p. 88].

It is also worth noting that within physics, it is well known that at the quantum scale the present can be influenced by the future. As described by Greene, referring to the delayed choice experiment in quantum mechanics:

> By any classical-common sense-reckoning, that's, well, crazy. Of course, that's the point: classical reckoning is the wrong kind of reckoning to use in a quantum universe [64, p. 875].

This retrocausal effect, first proposed as a thought experiment by physicist John Wheeler, has been experimentally demonstrated to high degrees of confidence in physics labs around the world [65–67]. A critic might respond by saying that time reversal might exist at microscopic scales, but that is irrelevant for understanding unorthodox forms of anticipation at the human scale because the special state of quantum coherence—required to sustain these strange effects—is fragile and rapidly washed out within the hot, wet environment of the brain. This was the prevailing view for many years [68]. But today, with rapid theoretical and experimental advancements in quantum biology [69], there is good reason to suspect that living systems, by their nature, take advantage of quantum effects in nontrivial ways, including "harnessing quantum coherence on physiologically important timescales" [70, p. 10].

In addition, with new evidence indicating that individual neurons are associated with cognitive tasks such as memory, learning, and reaction to stimulus novelty, it appears increasingly likely that quantum effects in the brain at the level of individual neurons may cause cascades that can influence unconscious processes, occasionally rising even to the level of conscious awareness [71]. This line of reasoning presents a new explanatory approach toward understanding unorthodox anticipatory phenomena. It also indicates that previous assertions that such effects are impossible are no longer tenable.

Beyond theoretical challenges in modeling these phenomena, the philosophical, and especially the epistemological, consequences of unorthodox forms of anticipation are far from settled. One disconcerting implication is that it may not be possible to prevent time-reversed influences in experiments, at least not through any currently known methods. Indeed, if the gold-standard, double-blind, randomized protocols used to demonstrate these effects continue to repeatedly support the existence of time-reversed effects in human experience, then we must be prepared to reconsider the possibility of retrocausation and—an even greater heresy—teleological pressure from the future [72, 73]. Indeed, at our current level of understanding, the idea that the present depends on both the past and the future is so remote from engrained ways of thinking that the first reaction to the evidence presented here is that it must be wrong. The second reaction, after a closer consideration of trends in quantum biology [65, 69, 74], may be surprise that a rational explanation for unorthodox forms of anticipation may be on the horizon. In his overview of some theories of anticipation expression, Nadin [75, 76] offers a number of such rational explanations.

Acknowledgments Portions of the studies mentioned in this article were supported by generous grants from the Bial Foundation, of Porto, Portugal, the Institut für Grenzgebiete der Psychologie und Psychohygiene, of Freiburg, Germany, and by the donors and members of the Institute of Noetic Sciences of Petaluma, California, USA.

References

1. Nadin, M.: Anticipation and computation. is anticipatory computing possible? In: Nadin, M. (ed.) Anticipation Across Disciplines, Cognitive Science Monographs. vol. 26, pp. 163–257. Springer, Cham (2015)
2. Mann, D.L., Spratford, W., Abernethy, B.: The head tracks and Gaze Predicts: How the world's best batters hit a ball. PLoS ONE **8**, e58289 (2013)
3. Friedman, H.S., Silver, R.C.: Foundations of Health Psychology. Oxford University Press, Oxford/New York (2007)
4. Simons, D.J.: Monkeying around with the Gorillas in our midst: familiarity with an inattentional-blindness task does not improve the detection of unexpected events. Iperception **1**, 3–6 (2010)
5. Ramachandran, V.S.: The neurology and evolution of humor, laughter, and smiling: the false alarm theory. Med. Hypotheses **51**, 351–354 (1998)
6. Seligman, M.E.P., Railton, P., Baumeister, R.F., Sripada, C.: Navigating into the future or driven by the past. Persp. Psych. Sci. **8**, 119–141 (2013)
7. Herbert, N.: Quantum Reality: Beyond the New Physics. Anchor Press/Doubleday, Garden City, N.Y. (1985)
8. Radin, D.I.: Entangled Minds: Extrasensory Experiences in a Quantum Reality. Paraview Pocket Books, New York (2006)
9. Briggs, G.A., Butterfield, J.N., Zeilinger, A.: The Oxford questions on the foundations of quantum physics. Proc. Math. Phys. Eng. Sci. 469 (2013)
10. Weihs, G.: Quantum mechanics: the truth about reality. Nature **445**, 723–724 (2007)
11. Honorton, C., Ferrari, D.C.: "Future telling": a meta-analysis of forced-choice precognition experiments, 1935–1987. J. Parapsych. **53**, 281–308 (1989)
12. Hedges, L.V., Olkin, I.: Statistical Methods for Meta-Analysis. Academic Press, Orlando (1985)
13. Storm, L., Tressoldi, P.E., Di Risio, L.: Meta-analysis of free-response studies, 1992–2008: assessing the noise reduction model in parapsychology. Psychol. Bull. **136**, 471–485 (2010)
14. Rhine, L.E.: Spontaneous Psi experiences. J. Miss. State Med. Assoc. **6**, 375–377 (1965)
15. Utts, J.: An assessment of the evidence for psychic functioning. J. Sci Explor. **10**, 3–30 (1996)
16. Dunne, B.J., Jahn, R.G.: Information and uncertainty in remote perception research. J. Sci. Expl. **17**, 207–241 (2003)
17. Ullman, M., Krippner, S., Vaughan, A.: Dream telepathy: scientific experiments in nocturnal extrasensory perception. Hampton Roads Pub, Charlottesville, VA (2001)
18. Dossey, L.: The power of premonitions: how knowing the future can shape our lives. Dutton, New York, NY (2009)
19. Krippner, S.: The Paranormal Dream and Man's Pliable Future. Psychoanal. Rev. **56**, 28–43 (1969)
20. Krippner, S., Ullman, M.: Telepathy and dreams: a controlled experiment with electroencephalogram-electro-oculogram monitoring. J Nerv Ment Dis. **151**, 394–403 (1970)

21. Watt, C., Wiseman, R., Vuillaume, L.: Dream precognition and sensory oncorporation: a controlled sleep laboratory study. J. Consc. Stud. **22**, 172–190 (2015)
22. Luke, D., Zychowicz, K., Richterova, O., Tjurina, I., Polonnikova, J.: A sideways look at the neurobiology of Psi: precognition and Circadian Rhythms. Neuroquantology **10**, 580–590 (2012)
23. Sherwood, S., Roe, C., Simmons, C., Biles, C.: An exploratory investigation of dream precognition using consensus judging and static targets. J. Soc. Psychical Res. **66**, 22–28 (2002)
24. Delplanque, S., Coppin, G., Bloesch, L., Cayeux, I., Sander, D.: The mere exposure effect depends on an odor's initial pleasantness. Front Psychol. **6**, 911 (2015)
25. Bem, D.J.: Feeling the future: experimental evidence for anomalous retroactive influences on cognition and affect. J. Pers. Soc. Psychol. **100**, 407–425 (2011)
26. Bem, D., Tressoldi, P.E., Rabeyron, T., Duggan, M.: Feeling the future: a meta-analysis of 90 experiments on the anomalous anticipation of random future events. F1000Research, 4:1188 (2016)
27. Kahneman, D.: Thinking, Fast and Slow. Farrar, Straus and Giroux, New York (2013)
28. Radin, D.I.: Unconscious perception of future emotions: an experiment in presentiment. J. Sci. Explor. **11**, 163–180 (1997)
29. Ellenberger, H.F.: The Discovery of the Unconscious: The History and Evolution of Dynamic Psychiatry. Basic Books, New York (1970)
30. Carpenter, J.: First sight: ESP and Parapsychology in Everyday Life. Rowman & Littlefield Publishers Inc, Lanham (2012)
31. Levin, J., Kennedy, J.: The relationship of slow cortical potentials to Psi information in man. J. Parapsychol. **39**, 25–26 (1975)
32. Hartwell, J.: Contingent negative variation as an index of precognitive information. Euro. J. Parapsychol. **2**, 83–103 (1978)
33. Hartwell, J.: An extension to the CNV study and an evaluation. Euro. J. Parapsych. **2**, 358–364 (1979)
34. Vassy, Z.: Method for measuring the probability of 1 bit extrasensory information transfer between living organisms. J. Parapsychol. **42**, 158–160 (1978)
35. Bierman, D.J., Radin, D.I.: Anomalous anticipatory response on randomized future conditions. Percept. Mot. Skills **84**, 689–690 (1997)
36. Radin, D.: Electrodermal presentiments of future emotions. J. Sci. Expl. **18**, 253–273 (2004)
37. Radin, D., Lobach, E.: Toward understanding the placebo effect: investigating a possible retrocausal factor. J. Alt. Complement. Med. **13**, 733–739 (2007)
38. Hinterberger, T., Studer, P., Jäger, M., Haverty-Stacke, C., Walach, H.: Can a slide-show presentiment effect be discovered in brain electrical activity? J. Soc. Psychical Res. **71**, 148–166 (2007)
39. Bierman, D.J., van Ditzhuyzen, J.: Anomalous Slow Cortical Components in a Slotmachine Task. Paper presented at: Proceedings of the 49th Annual Parapsychological Association. (2006)
40. Bierman, D.J., Scholte, H.S.: Anomalous Anticipatory Brain Activation Preceding Exposure of Emotional and Neutral pictures. In Tucson IV: Toward a Science of Consciousness (2002)
41. Bierman, D.J., Radin, D.I.: Anomalous Unconscious Emotional Responses: Evidence for a Reversal of the Arrow of Time. In Tuscon III: Toward a Science of Consciousness (1998)
42. Radin, D.I., Vieten, C., Michel, L., Delorme, A.: Electrocortical activity prior to unpredictable stimuli in meditators and nonmeditators. Explore **7**, 286–299 (2011)
43. Bierman, D.J.: Presentiment in a fMRI Experiment with Meditators. Euro-Parapsychological Association Convention, Paris (2007)
44. Broughton, R.: Exploring the reliability of the "presentiment" effect. Paper presented at: proceedings of the 47th convention of the parapsychological association (2004)
45. Wildey, C.: Impulse Response of Biological Systems. Department of Electrical Engineering, University of Texas, Arlington (2001)

46. McCraty, R., Atkinson, M., Bradley, R.T.: Electrophysiological evidence of intuition: Part 2. A system-wide process? J. Alt. Complement. Med. 10, 325–336 (2004)
47. McCraty, R., Atkinson, M., Bradley, R.T.: Electrophysiological evidence of intuition: Part 1. The surprising role of the heart. J. Alt. Complement. Med. 10, 133–143 (2004)
48. Bradley, R.T., Gillin, M., McCraty, R., Atkinson, M.: Non-local intuition in entrepreneurs and non-entrepreneurs: results of two experiments using electrophysiological measures. Int. J. Entrepreneurship Small Bus. 12, 343–372 (2011)
49. La Pira, F., Gillin, M., McCraty, R., Bradley, R.T., Atkinson, M., Simpson, D.: Validating nonlocal intuition in repeat entrepreneurs: a multi-method approach. J. Behav. Stud. Bus. 6, 1–22 (2013)
50. May, E.C., Paulinyi, T., Vassy, Z.: Anomalous anticipatory skin conductance response to acoustic stimuli: experimental results and speculation about a mechanism. J. Alt. Complement. Med. 11, 695–702 (2005)
51. Tressoldi, P., Martinelli, M., Massaccesi, S., Sartori, L.: Heart rate differences between targets and non-targets in intuitive tasks. Human Physiol. 31, 646–650 (2005)
52. Tressoldi, P., Martinelli, M., Zaccaria, E., Massaccesi, S.: Implicit intuition: how heart rate can contribute to prediction of future events. J. Soc. Psychical Res. 73, 1–16 (2009)
53. Tressoldi, P., Martinelli, M., Scartezzini, L., Massaccesi, S.: Further evidence of the possibility of exploiting anticipatory physiological signals to assist implicit intuition of random events. J. Sci. Explor. 24, 411–424 (2010)
54. Spottiswoode, S.J.P., May, E.: Skin conductance prestimulus response: analyses, artifacts and a pilot study. J. Sci. Explor. 17, 617–641 (2003)
55. Schönwetter, T., Ambach, W., Vaitl, D.: Does autonomic nervous system activity correlate with events conventionally considered as unpreceivable? Using a guessing task with physiological measurement. J. Parapsychol. 75, 327–328 (2011)
56. Don, N.S., McDonough, B.E., Warren, C.A.: Event-related brain potential (ERP) indicators of unconscious psi: a replication using subjects unselected for Psi. J. Parapsychol. 62, 127–145 (1998)
57. Sartori, L., Massaccesi, S., Martinelli, M., Tressoldi, P.: Physiological correlates of ESP: heart rate differences between targets and nontargets. J. Parapsychol. 68, 351–360 (2004)
58. Tressoldi, P.E.: Extraordinary claims require extraordinary evidence: The case of non-local perception, a classical and Bayesian review of evidences front. Psychology 2 (2011)
59. Jeffreys, H.: The Theory of Probability, 3rd edn. Oxford University Press, Oxford, UK (1961)
60. Mossbridge. J., Tressoldi, P., Utts, J.: Predictive physiological anticipation preceding seemingly unpredictable stimuli: A meta-analysis. Front. Psychol. 3 (2012)
61. Greene, B.: The Fabric of the Cosmos: Space, Time, and the Texture of Reality, 1st edn. A.A. Knopf, New York (2004)
62. Pearl, J.: Causality: Models, Reasoning, and Inference. Cambridge University Press, New York (2000)
63. Rosen, R.: Essays on Life Itself. Columbia University Press, New York (1999)
64. Kim, Y.H., Yu, R., Kulik, S.P., Shih, Y., Scully, M.O.: Delayed choice quantum eraser. Phys. Rev. Lett. 84, 1–5 (2000)
65. Aharonov, Y., Zubairy, M.S.: Time and the quantum: erasing the past and impacting the future. Science 307, 875–879 (2005)
66. Ma, X.-S., Zotter, S., Kofler, J., et al.: Experimental delayed-choice entanglement swapping. Nat. Phys. 8, 479–484 (2012)
67. Ma, X.-S., Kofler, J., Qarry, A., et al.: Quantum erasure with causally disconnected choice. Proc. Natl. Acad. Sci. U S A. 110, 1221–1226 (2013)
68. Tegmark, M.: Importance of quantum decoherence in brain processes. Phys. Rev. E Stat. Phys. Plasmas Fluids Relat. Interdiscip. Topics. 61, 4194–4206 (2000)
69. Melkikh, A.V., Khrennikov, A.: Nontrivial quantum and quantum-like effects in Biosystems: Unsolved questions and paradoxes. Prog. Biophys. Mol. Biol. 119, 137–161 (2015)
70. Lambert, N., Chen, Y.-N., Cheng, Y.-C., Li, C.-M., Chen, G.-Y., Nori, F.: Quantum biology. Nat. Phys. 9, 10–18 (2013)

71. Rutishauser, U., Ye, S., Koroma, M., et al.: Representation of retrieval confidence by single neurons in the human medial temporal Lobe. Nat. Neurosci. **18**, 1041–1050 (2015)
72. Wolfe, C.T.: Teleomechanism redux? Functional physiology and hybrid models of life in early modern natural philosophy. Gesnerus **71**, 290–307 (2014)
73. Ribeiro, M.G., Larentis, A.L., Caldas, L.A., et al.: On the debate about teleology in biology: The notion of "teleological obstacle". Hist. Cienc. Saude Manguinhos. **22**, 1321–1333 (2015)
74. Tonello, L., Cocchi, M., Gabrielli, F., Tuszynski, J.A.: On the possible quantum role of serotonin in consciousness. J. Integr. Neurosci. **14**, 295–308 (2015)
75. Nadin, M.: Anticipation and dynamics: Rosen's anticipation in the perspective of time. Int. J. Gen. Syst. (special issue), vol. 39:1, pp. 3–33. Taylor and Blackwell, London (2010)
76. Nadin, M.: Variability by another name: "repetition without repetition." In: Nadin, M. (ed.) Learning from the Past. Early Soviet/Russian contributions to a science of anticipation. Cognitive Science Monographs vol. 25, pp. 329–340, Springer, Cham (2015)

Anticipating Diabetes, Obesity and Polycystic Ovarian Syndrome and Applying Integrative Techniques Using Functional and Oriental Medicine

Dagmar Ehling

Abstract Metabolic syndrome is rampant in our society with increases of epidemic proportions in obesity and diabetes. This paper provides an overview of integrative medical diagnostic and treatment options that favor proactive interventions to prevent the long-term consequences of these maladies. Topics include Oriental medicine, blood sugar balance, adrenal health, immune system, hormones, brain health, dietary modifications, including their application for patients with polycystic ovarian syndrome (PCOS), a major cause of infertility. This approach offers useful and safe alternatives to drug therapy, bariatric surgery, and in the case of PCOS, assisted reproductive technologies, which have known risks, complications, and high costs. This paper suggests new possibilities for managing these issues proactively and effectively. Anticipation of outcomes of dysfunctional lifestyle choices offers the opportunity to prevent disease, i.e., "treat" it, before it occurs.

Keywords Diabetes · Obesity · Acupuncture · Functional medicine · Polycystic ovarian disease · Oriental medicine · Paleolithic diet

1 Introduction

Obesity is defined by the National Institutes of Health as having a body mass index (BMI) of 30 or higher, which is the equivalent of being overweight by 30 lb [1]. Diabetes is defined as having a fasting blood sugar level higher than 126 mg/dL on two different occasions or having a Hemoglobin A1c blood test of 6.5 % or higher [2].

The Centers for Disease Control (CDC) began tracking obesity and diabetes rates in 1994. Since that time, accumulated data show that the United States has experienced a staggering increase in those rates [3]. According to the CDC, as of

D. Ehling (✉)
Oriental Health Solutions, LLC, Durham, NC, USA
e-mail: d.ehling@yahoo.com

D. Ehling
Forschungsgruppe Akupunktur, Düsseldorf und Grafing, Germany

© Springer International Publishing Switzerland 2017
M. Nadin (ed.), *Anticipation and Medicine*, DOI 10.1007/978-3-319-45142-8_18

2010 more than 26 % of US Adults aged 18 years or older are obese, and more than 9 % or 1 in 10 adult Americans have type 2 diabetes [4]. By the year 2030, the incidence of diabetes is expected to affect 7.7 % of the world's population or 439 million adults. This means that between 2010 and 2030, there will be a 69 % increase for developing countries in adults with diabetes and for developed countries a 20 % increase [5]. The CDC estimates that by 2050 1 in 3 to 1 in 5 USA adults could have diabetes [6]. Linear forecasts for obesity indicate that by 2030, 51 % of the USA population will be obese [7].

In the last sixty years human diets have changed dramatically [8]. In 1977 the Senate Select Committee on Nutrition and Human Needs recommended modified Dietary Goals for Americans. It encouraged "increasing carbohydrates to 55–60 % of calories; decreasing dietary fat to no more than 30 % of calories; decreasing cholesterol to 300 mg per day; decreasing sugar intake to 15 % of calories; and decreasing salt intake to 3 g per day" [9]. Such goals testify to an anticipatory approach, provided that they are scientifically proven to be correct. We shall see that this is not the case.

It is worth noting that obesity rates began to significantly increase that same year. Prior to 1977, obesity rates of Americans were less than 15 % [10]. A study by Zhou et al. [11] shows charts in which obesity rates directly correlate with the increase in grains and sugars at the same time as meat consumption declined. The graphs show the mid-seventies as a major turning point. Various versions of the Food Guides were published; the most notable was the Food Pyramid in 1992. It recommended 6–11 servings of carbohydrates (bread, cereal, rice, and pasta) while limiting fats and dairy to 2–3 servings respectively per day [12]. *MyPlate*, introduced in 2011, recommends consuming more than 50 % fruits, grains, and dairy [13]. Most dairy sold in the US is low-fat dairy; therefore, because it is unopposed by sufficient fat, it is relatively higher in lactose, a known sugar. When carbohydrates are combined with fat or protein, the blood sugar spike after a meal is lowered. Based on outcomes, it appears that the governmental recommendations of lowering dietary fat and increasing carbohydrates—thereby increasing overall sugar intake—are a direct cause of the diabetes and obesity epidemic.

1.1 Costs Associated with Obesity and Type 2 Diabetes

According to the American Diabetes Association, the cost of diabetes in 2012 was US$245 billion, which includes US$176 billion in medical costs and US$69 billion in lost or reduced productivity [14]. The Gallup estimates indicate that costs associated with obesity hover around US$50 million per 100,000 residents annually [15]. Costs due to absenteeism from obesity between 2000 and 2004 are about US$4.3 billion per year in the USA [16].

Many women with PCOS undergo assisted reproductive therapies (ART) in order to conceive a child. Those costs can range from US$10,000 to US$100,000 depending on the need for ART. This can be compared to managing someone's diet, reversing their IR and enhancing the body's physiology to function on its own where the costs are around US$2000 [17].

1.2 Complications and Risk Factors

Complications and risk factors of diabetes are as follows: diabetic retinopathy, which may lead to blindness; diabetic neuropathy, which may lead to leg pain, numb toes (possibly requiring amputation); stroke; heart disease; kidney disease; high blood pressure; skin infections; mood changes such as depression [18]. Neuro-degeneration, also known as type 3 diabetes [19], may contribute to some forms of dementia. Type 2 diabetes and high insulin levels appear to contribute to cancer [20] and diminish cancer survival rates [21, 22]. The risk factors of obesity are heart disease, stroke, high blood pressure, type 2 diabetes, liver and gallbladder disease, degeneration of cartilage and underlying bone within a joint, infertility, and mental health problems [23].

2 Associated Physiological Phenomena in Patients with Insulin Resistance and Reactive Hypoglycemia

Many conditions accompany insulin resistance (IR), a precursor to type 2 diabetes: pain; joint degeneration (e.g., osteoarthritis); chronic muscle pain; digestive complaints; systemic inflammation due to blood sugar fluctuations; non-healing skin conditions (e.g., eczema, chronic pruritus); food sensitivities; polycystic ovarian disease (PCOS); premature andropause; and others [24]. It is not uncommon to see blood sugar fluctuations in IR patients; many, in fact, will fluctuate between reactive hypoglycemia (RHG) and insulin resistance throughout the day. Table 1 summarizes the most common symptoms of many IR and RHG patients.

Over the past few years the term "Type 3 diabetes" has been used to describe a type of dementia similar to that of Alzheimer's disease (AD) [26]. High blood sugar may be one of many contributing factors to the epidemic of dementia in our elderly population. It has been documented that an HbA1c level of greater than 5.9 represents an annual loss of brain mass of 0.5 %. This can be observed with standard brain MRI [27]. Type 2 diabetes is associated with hippocampal and amygdalar atrophy, regardless of vascular pathology [28].

Long-term blood sugar problems affect the adrenal glands. The adrenal cortex secretes three hormones: aldosterone, cortisol, and androgens. Table 2 summarizes the most common symptoms associated with compromised adrenal health.

Table 1 [25]

Reactive hypoglycemia	Insulin resistance
Sweet cravings when fatigued	Sugar cravings after meals
Irritable/lightheaded when meals are missed	Eating sweets does not relieve sugar cravings
Dependence on coffee to keep motivated	Fatigue after meals
Fatigue is relieved by eating	Frequent urination, thirst, increase appetite
Poor memory/forgetfulness/brain fog	Difficulty losing weight
Waking tired, insomnia, heart palpitations	Apple or pear body shape
Low blood pressure/orthostatic hypotension	If chronic: in males: gynecomastia; in females: polycystic ovaries, hirsutism
Skipping breakfast or other meals	
Mood swings, negativity	
Fasting serum glucose: <85 mg/dL	Fasting serum triglycerides >100 mg/dL

Table 2 [25]

Adrenal health	
Cannot stay asleep	Cannot fall asleep
Craves salt	Perspires easily
Slow starter in the morning	Experiences high amount of stress
Afternoon fatigue	Weight gain when under stress
Dizziness when standing up quickly	Feels not rested even after 6 or more hours of sleep
Headaches with exertion or stress	Inner tension

Blood sugar irregularities can also affect the circadian rhythm of cortisol, which can impact sleep. The circadian rhythms of cortisol (controlled by the hippocampus of the brain) and of melatonin (controlled by the pineal gland) are inverse to each other. When cortisol does not descend properly at nighttime the patient will be wide awake. Likewise, as cortisol does not rise at sunrise, it might be hard to get out of bed in a timely fashion, or the person may have no appetite at breakfast time. Shift workers and people who stay up late at night are particularly affected by this phenomenon. Over time an abnormal circadian rhythm of cortisol can lead to shrinkage of the hippocampus, impairing short-term to long-term memory conversion [29]. Chronic insomnia may stimulate pro-inflammatory proteins [30].

Rapid blood sugar fluctuations lead to abnormal cortisol secretions by the adrenal glands, stressing the body. Chronic stress or post-traumatic stress disorder (PTSD) can also promote inflammatory signaling [31]. The inflammation can manifest in various parts of the body: in joints, muscles, organs or brain. Long-term inflammation can lead to auto-immune diseases as the immune system loses its ability to discern self from non-self. Regulatory T-cells, however, can prevent or dampen the expression of autoimmune disease [32]. Functional medicine in particular can enhance regulatory T-cells by supplementation of certain nutrients; in particular, L-glutathione and vitamin D3 appear to modulate the immune system and therefore may help to dampen pro-inflammatory activities [33, 34]. Inflammation may affect gut permeability in which the proteins of the enterocytes become unstable allowing partially digested food particles to re-enter the blood stream, which further triggers the immune system [35]. Factors that may compromise gut permeability can be life stressors [36], traumatic brain injury [37], hypothyroidism [38], celiac disease and other autoimmune diseases [39]. Gut permeability problems are seen in patients with multiple food sensitivities and allergies, ulcerative colitis, or irritable bowel disease. Enhancing regulatory T-cells via supplementation of vitamin D3 appears to be helpful here as well [40]. Patients who consume a lot of alcohol experience a greater probability of gut wall degradation [41].

Insulin resistance in particular contributes to polycystic ovarian disease (PCOS) [42] in females, whereas in males IR can promote male menopause or andropause [43, 44]. Women of child-bearing age with PCOS commonly experience anovulation, irregular periods or amenorrhea, multiple unripe follicles on their ovaries, infertility, hair loss, hirsutism, and weight gain. PCOS patients have a greater propensity toward non-alcoholic fatty liver disease [45, 46]. Many of these women are more prone to develop hypothyroidism and subsequently, Hashimoto's thyroiditis [47–49]. Reasons for this are not very well understood, but clearly, women with PCOS experience more inflammation [50]. Insulin resistance in women stimulates the hypersecretion of androgens, which promotes all the above mentioned symptoms. (Not all women with PCOS experience amenorrhea or anovulation or obesity; many manifestations can be seen in women who are slender and have irregular periods yet have IR.) The continuum of PCOS ranges from mild IR (mildly irregular menses with normal ovulation) to severe IR (anovulation and amenorrhea). In males with IR, gynecomastia, low libido, erectile dysfunction, and adipose tissue around the waist can be observed. Insulin resistance in males promotes the aromatization of testosterone into estrogen [51].

3 Functional and Oriental Medicine

Functional medicine (FM) focuses on improving physiological function as a primary method for improving health in patients with chronic diseases [52]. It is a systems-approach versus a disease-oriented approach. That is, it addresses the whole human being; instead of treating him like a machine that needs a part replaced [53–55]. Oriental Medicine (OM) includes treatment modalities such as Chinese herbal medicine, acupuncture, diet, *tui na* and *shiatsu* massage, and *qi gong* exercises. Major diagnostic theories include: Yin/Yang theory, the Five Elements, Six Divisions (Shang Han Lun), the human body meridian/channel system, among others. OM's goal is to balance the body by adjusting the energy of organs and meridians thereby generating healthy functioning of the body. Great emphasis is placed on nourishing the body so that existing physiology is enhanced and improved [56, 57]. Acupuncture and Chinese herbal medicine have a history of thousands of years of treating a variety of disorders, including associated complications of obesity and type 2 diabetes [17, 56–58].

Various styles of acupuncture exist: traditional Chinese medicine, Worsley 5 element, Korean acupuncture, a number of different Japanese styles, and "medical" acupuncture, which has its foundation in French acupuncture [59]. The primary goal of all styles is to achieve harmony and wellness by modulating the patient's meridian system and energy flow. Disease manifests when the meridian system gets stuck or energy stagnates [56]. A skilled practitioner can detect early stages of blood sugar-related conditions and correct them using acupuncture and herbs, and counsel patients on diet.

Both FM and OM view health as a web that is interconnected between organ, nervous, and endocrine systems which all work synergistically, rather than individual systems functioning autonomously and without any effect on each other. Both treatment systems offer strategies that are personalized and anticipatory, and empower patients to express themselves optimally in body, mind, and spirit.

Figure 1 attempts to show the synergistic ideas of both forms of medicine within the Five Element system. The element Fire, ruled by the Heart and Small Intestine affects (via the Sheng cycle shown as the partial circle connecting Fire and Earth) Earth, which is about digestion, transportation and transformation of nutrients, and blood sugar metabolism. Likewise, Earth feeds into Metal (Lung and Large Intestine) which affects breath, immunity and the barrier system (gut, blood brain

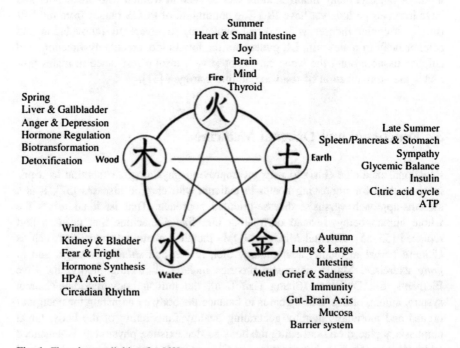

Fig. 1 Five element linking fist [60]

barrier, lung, endothelial, and the skin barrier). The barrier system is attributed to Metal since it rules the mucosa. Metal then feeds into Water, which rules bones, teeth and marrow, manages water, and deals with hormone synthesis, while Wood deals with hormone regulation. Fire regulates Metal, thereby influencing the brain-gut axis (via the Ko cycle shown as the straight line connecting Fire and Metal). Figure 1 shows possible interactions between OM and FM; these still need to be explored further.

Integration between FM, western medicine, nutrition and OM is happening in a variety of places: The Cleveland Clinic has hired a director of Functional Medicine. Dr. Mark Hyman, MD (the author of *The Blood Sugar* Solution [61]) has taught extensively at The Institute of Functional Medicine (IFM), which is an organization that trains physicians and allied health practitioners in functional medicine. IFM and The Cleveland Clinic have formed a collaborative relationship [62]. Eric Westman, MD, who does ongoing research at Duke University, is a major proponent of the low-carb, high-fat diet. His clinical research and clinical care focuses on lifestyle treatments for obesity and diabetes. (He has published over ninety peer-reviewed publications and several books [63].) Chris Kresser, another specialist in FM has a major online presence with his website and blog containing a lot of useful information for the public and physicians alike [64]. Diane Sanfilippo, author of *Practical Paleo*, has been on the forefront on helping consumers define which low-carb foods are optimal to manage blood sugar and other conditions [65]. The *Paleo Mom* (aka Sarah Ballantyne, PhD) has written about auto-immune disease and published two books on how a patient might include the paleo diet to address his/her auto-immune disease [66–68].

4 Preventing Obesity and Type 2 Diabetes—Anticipatory Approaches

The human food supply has changed rapidly in the past sixty years and therefore, based on the historic outline mentioned above, it seems appropriate to consider the possibility that food plays an important role in disease outcomes. A hundred years ago obesity, diabetes, and PCOS were unheard of. The USDA recommendations of reducing saturated fat and increasing carbohydrates in 1977 led to a steady increase of obesity and diabetes. It's time to reverse this trend. If current epidemiological trends continue, the expense and lack of quality of life that these conditions provoke will be unsustainable. The USA and many other countries in a similar situation face healthcare expenses and economic hardship of unprecedented proportion if this issue is not remedied.

4.1 Diet

Conventional methodologies have not worked. Drug therapies offer limited results. In addition, drug therapy as a treatment strategy is *reactive*, rather than *proactive* or *anticipatory*, as it is only implemented in patients who already have diabetes.

Anticipatory dietary recommendations would include a low-carbohydrate diet, in particular a low-grain diet, or even a no-grain paleolithic diet. This essentially means consuming a diet of mainly seafood, meat, poultry, eggs, along with lots of vegetables, fermented vegetables, and low-glycemic fruits, preferably all freshly prepared. Such a diet will have positive effects on lowering insulin levels and prevent inflammation. Even short-term consumption of a paleo-style diet improved cardiovascular risk factors in those with metabolic syndrome [69]. The paleo diet can achieve significant and sustained weight-loss [70, 71]. In some circumstances, a ketogenic, very low-carbohydrate diet has reversed type 2 diabetes completely [72, 73]. PCOS may be completely reversed with a paleo diet [17, 74]. A reduced-carbohydrate diet affects body composition and fat distribution favorably in women with PCOS [75].

For decades, consumers have learned through advertising to purchase processed and high carbohydrate-containing foods. Unlearning this past marketing effort takes patience and persistence. One suggestion to alter the current trend is education: teach children and adults about eating well and avoiding fast food restaurants, avoiding foods with additives, artificial sweeteners, coloring agents, and giving up soft drinks and pre-packaged "convenience" foods of questionable nutrient content. Artificial sweeteners, in particular, have been associated to alter the intestinal microbiota, which results in glucose intolerance [76].

It would also be advisable to improve school lunches in the USA [77, 78]. Government agencies must also change the *MyPlate* by reducing carbohydrates and sugars and increasing fat and protein consumption. This will lead to increased satiety that decreases carbohydrate intake. These basic steps would seem like some of many logical actions for preventing or at least subduing the expected economic tsunami of metabolic syndrome and obesity and their associated health care costs.

4.2 Oriental Medicine

Blood sugar imbalances can lead to chronic pain. In Oriental medicine, the flavor "sweet" is controlled by the element "Earth" within the Five Element Theory. The Earth element, among other tissues, rules the "muscles." Overconsumption of sweets, as in excess carbohydrates, alcohol, fructose, and sugars can lead to chronic muscle pain. Likewise, skipping meals and starving oneself lead to the same [25]. In Oriental medicine, the Spleen and Stomach promote transformation and transportation of food and liquids. Therefore, optimal absorption of nutrients and blood sugar modulation are important for healthy muscle and other bodily functions.

One acupuncture style in particular has been very useful in confirming blood sugar and other health concerns. Kiiko Matsumoto, a Japanese practitioner (residing in Boston, Massachusetts) teaches her particular style to practitioners in the USA and Europe. (Her methods are summarized in her books Clinical Strategies Volume 1 and 2 [79].) She regularly goes back to Japan to study with expert practitioners there. This style incorporates both western and eastern physiology. It is a palpatory method in which certain reflex zones that exhibit pressure pain upon palpation signify specific imbalances in individual patients. A practitioner can palpate specific reflexes such as adrenal, blood sugar, stomach, liver, circulation, ovarian, prostate, thyroid, or autonomic nervous system, etc. These are micro-systems of the body or "homunculi" that are accessed via palpation. When pressure pain is elicited, a practitioner can, along with a thorough history and blood work, surmise diagnostic conclusions. Distal acupuncture points are then selected to ameliorate the pressure pain in affected reflex zones.

Treating the micro-system has impact on the macro-system. A patient with IR and PCOS may exhibit pressure pain in the left lower costal region and the left area of the second and third cervical vertebrae. Possible points that will ameliorate this pressure pain are Spleen 3 (*taibai*) and/or Spleen 9 (*yinlingquan*) and Oddi point. As mentioned above, blood sugar imbalances will stress the adrenals. Therefore, points to address adrenal function are chosen as well. Blood sugar imbalances can contribute to inflammation; in this case, the fire point on one or several meridians can exhibit pressure pain; if so, the metal/water points on the affected respective meridian may be needled.

Oriental medicine diagnosis is based on patterns of disharmony. A person with poor blood regulation, digestive disorders such as loose stools and muscle aches may have a pattern called Spleen qi deficiency. Likewise a patient with pedal edema, low back pain and weakness in the knees may have a pattern called Kidney qi deficiency. A thorough history, assessment of the tongue and pulse (there are 28 pulse qualities to be considered), visual and palpatory assessment guides the practitioner toward an Oriental medical diagnosis or patterns of disharmony. The most common patterns can be reviewed in The Foundations of Chinese Medicine [80]. After a pattern of disharmony is determined a Chinese herbal formula (CHF) may be prescribed.

To illustrate this here is a case of a 66 year old male with type 2 diabetes which was diagnosed five years ago: he is 35 lbs overweight and recently was diagnosed with gallstones and non-alcoholic fatty liver disease. He complained of right upper quadrant pain, hypertension, benign prostatic hypertrophy, depression, mood swings, fatigue, frequent nighttime urination, and low libido. His medication list includes Glucotrol (Glipizide) 10 mg, Metformin 500 mg twice daily, Benicar HCT (olmesartan medoxomil-hydrochlorothiazide) 20/12.5 mg, atorvastatin (Lipitor) 110 mg, and Flomax (Tamsulosin Hydrochloride) 0.4 mg. His blood work was remarkable for the following: HbA1c 7.9 %, fasting glucose 156 mg/dL, CRP 1.8 mg/L, Homocysteine 14.7 μmol/L; his complete metabolic panel and complete blood count with differential were normal. His diet revealed the following meals: breakfast included oatmeal with herbal tea. On the weekends he had Cheerios™

with non-fat milk, English muffin with marmalade, or an occasional omelet, bacon, with green onions and tomatoes. Lunch included a salad with chicken breast, low-fat cheese and commercial salad dressing. The afternoon snack consisted of a banana, or an apple. Dinner usually was chicken, fish or beef, veggies, with organic whole-wheat bread, brown rice or potatoes. He liked a glass of wine or two with dinner. Dessert was a portion of low-fat ice cream. His fluid intake is mostly water with ice.

His tongue was swollen, pink with slightly red sides, and a thin white coat; his pulse was slippery with a slight wiry quality. Upon palpation he showed pressure pain in the right upper quadrant (RUQ) and also in the lower left quadrant. Overall his abdomen and epigastric region felt tight and mildly cold.

His primary Oriental medical diagnosis was a *Shaoyang* stage disorder, Liver *qi* stagnation with Spleen dampness, and mild Heat in the Liver/Gallbladder. His treatment principle was to harmonize and release the *Shaoyang* disorder, course Liver *qi*, resolve dampness in the Spleen, and cool Heat in the Liver/Gallbladder. The chosen formula was Minor Bupleurum Decoction (*Xiao Chai Hu Tang*) consisting of Radix Bupleuri (*Chai Hu*), Radix Scutellariae (*Huang Qin*), Rhizoma Pinelliae (*Ban Xia*), Rhizoma Zingiberis (*Sheng Jiang*), Radix Codonopsis (*Dang Shen*), Honey-Fried Radix Glycyrrhizae (*Zhi Gan Cao*), Fructus Jujubae (*Da Zao*). The formula was modified by adding the following herbs: Spora Lygodii (*Hai Jin Sha*), Rhizoma Curcumae Longae (*Jiang Huang*), Rhizoma Cyperi (*Xiang Fu*), Fructus Meliae Toosendan (*Chuan Lian Zi*), Pericarpium Citri Reticulate Viride (*Qing Pi*), and Rhizoma Atractylodis Macrocephalae (*Bai Zhu*). This was given as a granule herbal formula at 3 g twice daily in boiling water [81–83]. The main purpose of this formula was to treat his RUQ pain and improve digestive function and blood sugar regulation.

In addition, he was instructed to consume a paleo-type diet. He was to have a warm breakfast that contained both fat and protein. It was recommended that for the remaining meals he was to have some meat, poultry or fish with lots of vegetables, and eliminate the starches, wine, and dessert. Since there was no obvious reason to eliminate dairy completely (such as lactose or casein intolerance) he was instructed to consume full-fat dairy in very small amounts. Too much dairy can aggravate dampness in the Spleen but it was felt small amounts would be fine, especially since his tongue coating was thin and not thick. In addition, using ice with liquids furthers the dampness in the Spleen; therefore, it was suggested that he consume room temperature or warm liquids.

Lastly, he was instructed to consider regular exercise with a high intensity interval training component. Exercise lowers blood glucose [84] and lowers blood pressure [85].

After 4 months of treatment equaling three herbal/nutritional consults (no acupuncture was performed due to patient's preference) his HbA1c was reduced to 6.1; he had lost 25 lbs; his energy level was great; his depression was gone and his mood swings were much improved; his RUQ pain was diminished. Obviously, more work is needed but this case illustrates the possibilities.

4.3 Other Anticipatory Approaches

Utilizing medical data technology can facilitate implementation and follow-through of dietary changes. Many such devices exist today: Fitbit™ tracks daily steps taken; Jawbone™ tracks sleep changes; and apps such as MyFitnessPal™ can track carbohydrate, protein, fiber, and fat consumption in grams per day. Devices such as wearables or watches may facilitate motivation to follow-through with dietary recommendations. Additionally, computer games can help to enhance motivation, record progress and often, can be personalized. These devices are being designed by a number of biomedical engineers, some of whom presented their developments at the International Conference: Anticipation in Medicine in Delmenhorst, Germany [86]. The multidisciplinary academic study of anticipation sheds light on expectations, premonition, guessing, forecasting, and prediction [87].

Organizations such as The European Association for Predictive, Preventative and Personalised Medicine [88] take an anticipatory perspective to medicine: preventive programs rather than reactive medical treatments. Anticipation is a much needed approach if human beings intend to thrive in a cost-effective, safe, and personalized health care system. Educating health care practitioners and patients alike about the anticipation of disease—in its pre-symptomatic phase, before the symptomatic stage when frequently treatment is less effective and more costly—is a first step.

References

1. Centers for Disease Control and Prevention. Defining Overweight and Obesity. http://www.cdc.gov/obesity/adult/defining.html
2. American Diabetes Association. Diagnosing Diabetes and Learning About Prediabetes. http://www.diabetes.org/diabetes-basics/diagnosis/
3. Centers for Disease Control and Prevention. PowerPoint Slides on Diabetes. http://www.cdc.gov/diabetes/data/center/slides.html
4. Centers for Disease Control and Prevention. Press Release October 22, 2010. http://www.cdc.gov/media/pressrel/2010/r101022.html
5. Shaw, J.E., Sicree, R.A., Zimmet, P.Z.: Global estimates of the prevalence of diabetes for 2010 and 2030. Diabetes Res. Clin. Pract. **87**, 4–14 (2010)
6. Medline Plus. Type 2 Diabetes: Causes, Symptoms, Exams and Tests, Treatment. http://www.nlm.nih.gov/medlineplus/ency/article/000313.htm
7. Finkelstein, E.A., Khavjou, O.A., Thompson, H., Trogdon, J.G., Pan, L., Sherry, B., Dietz, W.: Obesity and severe obesity forecasts through 2030. Am. J. Prev. Med. **42**(6), 563–570 (2012)
8. United States Department of Agriculture. Agriculture factbook 2001–2002: profiling food consumption in America. (2002). http://www.diet.com/g/usda-food-guide-pyramid-mypyramid
9. United States Department of Agriculture. Nutrition and Your Health: Dietary Guidelines for Americans, Appedix G-5: History of the Dietary Guidelines for Americans. http://www.health.gov/dietaryguidelines/dga2005/report/html/G5_History.htm

10. Ogden, C.L., Carroll, M.D.: Division of Health and Nutrition Examination Surveys; Prevalence of Overweight, Obesity, and Extreme Obesity Among Adults: United States, Trends 1960–1962 Through 2007–2008. www.cdc.gov/NCHS/data/hestat/obesity_adult_07_08/obesity_adult_07_08.pdf

11. Zhou, S.S., Li, D., Zhou, Y.M., Sun, W.P., Liu, Q.G.: B-vitamin consumption and the prevalence of diabetes and obesity among the US adults: population based ecological study. BMC Public Health 10, 746 (2010). http://openi.nlm.nih.gov/detailedresult.php?img=3014900_1471-2458-10-746-8&req=4

12. United States Department of Agriculture. A Brief History of USDA Food Guides. Center for Nutrition Policy and Promotion, June 2011. http://fnic.nal.usda.gov/dietary-guidance/myplate-and-historical-food-pyramid-resources/past-food-pyramid-materials

13. United States Department of Agriculture. Center for Nutrition Policy and Promotion. MyPlate (2010). http://www.cnpp.usda.gov/MyPlate

14. American Diabetes Association. The Cost of Diabetes. http://www.diabetes.org/advocacy/news-events/cost-of-diabetes.html

15. Witters, D., Harter, J., Bell, K., Ray, J.: The cost of obesity to US cities, gallup business journal, gallup-healthways well-being index. http://businessjournal.gallup.com/content/145778/cost-obesity-cities.aspx#1

16. Cawley, J., Rizzo, J.A., Haas, K.: Occupation-specific absenteeism costs associated with obesity and morbid obesity. J. Occup. Environ. Med. 49(12), 1317–1324 (2007)

17. Ehling, D.: Integrative techniques using acupuncture, Chinese herbal medicine, diet, and supplements for polycystic ovary syndrome: a case report. J. Integr. Med. 11(6), 422–427 (2013)

18. Centers for Disease Control and Prevention. Diabetes Public Health Resource. Diabetes Complications. http://www.cdc.gov/diabetes/statistics/complications_national.htm

19. De la Monte, S., Wands, J.R.: Alzheimer's disease is type 3 diabetes—evidence reviewed. J. Diabetes Sci Technol. 2(6), 1101–1113 (2008)

20. Emerging Risk Factors Collaboration. Seshasai, S.R., Kaptoge, S., Thompson, A., Di Angelantonio, E., Gao, P., Sarwar, N., Whincup, P.H., Mukamal, K.J., Gillum, R.F., Holme, I., Njølstad, I., Fletcher, A., Nilsson, P., Lewington, S., Collins, R., Gudnason, V., Thompson, S.G., Sattar, N., Selvin, E., Hu, F.B., Danesh, J.: Diabetes mellitus, fasting glucose, and the risk of cause-specific death. N. Engl. J. Med. 364, 829–841 (2011)

21. Lee, J.S., Cho, S.I., Park, H.S.: Metabolic Syndrome and cancer-related mortality among Korean men and women. Ann. Oncol. 21(3), 640–645 (2010)

22. Goodwin, P.J., Ennis, M., Pritchard, K.I., Trudeau, M.E., Koo, J., Madamas, Y., Hartwick, W., Hoffman, B., Hood, N.: Fasting insulin and outcome in early-stage breast cancer: results of a prospective cohort study. J. Clin. Oncol. 20(1), 42–51 (2002)

23. Centers for Disease Control and Prevention. Vital Signs, Medical Complications of Obesity. http://www.cdc.gov/vitalsigns/AdultObesity/Risk-large.html

24. American Diabetes Association. Diabetes Symptoms. http://www.diabetes.org/diabetes-basics/symptoms/

25. Kharrazian, D.: Metabolic Assessment Form 2012

26. De la Monte, S., Wands, J.R.: Alzheimer's disease is type 3 diabetes—evidence reviewed. J. Diabetes Sci. Technol. 2(6), 1101–1113 (2008)

27. Enzinger, C., Fazekas, F., Matthews, P.M., Ropele, S., Schmidt, H., Smith, S., Schmidt, R.: Risk factors for progression of brain atrophy in aging: six-year follow-up of normal subjects. Neurology 64(10), 1704–1711 (2005)

28. den Heijer, T., Vermeer, S.E., van Dijk, E.J., Prins, N.D., Koudstaal, P.J., Hofman, A., Breteler, M.M.: Type 2 diabetes and atrophy of medial temporal lobe structures on brain MRI. Diabetologia 46(12), 1604–1610 (2003)

29. McAuley, M.T., Kenny, R.A., Kirkwood, T.B., Wilkinson, D.J., Jones, J.J., Miller, V.M.: A mathematical model of aging-related and cortisol induced hippocampal dysfunction. BMC Neurosci. 25(10), 26 (2009). doi:10.1186/1471-2202-10-26

30. Frey, D.J., Fleshner, M., Wright Jr., K.P.: The effects of 40 hours of total sleep deprivation on inflammatory markers in healthy young adults. Brain Behav. Immun. **21**(8), 1050–1057 (2007). Epub 2007 May 23

31. Gill, J., Vythilingam, M., Page, G.G.: Low cortisol, high DHEA, and high levels of stimulated TNFα, and IL-6 in women with PTSD. J. Trauma. Stress **21**(6), 530–539 (2008). doi:10.1002/jts.20372

32. Yan, Z., Banerjee, R.: Redox remodeling as an immunoregulatory strategy. Biochemistry **49** (6), 1059–1066 (2010). doi:10.1021/bi902022n

33. Fraternale, A., Paoletti, M.F., Casabianca, A., Oiry, J., Clayette, P., Vogel, J.U., Cinatl Jr., J., Palamara, A.T., Sgarbanti, R., Garaci, E., Millo, E., Benatti, U., Magnani, M.: Antiviral and immunomodulatory properties of new pro-glutathione (GSH) molecules. Curr. Med. Chem. **13**(15), 1749–1755 (2006)

34. Lucas, R.M., Gorman, S., Geldenhuys, S., Hart, P.H.: Vitamin D and immunity. F1000Prime Rep. **1;6**, 118. doi:10.12703/P6-118 (2014). eCollection 2014

35. Gareau, M.G., Silva, M.A., Perdue, M.H.: Pathophysiological mechanisms of stress-induced intestinal damage. Curr. Mol. Med. **8**(4), 274–281 (2008)

36. Alonso, C., Guilarte, M., Vicario, M., Ramos, L., Ramadan, Z., Antolín, M., Martínez, C., Rezzi, S., Saperas, E., Kochhar, S., Santos, J., Malagelada, J.R.: Maladaptive intestinal epithelial responses to life stress may predispose healthy women to gut mucosal inflammation. Gastroenterology. **135**(1), 163–172.e1 (2008). doi:10.1053/j.gastro.2008.03.036. Epub 2008 Mar 22

37. Bansal, V., Costantini, T., Kroll, L., Peterson, C., Loomis, W., Eliceiri, B., Baird, A., Wolf, P., Coimbra, R.: Traumatic brain injury and intestinal dysfunction: uncovering the neuro-enteric axis. J. Neurotrauma **26**(8), 1353–1359 (2009). doi:10.1089/neu.2008-0858

38. Koyuncu, A., Aydintu, S., Koçak, S., Aydin, C., Demirer, S., Topçu, O., Kuterdem, E.: Effect of thyroid hormones on stress ulcer formation. ANZ J. Surg. **72**(9), 672–675 (2002)

39. Visser, J., Rozing, J., Sapone, A., Lammers, K., Fasano, A.: Tight junctions, intestinal permeability, and autoimmunity: celiac disease and type 1 diabetes paradigms. Ann. N. Y. Acad. Sci. **1165**, 195–205 (2009). doi:10.1111/j.1749-6632.2009.04037.x

40. Li, Y.C., Chen, Y., Du, J.: Critical roles of intestinal epithelial vitamin D receptor signaling in controlling gut mucosal inflammation. J. Steroid Biochem. Mol. Biol. **17**. pii: S0960-0760 (15)00013-8 (2015). doi:10.1016/j.jsbmb.2015.01.011

41. Tang, Y., et al.: Nitric oxide-mediated intestinal injury is required for alcohol-induced gut leakiness and liver damage. Alcohol. Clin. Exp. Res. **33**(7), 1220–1230 (2009). (Epub 2009 Apr 9)

42. Spritzer, P.M., Lecke, S.B., Satler, F., Morsch, D.M.: Adipose tissue dysfunction, adipokines and low-grade chronic inflammation in PCOS. Reproduction. (2015) pii: REP-14-0435

43. Zitzmann, M.: Testosterone deficiency, insulin resistance and the metabolic syndrome. Nat. Rev. Endocrinol. **5**(12), 673–681 (2009). doi:10.1038/nrendo.2009.212. (Epub 2009 Oct 27)

44. Dandona, P., Dhindsa, S., Chaudhuri, A., Bhatia, V., Topiwala, S.: Hypogonadotrophic hypogonadism in type 2 diabetes. Aging Male. **11**(3), 107–117 (2008). doi:10.1080/13685530802317934

45. Ramezani-Binabaj, M., Motalebi, M., Karimi-Sari, H., Rezaee-Zavareh, M.S., Alavian, S.M.: Are women with polycystic ovarian syndrome at a high risk of non-alcoholic Fatty liver disease; a meta-analysis. Hepat Mon. **14**(11):e23235 (2014). doi:10.5812/hepatmon.23235 (eCollection 2014)

46. Vassilatou, E.; Nonalcoholic fatty liver disease and polycystic ovary syndrome. World J. Gastroenterol. **20**(26), 8351–8363 (2014). doi:10.3748/wjg.v20.i26.8351

47. Gaberšček, S., Zaletel, K., Schwetz, V., Pieber, T., Obermayer-Pietsch, B., Lerchbaum, E.: Mechanisms in endocrinology: thyroid and polycystic ovary syndrome. Eur. J. Endocrinol. **172**(1), R9–R21 (2015). doi:10.1530/EJE-14-0295

48. Janssen, O.E., et al.: High prevalence of autoimmune thyroiditis in patients with polycystic ovary syndrome. Eur. J. Endocrinol. **150**(3), 363–369 (2004)

49. Garelli, S. et al.: High Prevalence in chronic thyroiditis in patients with polycystic ovarian syndrome. Eur. J. Obstet. Gynecol. Reprod. Biol. (2013). pii: S0301-2115(13)00116-4 (Epup ahead of print)
50. Gonzales, F., et al.: Elevated serum levels of tumor necrosis factor alpha in normal-weight women with polycystic ovary syndrome. Metabolism **48**(4), 437–441 (1999)
51. Fukui, M., Tanaka, M., Toda, H., Okada, H., Ohnishi, M., Mogami, S., Kitagawa, Y., Hasegawa, G., Yoshikawa, T., Nakamura, N.: Andropausal symptoms in men with Type 2 diabetes. Diabet. Med. **29**(8), 1036–1042 (2012). doi:10.1111/j.1464-5491.2012.03576.x
52. Pal, S.K.: Complementary and alternative medicine: an overview. Current Sci. **82**(5), 518–524 (2002)
53. Nadin, M.: Medicine: the decisive test of anticipation. In: Nadin, M. (ed.) Anticipation and Medicine, pp. 1–25. Springer, Cham (2016)
54. Gatherer, D.: So what do we really mean when we say that systems biology is holistic? BMC Syst. Biol. **4**, 22 (2010) doi:10.1186/1752-0509-4-22. http://bmcsystbiol.biomedcentral.com/articles/10.1186/1752-0509-4-22
55. Wang, M., Lamers, R.J., Korthout, H.A., van Nesselrooij, J.H., Witkamp, R.F., Heijden, R. van der, Voshol, P.J., Havekes, L.M., Verpoorte. R., Greef, J. van der.: Metabolomics in the context of systems biology: bridging traditional Chinese medicine and molecular pharmacology. Phytother Res. **19**, 173–182 (2005). doi:10.1002/ptr.162
56. Ehling, D.: Oriental medicine: an introduction. Altern. Ther. Health Med. **7**(4), 71-82 (2001)
57. University of California San Francisco Medical Center. Acupuncture. http://www.ucsfhealth.org/treatments/acupuncture/
58. Vickers, A., Zollman, C.: Acupuncture. BMJ **319**(7215), 873–976 (1999)
59. Joswick, D.: What are the different styles of acupuncture? Acufinder.com https://www.acufinder.com/Acupuncture+Information/Detail/What+are+the+different+styles+of+acupuncture+
60. Five Element Linking Fist. https://theory.yinyanghouse.com/theory/chinese/five_element_acupuncture_theory. [See also http://www.axelradclinic.com/]. Five Element associations adapted and modified from Chris Axelrad LAc. Lecture notes Integrative Fertility Symposium. 2015. Vancouver, Canada. Please note that OM terms are capitalized to distinguish them from Western meanings. For example, kidneys, two anatomical organs located within the posterior thoracic cavity, is different from Kidney, and Organ that, among other things, rules the bones, teeth and marrow.
61. Hyman, M.: The Blood Sugar Solution. Little, Brown and Company (2012)
62. Cleveland Clinic. Center for Functional Medicine. http://my.clevelandclinic.org/services/center-for-functional-medicine
63. Westman, E.: A Low Carbohydrate, Ketogenic Diet Manual. No Sugar, No Starch Diet. 2013. CreateSpace Independent Publishing Platform
64. Let's take back your health. http://chriskresser.com/
65. Sanfilippo, D.: Practical Paleo. 30-Day Meal Plans. 2012. Victory Belt Publishing, Inc. http://balancedbites.com/practicalpaleo
66. http://www.thepaleomom.com/about-sarah-2
67. Dr. Sarah Ballantyne and the Paleo Diet. http://www.thepaleomom.com
68. Ballantyne, S.: The Paleo Approach Cookbook: A Detailed Guide to Heal Your Body and Nourish Your Soul. 2014. Victory Belt Publishing, Inc. http://www.thepaleomom.com/about-the-paleo-mom/the-paleo-approach-cookbook
69. Boers, I., Muskiet, F.A., Berkelaar, E., Schut, E., Penders, R., Hoenderdos, K., Wichers, H.J., Jong, M.C.: Favourable effects of consuming a Palaeolithic-type diet on characteristics of the metabolic syndrome: a randomized controlled pilot-study. Lipids Health Dis. **11**(13), 160 (2014). doi:10.1186/1476-511X-13-160
70. Stomby, A., Simonyte, K., Mellberg, C., Ryberg, M., Stimson, R.H., Larsson, C., Lindahl, B., Andrew, R., Walker, B.R., Olsson, T.: Diet-induced weight loss has chronic tissue-specific effects on glucocorticoid metabolism in overweight postmenopausal women. Int. J. Obes. (Lond). (2014). doi:10.1038/ijo.2014.188

71. Masharani, U., Sherchan, P., Schloetter, M., Stratford, S., Xiao, A., Sebastian, A., Nolte Kennedy, M., Frassetto, L.: Metabolic and physiologic effects from consuming a hunter-gatherer (Paleolithic)-type diet in type 2 diabetes. Eur. J. Clin. Nutr. doi:10.1038/ejcn.2015.39 (2015). http://www.ncbi.nlm.nih.gov/pubmed/25828624
72. Yancy, Jr, W.S., Foy, M., Chalecki, A.M., Vernon, M.C., Westman, E.C.: A low-carbohydrate, ketogenic diet to treat type 2 diabetes. Nutr. Metab. **2**, 34 (2005). doi:10.1186/1743-7075-2-34
73. Westman, E.C., Yancy Jr., W.S., Mavropoulos, J.C., Marquart, M., McDuffie, J.R.: The effect of a low-carbohydrate, ketogenic diet versus a low-glycemic index diet glycemic control in type 2 diabetes mellitus. Nutr. Metab. (Lond). **18**(5), 36 (2008). doi:10.1186/1743-7075-5-36
74. Gower, B.A., Goss, A.M.: A lower-carbohydrate, higher-fat diet reduces abdominal and intermuscular fat and increases insulin sensitivity in adults at risk of type 2 diabetes. J. Nutr. **145**(1), 177S–183S (2015). doi:10.3945/jn.114.195065. (Epub 2014 Dec 3)
75. Goss, A.M., Chandler-Laney, P.C., Ovalle, F., Goree, L.L., Azziz, R., Desmond, R.A., Wright Bates, G., Gower, B.A.: Effects of a eucaloric reduced-carbohydrate diet on body composition and fat distribution in women with PCOS. Metabolism. **63**(10), 1257–1264 (2014). doi:10.1016/j.metabol.2014.07.007 (Epub 2014 July 18)
76. Suez, J., Korem, T., Zeevi, D., et al.: Artificial sweeteners induce glucose intolerance by altering the gut microbiota. Nature **514**(7521), 181–186 (2014)
77. Plantier, R.: What French kids eat for school lunch. http://www.mindbodygreen.com/0-14845/what-french-kids-eat-for-school-lunch-it-puts-americans-to-shame.html
78. Godoy, M.: This is What America's School Lunches Really Look like. http://www.npr.org/blogs/thesalt/2013/11/20/246400702/this-is-what-america-s-school-lunches-really-look-like
79. Matsumoto, K., Euler, D.: Kiiko Matsumot's clinical strategies vols. 1 and 2. 2004 and 2008, Kiiko Matsumoto International, Natick, MA
80. Maciocia, G.: The Foundations of Chinese Medicine. Churchill Livingstone Inc. (1989)
81. Bensky, D., Barolet, R.L.: Chinese Herbal Medicine—Formula & Strategies. Eastland Press Inc. (1990)
82. Huang, H.: Ten Key Formula Families in Chinese Medicine. Trans. Max M. Eastland Press Inc. (2009)
83. Ehling, D., Swart, S.: The Chinese Herbalist's Handbook, 2nd edn. In Word Press (1996)
84. Marcinko, K., Sikkema, S.R., Smaan, M.C., Kemp, B.E., Fullerton, M.D., Steinberg, G.R.: High Intensity interval training improves liver and adipose tissue insulin sensitivity. Mol. Metb. **4**(12), 903–915 (2015). doi:10.1016/j.molmet.2015.09.006. eCollection 2015. http://www.ncbi.nlm.nih.gov/pubmed/26909307
85. Lemes, I.R., Ferreira, P.H., Linares, S.N., Machado, A.F., Pastr, Cm, N.J. Jr.: Resistance training reduces systolic blood pressure in metabolic syndrome: a systematic review and meta-analysis of randomised controlled trials. Br. J. Sports Med. Pii: bjsports-2015-094715. doi:10.1136/bjsports-2015-094715. [Epub ahead of print] http://www.ncbi.nlm.nih.gov/pubmed/26964146
86. International Conference: Anticipation and Medicine, Hanse Wissenschaftskolleg, Delmenhorst, Germany, September 28–30 (2015)
87. Nadin, M. (ed.): Anticipation across disciplines. cognitive systems monographs, vol. 29. Springer, Cham CH (2015) http://www.springer.com/us/book/9783319225982
88. The European Association for Predictive, Preventative & Personalized Medicine. http://www.epmanet.eu

Anticipating the Diagnosis of Breast Cancer: Screening Controversies and Warning Dreams

Larry Burk

Abstract Mass population screening programs such as mammography represent an attempt to use technology to forecast the presence of disease in asymptomatic individuals. Unfortunately, sometimes the benefits outweigh the risks when there are too many false positives that require further invasive diagnostic procedures or when detection of carcinoma in situ leads to aggressive treatment of disease that may have never become clinically relevant. There are options of using other imaging approaches such MRI, ultrasound, and thermography, as well as offering genetic screening. It is also a timely opportunity to investigate an intuitive approach to self-care using dreams as a supplement to breast self-examination. A recent survey of 18 women, from around the world, who had life-changing warning dreams of breast cancer prior to symptoms that prompted medical attention leading directly to diagnosis will be reviewed.

Keywords Breast cancer screening · Dreams · Precognition · Magnetic resonance imaging · Premonition · Mammography · Ultrasound · BRCA genetic testing · Thermography

1 Introduction

Prior to the development of radiological imaging techniques, having a positive family history was the only way to anticipate a diagnosis of breast cancer. The National Cancer Institute (NCI) estimates that in 2015 there will be 232,000 women newly diagnosed with breast cancer and 40,000 breast cancer deaths with 89 % surviving five years [1]. Mass population screening mammography was introduced in the 1960s in an attempt to detect cancer in women prior to the onset of symptoms. This approach has been called into question through an ongoing debate about the risks and benefits of mammography [2]. There are other imaging modalities

L. Burk (✉)
Healing Imager, P.C, Durham, NC, USA
e-mail: burk0001@yahoo.com

© Springer International Publishing Switzerland 2017
M. Nadin (ed.), *Anticipation and Medicine*, DOI 10.1007/978-3-319-45142-8_19

available as alternatives to mammography, such as MRI, ultrasound, and thermography, which also have advantages and disadvantages. In the past two decades genetic screening has been developed to identify women at high risk of breast cancer who carry the BRCA1 and BRCA2 genes, with the result that some women elect to have prophylactic bilateral mastectomies [3]. In the past two years, a survey was completed of asymptomatic women who reported warning dreams that prompted imaging studies leading directly to the diagnosis of breast cancer [4]. With all of these options available to women, it is important that they be well informed about each of these ways to maintain the health of their breasts.

2   Benefits of and Problems with Current Screening Methods

2.1   Mammography

Screening mammography refers to annual or biannual routine examinations performed in asymptomatic women. The breast is individually compressed between two plates to flatten the tissue and improve image quality, which can be quite uncomfortable for some women. For screening, two orthogonal views are taken and reviewed by a radiologist for the presence of any suspicious lesions or calcifications. If the radiologist is on-site, any necessary extra views can be obtained immediately; but if there is no radiologist available for real time evaluation, the woman may need to be called back at a later date for more views. Criteria from the BI-RADS Atlas by the American College of Radiology (ACR) are used in the evaluation to determine the likelihood of malignancy on a scale of 0–6 with 4 or greater indicating the need for a biopsy [5]. Positive biopsies are found in about 30 % of those with a score of 4. A score of 3 may generate a recommendation to return for a follow-up exam at a shorter 6-month interval.

The first randomized control trial of screening mammography was begun in 1963. Based on its results showing a 30 % reduction in death rate, the NCI started an annual screening program in 1973, excluding women under 50 due to lack of proven benefit [6]. By 1988 the starting age was reduced to 40 by the American Cancer Society (ACS) due to a reanalysis of the study results and a reduction in radiation dose from improved radiology equipment. Eight randomized clinical trials were completed by 1992, and the results were reviewed at an international NCI workshop reaffirming the apparent lack of benefit for women under 50 [7]. Despite this lack of scientific support, the ACS has maintained the recommendation of 40 as the starting age. However, in 2009, the U. S. Preventive Services Task Force (USPSTF) published recommendations for routine screening every two years starting at age 50 [8] in conflict with the current ACR recommendations for screening every year starting at age 40 [9]. The Cochrane Collaboration review in 2011 of seven of the randomized controlled trials concluded that, "the studies which provided the most reliable information showed that screening did not reduce breast cancer mortality" [10].

Detractors cite the above Cochrane review:

> If we assume that screening reduces breast cancer mortality by 15 % and that over-diagnosis and over-treatment is at 30 %, it means that for every 2000 women invited for screening lasting a period of 10 years, one will avoid dying of breast cancer and 10 healthy women, who would not have been diagnosed if there had not been screening, will be treated unnecessarily. Furthermore, more than 200 women will experience important psychological distress including anxiety and uncertainty for years because of false positive findings [10].

Over-diagnosis includes many women who have low-grade cancer, such as ductal carcinoma in situ (DCIS), that likely would never cause death or even any clinical symptoms [11]. In fact, there is even evidence that some cancers may regress over time [12]. Yet women who are treated for DCIS found on a screening exam feel their lives have been saved. However, autopsy series of women who die of other causes reveal that up to 17 % of these women had pathological evidence of breast cancer [13]. That means that the harder we look for cancer with higher and higher resolution imaging studies, the more we will find. It also means that we are exposing one of the most radiation sensitive organs in the body to an annual dose of radiation with unknown consequences, perhaps, or probably, inducing some new cancers in the process [14]. This controversy has raged on to the present day [15].

The proponents of mammography note that a 2014 study [16], which suggested that mammography does not reduce mortality beyond that of physical examination, was performed without state-of-the-art technology [17]. They also calculate that the number of women needed to be screened to save one woman from a breast cancer death is less than other estimates: on the order of 465 rather than 2000 [18]. There is evidence in one study that 70 % of the women who died from breast cancer were among the 20 % of women who were not participating in screening [19]. Technological improvements such as digital breast tomosynthesis also show promise in improving accuracy while decreasing radiation dose [20]. However, most of the promoters of mammography are radiologists, so there is some question of conflict of interest regarding interpretation of the studies. In 2010, two prominent radiologists observed:

> … the response of the radiologic medical community to the new USPSTF guidelines for screening mammography was needlessly confrontational and not in the best interest of everyone's fight against breast cancer. Because these critical views are not shared by the bulk of the medical community, we fear an unwarranted backlash against our specialty. Controversies regarding medical screening and many other cost-benefit health care decisions are increasingly societal issues rather than purely scientific ones, and therefore, open-minded public discussion and education should be welcome [21].

2.2 Magnetic Resonance Imaging

In high-risk women with the BRCA 1 and 2 genes, magnetic resonance imaging (MRI) is recommended as an additional screening method. MRI is an order of magnitude more expensive than mammography, so could never be justified for routine general screening. Both breasts are scanned at the same time without compression or ionizing radiation. Intravenous chelated gadolinium contrast is administered which has less than a 0.5 % risk of allergic reaction [22]. Rare cases of deadly nephrogenic systemic fibrosis (NSF) have been reported after administration of gadolinium contrast in patients with significant renal insufficiency since 2006 [23]. Guidelines have been established to screen for decreased renal function prior to MRI, due to concerns about delayed clearance of gadolinium from tissues, as transmetallation can occur resulting in release of toxic free gadolinium from the contrast agent. However, in the past three years, there have been reports of long-term gadolinium deposition in the brain after contrast administration in patients with normal renal function, which is of uncertain clinical significance regarding potential neurotoxicity [24].

In a 2005 study, MRI was shown to be much more sensitive (100 %) than mammography (25 %) in screening women at high risk for breast cancer with a comparable specificity [25]. The need for this alternative is readily apparent as no trials of mammography have shown a decrease in mortality in women with a positive family history, and MRI is particularly useful in women with dense breasts where mammography has a difficult time finding cancer.

Unfortunately, over-diagnosis may actually increase with MRI in this population where more cases of DCIS are discovered [26]. The National Comprehensive Cancer Network guidelines currently recommend annual MRI alone in high-risk women aged 25 to 30 years, with annual mammography in addition after the age of 30 years [27]. About 18 % of women may not be able to tolerate the MRI environment due to claustrophobia, preexisting metal implants, or contrast agent issues [28]. Despite these concerns, MRI is considered safer than mammography as it uses non-ionizing radiation in the form of radiofrequencies in the megahertz range similar to regular television broadcasts.

2.3 Ultrasound

Ultrasound is predominantly used as a problem solving technique in differentiating solid from cystic masses discovered on physical examination or through mammography. It is the simplest and safest approach without the use of ionizing radiation, but does require considerable operator expertise. A handheld ultrasound probe is applied directly to the breast and images are obtained in real-time. In most cases it is relatively straightforward in determining whether a lesion is a simple cyst or not. Although ultrasound has lower sensitivity and specificity for detecting cancer than the other radiological tests, it does have value as an adjunct screening method to mammography and is less expensive than MRI [29]. It can be effective in finding cancers in women under the age of 50 with dense breasts and negative mammograms [30]. In women with a positive family history who cannot undergo MRI, ultrasound may be also be used as a supplemental screening test.

2.4 Thermography

Thermograms are performed noninvasively using a digital infrared camera, which does not touch the breasts. The images are created from the heat that radiates from the surface of the breast providing a physiologic indicator of relative vascularity and blood flow. Thermography does not directly image breast cancer anatomically, but visualizes alterations in the skin temperature, which reflect autonomic nervous system mediated changes in blood flow in the breast related to malignancy. However, changes in skin temperature are non-specific and can also be seen in benign fibrocystic disease. Thermography was judged ineffective by radiologists in 1983 [31], but has been gaining in popularity in the holistic medicine community with improvements in technology since the earlier studies [32]. Thermography may detect some cancers before mammography, but may also miss cancers detected by mammography making the two approaches potentially synergistic in screening. Normal breast thermograms are symmetrical and stable over time on follow up examinations, and the development of a positive asymmetric pattern is considered to be a high risk factor for breast cancer [33].

2.5 Genetic Screening

The first genetic testing for breast cancer began in 1994, and in 2013 the Supreme Court invalidated the patents that gave Myriad Genetics a monopoly, thus allowing the development of BRCA Share, a blood testing collaboration between LabCorp and Quest Diagnostics. In addition, Color Genomics is also beginning to offer a saliva test for the BRCA 1 and BRCA 2 genes [34]. The tests detect mutations in these genes, which code for tumor suppressor proteins and which also impact the risk for ovarian cancer and other tumors. The 12 % lifetime risk of breast cancer in the general population is increased up to 45 % with BRCA 2 mutations, and up to 65 % with BRCA 1 mutations [35]. Up to 25 % of hereditary cases of breast cancer are due to a mutation in one these rare highly penetrant genes [36]. These mutations can be passed down from either parent and are most common in families with a history of breast cancer before age 50 years, bilateral breast cancer, presence of breast and ovarian cancer, presence of breast cancer in one or more male family members, multiple cases of breast cancer in the family, one or more family members with two primary types of BRCA-related cancer, and Ashkenazi Jewish ethnicity.

The USPSTF recommends several screening tools to determine the need for testing and genetic counseling [37]. Depending on the results, which may sometimes be equivocal or report variants of unknown significance, decisions can then be made regarding enhanced screening options, prophylactic bilateral mastectomies or oophorectomies and chemoprevention with tamoxifen or other agents [38].

3 Warning Dreams

After a review of all these high-tech imaging and genetic options, it is not inopportune to think about alternatives. An intuitive, low-tech approach to self-care could be acknowledgment of dreams—of course in addition to breast self-examination—and followed by a physician's evaluation. Dreams are on historic record; they were once used as diagnostic tools in the Asclepian temples of ancient Greece [39]. There is also a long tradition of shamanic dream interpretation and guidance in indigenous cultures [40]. In Russia, Kasatkin [41] published the first modern research in 1967 correlating dreams with physical illness, focusing mainly on neurological conditions without mention of breast cancer. A psychiatrist by profession, Kasatkin noted the following common dream features related to physical illness: (1) an increase in dream recall; (2) distressful, violent and frightening images; (3) occurrence preceding the first symptoms; (4) long duration and persistence; (5) content revealing the location and seriousness of the illness. (One of the first available English reports about his work was provided by Van de Castle [42].) Later, in England, Royston, also a psychiatrist, collected over 400 health-related dreams [43], including the dream of a woman named Nancy who self-diagnosed her own breast "malig-nancy," shouting the play-on-words name "Bad Nancy" at her dream self. This dream was described by Barasch along with this insightful observation by Royston, "These are not ordinary dreams, but big dreams, archetypal dreams, so laden with powerful emotional affect that the dreamer is forced to take them seriously" [44].

In 1996, Kinney published the first formal case report of a breast cancer warning dream in the nursing literature, which was confirmed by mammogram and subsequent biopsy [45]. She wrote a personal narrative of having a maternal history of breast cancer without symptoms and being awakened by an authoritative dream message, "Go make your appointment for your mammogram right now. Do not delay." The diagnosis was followed by successful treatment with double mastectomies and holistic therapies. Burch performed an informal retrospective study of 19 women from a breast cancer support group, which was included in 2013 as part of a literature review with case reports by Burk [46]. Ten of the women experienced breast cancer warning dreams featuring messages from deceased family members in all but one. Ten other case reports were included in the review, five of which described specific localization of the tumors confirmed on imaging studies and at surgery. Unfortunately, one of these cases was reported by the patient to her doctor but dismissed without appropriate follow-up, resulting in her untimely death. As she described it:

> I had a dream that I had cancer. I went to the G.P. complaining of a lump and spasm-like feelings on my sternum. The G.P. concluded it was normal breast tissue, and the feeling in my sternum was dismissed, a devastating mistake. A year later, a different doctor diagnosed stage 3 breast cancer [47].

Dossey [48] reported another example of tumor localization:

> A woman had a dream that she had breast cancer. Worried sick, she visited her physician the next morning. She pointed with one finger to a specific spot in her upper left breast where she'd seen the cancer in the dream. "It's right here," she said. She could not feel a lump, however, and neither could her physician. A mammogram was done, which was normal. When the physician reassured her that nothing was wrong and that they should take a wait-and-see approach with frequent exams, she was not satisfied. "This was the most vivid dream I've ever had," she protested. "I'm certain I have breast cancer at this exact spot." When she insisted on going further, the physician, against his better judgment, pressured a surgeon to do a biopsy. "But where? There's nothing there," the surgeon objected. "Look, just biopsy where she points," the physician said. In a few days the pathologist called the original doctor with the report. "This is the most microscopic breast cancer I've seen," he said. "You could not have felt it. There would have been no signs or symptoms. How did you find it?" "I didn't," he replied. "She did. In a dream."

Burch and Kanavos-O'Keefe both published books about their breast cancer warning dream experiences emphasizing the importance of dream messengers:

> My father appears and seems to be checking on me. Someone else is also in the dream: a man dressed in a medical coat. He tells me, almost shouting, that I have a malignant lump in my breast and that I must have my breast removed. He continues to shout, telling me that, no matter what I hear, it is not benign. He is now leading me out the door to the Mayo Clinic, where a doctor is shouting to me that I have a malignancy and I must act immediately [49].

> Enjoying my dream, it suddenly stops, much like what happens when a computer screen freezes, and a pop-up window appears, also similar to that of a computer. My spiritual guide/guardian angel, in the brown robe, rope belt, and leather sandals of a monk, steps through the window and says, "Come with me. We have something to tell you." I obediently follow him into a room I call the Room Between Realms, a place that is neither of the living nor the dead, yet both can visit to share information. It is a parallel universe of consciousness. A guide takes my hand, places it on my right breast, and says, "You have cancer right here. Feel it? Go back to your doctor tomorrow. Don't wait for an appointment" [50].

These reports inspired a retrospective survey of women with known breast cancer from the United States, Europe, and South America who described warning dreams. The study was sponsored by DreamsCloud, a social networking company with a database of over 2 million dreams [51]. It is a social platform and resource for people of all ages with an interest in exploring dreams and the unconscious mind. Through its website and mobile apps for all smart devices, DreamsCloud provides a community within which users can interact and explore dream meanings. It allows users to log their dreams, create dream journals, save them privately or share them publicly or anonymously, and receive personalized reflections on their

Table 1 Breast cancer warning dreams questionnaire, 18 responses

		#Yes	% Yes
1	Did you have any dreams warning about breast cancer before your diagnosis?	17	94
2	Do you keep a dream diary to record your dreams?	9	50
3	Did the first clues about the breast cancer come in your dreams?	17	94
4	Did you have more than one dream warning you about breast cancer?	5	28
5	Did the dreams increase in intensity, specificity or urgency with time?	4	22
6	Were the dream(s) more vivid, real or intense than your ordinary dreams?	15	83
7	Did the dream(s) contain an emotional sense of threat, menace or dread?	13	72
8	Were the specific words breast cancer/tumor used in the dream(s)?	8	44
9	Did the dream(s) localize the tumor to a specific breast location?	10	56
10	Did the dream(s) involve the sense of physical contact with your breast?	7	39
11	Did you receive a breast cancer dream message from a deceased family member?	3	17
12	Was there a sense of conviction about the importance of the dream(s)?	17	94
13	Did the dream(s) prompt you to seek medical advice and diagnostic testing?	13	72
14	Did you share the dream(s) with your doctor?	11	61
15	Did the dream(s) directly lead to the diagnosis being made?	10	56
16	Did you ignore the dream(s) until the diagnosis was made for another reason?	3	17
17	Did you forget about the dream(s) until after the diagnosis was made?	2	11
18	Did you overlook the significance of the dream(s) until after the diagnosis?	6	33
19	Did anyone else you know have dreams warning about your breast cancer?	3	17

dreams. Eighteen women completed all phases of the study, which included a 19-item Yes or No questionnaire and dream narratives. Their dream questionnaire data is presented in Table 1 [52].

The most common characteristics of warning dreams in descending order of frequency reported in the survey were: (1) a sense of conviction about the importance; (2) more vivid, real or intense than ordinary dreams; (3) an emotional sense of threat, menace or dread; (4) the use of the specific words breast cancer/tumor; and (5) the sense of physical contact with the breast.

The following case is illustrative of all five of these common features of warning dreams:

In March 2004 I had a vivid dream (unlike any before) in which I was lying on an operating table and a woman surgeon was operating on my left breast. At one point, she went to a microscope and looked through it and came back and told me that I have breast cancer. After hearing this news from the doctor, my daughter and former husband broke down and cried. I woke up. While I was startled, there was also a sense of calm at the same time, a knowing that I needed to get checked medically as soon as possible. I was scheduled for an appointment several months later for my annual mammogram and I called and moved the appointment up.

The mammogram was "normal" so the radiologist told me "all is well" and that I can go home. I told her that I wanted to have an ultrasound. She insisted that it was not procedure unless there was something unusual found on the mammogram, and since nothing was found on the mammogram she was not going to do an ultrasound. I insisted, she insisted it's not procedure. We went back and forth, and I think she got tired of me insisting. She finally said OK with a sense of frustration/irritation and she personally did the ultrasound. There it was on the screen black, with tentacles. The doctor literally went white and silent. And then she turned to me and asked me how I knew. I told her about the dream.

On April 9, 2004, I was lying on an actual operating table. A woman surgeon excised breast tissue which was then examined under a microscope and determined to be cancer. Shortly after waking up from the anesthesia and getting dressed to go home, the doctor came to tell me that I had breast cancer. At home, my former husband and my daughter cried with the news [52].

In more than half of the cases, the dreams prompted medical attention, provided the location of the tumors, were shared with doctors, and led directly to diagnosis. In a third or less of the cases, the dreams were forgotten, ignored, or the significance was overlooked until the diagnosis was made for another reason. Six of the women in the study disclosed a family history of breast cancer including two sisters who reported carrying the BRCA gene mutation. In a few of the cases, women reported that their cancers were also dreamed about by other family members. One woman had a dream about her mother's breast cancer a week before it was diagnosed along with a message that the dreamer also had breast cancer which was eventually diagnosed 5 years later on an annual mammogram. One woman without an initial warning dream who was not included in the study had a warning dream prior to a recurrence guiding her to find a new lesion in her axilla. Another woman with benign breast disease who was not included in the study reported a dream that had many of the features described above. She later had an adenoma removed after 18 months of therapy with energy healing.

This small retrospective survey without a control group or statistical evaluation serves as a limited pilot study to lay the groundwork for future research. The lack of specific clinical correlation leaves the claims of tumor localization open to skeptical criticism regarding the possibility of inaccurate recall of events that occurred prior to biopsy. Documentation of the fact that the dreams predated the diagnosis was only available in a few cases. Another potential pitfall is the possibility that warning dreams might be related to symbolic psycho-spiritual metaphors rather than actual physical illnesses, a scenario that may have been considered by the woman in the study whose diagnosis occurred five years after her dream. However, in most of the cases, the dreams led to immediate diagnosis, and in a few cases the dreamers were so convinced of the reality of the dream guidance that they persisted in seeking a confirmatory diagnosis even in the face of initial negative testing.

The specific localization of tumors by the warning dreams is a paradigm-shifting observation that raises questions about the mechanism of transfer into conscious-ness and the source of the information. The new academic interdisciplinary field of anticipation concerns itself with exploring such issues. Nadin describes an antici-patory system is described as "a system whose current state is determined not only by a past state, but also by possible future states" [53]. The most conservative explanation would be that the women already had experienced vague symptoms or detected subtle physical signs through self-examination, which were then translated into dream form and brought to conscious awareness. Another speculative option would be that there may be unknown psychophysiological mechanisms for transfer of information from the body to the brain. The most threatening position with respect to a materialistic worldview is that some of the warning dreams appear to be precognitive in nature suggesting a non-local, intuitive explanation [54]. The holistic nature of anticipatory processes (see Nadin[1]) must also be taken into consideration.

4 Conclusions

Attempts to anticipate the diagnosis of breast cancer utilizing technology have been evolving steadily over the past 50 years. Significant controversy in the field resulted in considerable confusion among both patients and healthcare practitioners. The debate about mammographic screening persists despite many large clinical trials that have been interpreted differently by the promoters and skeptics concerning the lives saved by early diagnosis and the lives harmed by over-diagnosis. Healthcare policy decisions hang in the balance with major financial implications for the delivery system and the insurance business that will impact millions of women [55]. Concerns are greatest among women with a positive family history and genetic risk factors, and MRI has emerged as a useful screening approach in this group. Ultrasound has a supplementary role to play, and thermography may also someday reemerge as another complementary noninvasive approach if the public demands that it be included in modern research protocols. As genetic testing matures, it will become less expensive and more accessible. Chances are good that considerations informed by the anticipation perspective could be introduced in genetic testing. It will be up to the ethicists and genetic counselors to determine how widespread its impact will be on screening.

Warning dreams of breast cancer invoke an ancient tradition that has been rediscovered and scrutinized in parallel with these technological developments over the last 50 years. These dreams frequently prompt women to seek medical attention, often leading to medical evaluations, breast imaging studies, biopsies, and surgery. The woman who died of breast cancer after her physician dismissed the warning provided by her dream and failed to initiate an appropriate workup is a cautionary tale regarding the potential significance of these dreams. The potential skepticism of physicians is part of the motivation for presenting the subject here and for pursuing further research in this area. A prospective study with imaging and pathological correlation to determine the predictive value of a warning dream in

---

[1]Nadin, M.: Anticipation and the Brain. In: Nadin, M.: (ed.) Anticipation and Medicine, pp. 135–162. Springer, Cham (2016).

comparison to a control group with benign disease would require participation by open-minded surgeons and holistic breast imagers, as well as support from the major breast cancer research organizations. This initial survey suggests that keeping a dream diary might be a useful adjunct to routine self-examination as part of a breast self-care program.

References

1. SEER Stat Fact Sheets: Breast Cancer. http://seer.cancer.gov/statfacts/html/breast.html
2. Biller-Andorno, N., Juni, P.: Abolishing mammography screening programs? A view from the Swiss Medical Board. NEJM **16** (2014) doi:10.1056/NEJMp1401875 http://www.nejm.org/doi/full/10.1056/NEJMp1401875
3. Rebbeck, T.R., Friebel, T., Lynch, H.T., et al.: Bilateral prophylactic mastectomy reduces breast cancer risk in BRCA1 and BRCA2 mutation carriers: the PROSE study group. J. Clin. Oncol. **22**, 1055–1062 (2004)
4. Burk, L.: Breast cancer warning dreams pilot project: characteristics of prodromal dreams. In: Proceedings of the 31st Annual International Association for the Study of Dreams Conference, Berkeley, CA (2014)
5. Eberl, M.M., Fox, C.H., Edge, S.B., Carter, C.A., Mahoney, M.C.: BI-RADS classification for management of abnormal mammograms. J. Am. Board Fam. Med. **19**, 161–164 (2006). doi:10.3122/jabfm.19.2.161
6. Shapiro, S.: Evidence on screening for breast cancer from a randomized trial. Cancer **39**(suppl 6), 2772–2782 (1977)
7. Fletcher, S., Black, W., Harris, R., Rimer, B., Shapiro, S.: Report of the international workshop on screening for breast cancer. J. Natl. Cancer Inst. **85**, 1621–1624 (1993)
8. U.S. Preventive Services Task Force. Screening for Breast Cancer: U.S. Preventive Services Task Force recommendation statement. Ann. Intern. Med. **17**, 151(10), 716–726, W-236 (2009). doi:10.7326/0003-4819-151-10-200911170-00008. http://www.guideline.gov/content.aspx?id=15429
9. Mainiero, M.B., et al.: ACR appropriateness criteria breast cancer screening. J. Am. Coll. Radiol. **10**(1); 11–14 (2013). doi:10.1016/j.jacr.2012.09.036. http://www.acr.org/~/media/ACR/Documents/AppCriteria/Diagnostic/BreastCancerScreening.pdf
10. Gøtzsche, P.C., Nielsen, M.: Screening for breast cancer with mammography. Cochrane Database Syst. Rev. **19**(1), CD001877. (2011) doi:10.1002/14651858.CD001877.pub4
11. Welch, H.G., Schwartz, L., Woloshin, S.: Overdiagnosed: Making People Sick in the Pursuit of Health. Beacon Press, Boston (2011)
12. Zackrisson, S., Andersson, I., Janzon, L., Manjer, J., Garne, J.P.: Rate of over-diagnosis of breast cancer 15 years after end of Malmö mammographic screening trial: follow-up study. BMJ **332**(7543), 689–692 (2006). Epub 3
13. Welch, H.G., Black, W.C.: Using autopsy series to estimate the disease "reservoir" for ductal carcinoma in situ of the breast: how much more breast cancer can we find? Ann. Intern. Med. **127**(11), 1023–1028 (1997)
14. Gofmann, J.W.: Preventing Breast Cancer: The Story of a Major, Proven, Preventable Cause of This Disease. Committee for Nuclear Responsibility, San Francisco (1996)
15. Hefti, R.: The Mammogram Myth: the independent investigation of mammography the medical profession doesn't want you to know about. Amazon Digital Services, Rolf Hefti (2013)
16. Miller, A.B., Wall, C., Baines, C.J., Sun, P., To, T., Narod, S.A.: Twenty five year follow-up for breast cancer incidence and mortality of the Canadian National breast screening study:

randomised screening trial. BMJ **348** (2014). doi:http://dx.doi.org/10.1136/bmj.g366 (Published 11 February 2014) http://www.bmj.com/content/348/bmj.g366

17. Kopans, D.B.: Arguments against mammography screening continue to be based on faulty science. Oncologist **19**(2), 107–112 (2014). doi:10.1634/theoncologist.2013-0184. https://theoncologist.alphamedpress.org/content/19/2/107.full.pdf+html

18. Tabár, L., Vitak, B., Yen, M.F., Chen, H.H., Smith, R.A., Duffy, S.W.: Number needed to screen: lives saved over 20 years of follow-up in mammographic screening. J. Med. Screen. **11**(3), 126–129 (2004)

19. Webb, M.L., Cady, B., Michaelson, J.S., et al.: A failure analysis of invasive breast cancer: most deaths from disease occur in women not regularly screened. Cancer **120**(18), 2839–2846 (2014)

20. Kopans, D.B.: Digital breast tomosynthesis from concept to clinical care. AJR **202**(2), 299–308 (2014). doi:10.2214/AJR.13.11520

21. Berlin, L., Hall, F.M.: More mammography muddle: emotions, politics, science, costs, and polarization. Radiology **255**(2), 311–316 (2010). doi:10.1148/radiol.10100056

22. Fakhran, S., Alhilali, L., Kale, H., Kanal, E.: Assessment of rates of acute adverse reactions to gadobenate dimeglumine: review of more than 130,000 administrations in 7.5 years. AJR **204**, 703–706. (2015) http://www.ajronline.org/doi/abs/10.2214/AJR.14.13430

23. Schlaudecker, J.D., Bernheisel, C.R.: Gadolinium-induced nephrogenic systemic fibrosis. Am. Fam. Physician **80**(7), 711–714. (2009) http://www.aafp.org/afp/2009/1001/p711.html

24. McDonald, R.J., McDonald, J.S., Kallmes, D.F., Jentoft, M.E., Murray, D.L., Thielen, K.R., Williamson, E.E., Eckel, L.J.: Intracranial gadolinium deposition after contrast-enhanced MR imaging. Radiology **5**, 150025 (2015). [Epub ahead of print] http://dx.doi.org/10.1148/radiol.15150025

25. Kuhl, C.K., et al.: Mammography, breast ultrasound, and magnetic resonance imaging for surveillance of women at high Familial risk for breast cancer. JCO **23**(33), 8469–8476 (2005). http://jco.ascopubs.org/content/23/33/8469.long

26. Berg, W.: How well does supplemental screening magnetic resonance imaging work in high-risk women? JCO **32**(21), 2193–2196 (2014)

27. National Comprehensive Cancer Network Genetics Screening. http://www.nccn.org/professionals/physician_gls/pdf/genetics_screening.pdf

28. Berg, W.A., Blume, J.D., Adams, A.M., et al.: Reasons women at elevated risk of breast cancer refuse breast MR imaging screening: ACRIN 6666. Radiology **254**, 79–87 (2010)

29. Berg, W.A., et al.: Combined screening with ultrasound and mammography vs. mammography alone in women at elevated risk of breast cancer. JAMA **299**(18), 2151–2163 (2008). http://jama.jamanetwork.com/article.aspx?articleid=181896

30. Corsetti, V., Houssami, N., Ferrari, A., et al.: Breast screening with ultrasound in women with mammography-negative dense breasts: evidence on incremental cancer detection and false positives, and associated cost. Eur. J. Cancer **44**, 539–544 (2008)

31. Moskowitz, M.: Screening for breast cancer: how effective are our tests? A critical review. CA-A Cancer J. Clin. **33**(1), 26–39 (1983). http://onlinelibrary.wiley.com/doi/10.3322/canjclin.33.1.26/pdf

32. Amalu, W.C., Hobbins, W.B., Head, J.F., Elliot, R.L.: Infrared imaging of the breast: an overview. In medical devices and systems. CRC/Taylor & Francis, Boca Raton, FL (2006)

33. Hudson, T.: Journey to Hope: Leaving the Fear of Breast Cancer Behind. Brush and Quill Productions, Naples, FL (2011)

34. Pollack, A.: New genetic tests for breast cancer hold promise. The New York Times, April 21 (2015). http://www.nytimes.com/2015/04/21/business/more-accurate-affordable-tests-for-detecting-breast-cancer-genes.html?_r=1

35. BRCA1 and BRCA2: Cancer risk and genetic testing. http://www.cancer.gov/cancertopics/causes-prevention/genetics/brca-fact-sheet#r27

36. Shiovitz, S., Korde, L.A.: Genetics of breast cancer: a topic in evolution. Ann. Oncol. (2015). doi:10.1093/annonc/mdv022

37. BRCA-related cancer: risk assessment, genetic counseling, and genetic testing. http://www.uspreventiveservicestaskforce.org/Page/Topic/recommendation-summary/brca-related-cancer-risk-assessment-genetic-counseling-and-genetic-testing
38. Phillips, K.A., Milne, R.L., Rookus, M.A., et al.: Tamoxifen and risk of contralateral breast cancer for BRCA1 and BRCA2 mutation carriers. JCO **31**(25), 3091–3099 (2013)
39. Sigerist, H.E.: A History of Medicine, Volume 2: early Greek, Hindu, and Persian Medicine, 1st edn. Oxford University Press, New York (1987)
40. Krippner, S.: Humanity's first healers: psychological and psychiatric stances on shamans and shamanism. Rev. Psiq. Clin. **34S**(1), 16–22 (2007)
41. Kasatkin, V.N.: Teoriya Snovidenii (Theory of Dreams). Meditsina, Leningrad (1967)
42. Van de Castle, R.L.: Our Dreaming Mind. Ballantine Books, New York (1994)
43. Royston, R., Humphries, A.: The Hidden Power of Dreams. Bantam Books, London (2006)
44. Barasch, M.I.: Healing Dreams: Exploring the Dreams That Can Transform Your Life. Riverhead Books, New York (2000)
45. Kinney, C.K.: Transcending breast cancer: reconstructing one's self. Issues Mental Health Nurs. **17**, 201–216 (1996)
46. Burk, L.: Prodromal dreams of breast cancer and clinical correlation. In: Proceedings of the IASD Psiber dreaming Conference (2013). http://www.letmagichappen.com/images/uploads/documents/pdc2013-burk.BreastCancerDreams.pdf
47. Lee-Shield, S.: My Appeal. July 4 (2013). http://soniaslifeappeal.wordpress.com/myappeal/
48. Dossey, L.: One Mind: How Our Individual Mind Is Part of a Greater Consciousness and Why It Matters. Hay House Inc, Carlsbad, CA (2013)
49. Burch, W.E.: She Who Dreams: A Journey Into Healing Through Dreamwork. New World Library, Novato, CA (2003)
50. O'Keefe-Kanavos, K.: Surviving Cancerland: Intuitive Aspects of Healing. Cypress House, Fort Bragg, CA (2014)
51. DreamsCloud, https://www.dreamscloud.com/
52. Burk, L.: Warning dreams preceding the diagnosis of breast cancer: a survey of the most important characteristics. Explore **11**(3), 193–198 (2015). doi:10.1016/j.explore.2015.02.008. Epub 2015 Feb 14
53. Nadin, M.: Anticipation across disciplines. Cognitive Systems Monographs, Vol. 29. Springer, Cham, CH (2015). http://www.springer.com/us/book/9783319225982
54. Dossey, L.: Nonlocal knowing: the emerging view of who we are. Explore **4**(1), 1–9 (2008). http://www.explorejournal.com/article/S1550-8307%2807%2900414-4/fulltext#sec5
55. Yee, K.M.: USPSTF mammo recs would cut coverage for 17M women. AuntMinnie.com. May 14 (2015). http://www.auntminnie.com/index.aspx?sec=sup&sub=imc&pag=dis&ItemID=110963&wf=6529

Anticipation in Traditional Healing Ceremonies: The Call from Our Past

Thomas Schack and Ellen Schack

Abstract A particular understanding of anticipation in human motor action is introduced in order to provide an example of how this research can be used to develop new technical devices to support anticipation in medicine. The second part serves as an introduction to certain traditional healing ceremonies, which may allow quite different reflections about the role of anticipation in human action and life. Ceremonies are described as technologies to improve personal and social anticipation at different dimensions.

Keywords Anticipation · Basic action concepts · Cognitive interaction technology · Healing ceremonies · Motor action

1 Introduction

Anticipation is the ability of a living organism and a biological system to construct and organize current system states in consideration of future oriented models, states or goals [1–3]. Anticipatory systems, such as the human motor system, are related to future states at different levels: perception, memory, motor organization, for example [4–6]. From a cultural-historical perspective, it is not only interesting how

T. Schack (✉)
Faculty of Psychology and Sport Sciences, Bielefeld University, Bielefeld, Germany
e-mail: thomas.schack@uni-bielefeld.de

T. Schack
Cognitive Interaction Technology, Center of Excellence, Bielefeld University, Bielefeld, Germany

T. Schack
CoR-Lab, Research Institute for Cognition and Robotics, Bielefeld University, Bielefeld, Germany

E. Schack
von Bodelschwinghschen Stiftungen Bethel, ProWerk, Bielefeld, Germany

© Springer International Publishing Switzerland 2017   323
M. Nadin (ed.), *Anticipation and Medicine*, DOI 10.1007/978-3-319-45142-8_20

many levels of action organization or interaction are anticipatory, but also what kind of cultural symbols and signs are involved in the anticipation process [7]. Also of interest is how far the topic of anticipation is considered at a social-communicative level in a culture or subculture for saving a particular level of expertise (e.g., sport, medicine, dance etc.).

Biological organisms are consciously or unconsciously always involved in a "game" of survival. From such a perspective, their stage of development and evolution depends on anticipatory capabilities and processes. As Nadin stated, "… the future in question is pertinent to the open-ended, ever-changing space of possibilities. Within this view … anticipation is a definitory characteristic of the living," [3, 624].

If anticipation is a characteristic of human life itself, we should not simply address this topic in a detached high-tech-performance frame of reference. Rather we should try to reflect this phenomenon form a holistic perspective, which involves basic components of human life. In this article we try to bridge a gap between new devices in technical supported modern medicine and the old methods of traditional healing ceremonies. In the first part, insight is provided into our research on assistive technologies useful for modern medicine. Because our research group investigates the neuro-cognitive nature of motor anticipation and interaction [8], we will first introduce a particular understanding of anticipation in human motor action in order to provide an example of how this research can be used to develop new technical devices to support anticipation in medicine. The second part serves as an introduction to certain traditional healing ceremonies, which may allow quite different reflections about the role of anticipation in human action and life. In recent research on traditional healing and shamanism [9, 10], the neurocognitive dimension of shamanic practices and the particular involvement of deeper neurophysiological levels of brain and motor organization in *altered states of consciousness* (ASC) have been discussed. Such a perspective makes it possible to view traditional healing from a neuro-motor and cognitive perspective and to develop particular reflections about anticipation in traditional healing ceremonies.

2 Anticipation in Human Motor Action

Motor anticipation and motion intelligence have been central dimensions of biological organisms since life began [11]. Important advances in evolution are often associated with the establishment of new functional links between the motor system, related memory structures, and perception. From an evolutionary perspective and from the cognitive sciences we now have strong evidence that the cognitive and motor processes underlying action are strongly interconnected. We view human motor actions not as isolated events with defined start and end points, but as built-upon evolved hierarchical structures consisting of different levels and modules. Cyclic movements for example—walking, swimming, cycling, etc.—are controlled by very old neurophysiological structures in our brains. In contrast, goal-directed manual actions and the anticipation required for tool use (such as turning a screw) are controlled by different brain structures that are much younger in evolutionary terms. It is our understanding that such distinctive goal-directed actions are performed on the basis of precise representations in motor memory [6].

Bernstein [1] was one of the first scientists who acknowledged the fundamental role of anticipation and sensory feedback processing in the control of voluntary movements to point out the goal-directed character of motor actions. This was acknowledged in former work [6, 12], but as well in more recent publications on anticipatory systems. Nadin [3] explicitly emphasized, in the same manner as Bernstein, the importance of anticipation in realizing any type of goal-directed motor act, explaining that any voluntary motor action cannot be initiated without a model of what should result from the planned action. This idea is expressed in Bernstein's "model of the desired future" (i.e., a model of what should be), which is supposed to play an important role in motor control. Consequently, such a model must possess the ability to form a representation of future events by integrating information from past (i.e., memory) and present (i.e., sensory) events in order to generate motor commands that transform the current state in the sensory environment into the desired state (i.e., achieving the action goal).

Building on this general idea and more recent research by Hoffmann [13, 14], Rosch [15], Prinz [16], Jeannerod [17], and others, Schack and colleagues have proposed a cognitive architecture model, which views the functional construction of actions on the basis of a reciprocal assignment of performance-oriented regulation levels and representational levels [4, 18, 19]. According to this view, basic action concepts (BACs) are thought to serve as major representation units for movement control. BACs are based on the cognitive chunking of body postures and movement events concerning common functions in realizing action goals. Taken together, such movement representations provide the basis for action anticipation and control by linking higher-level action goals with the lower-level perceptual effects in the form of cognitive reference structures [8, 20].

One interesting example for anticipation in motor planning is the so called "end-state comfort effect" (ESC) [21] setting forth that individuals are willing to transiently adopt uncomfortable initial limb positions as long as this leads to a comfortable position at the end of the movement. This sensitivity toward comfortable end postures has been taken as evidence that final body postures are represented in memory, and that these postures are specified before movements are initiated [22]. More importantly, the ESC effect clearly demonstrates that movements are planned, controlled and executed with respect to anticipated final positions. Weigelt and Schack [23] showed that the ESC effect develops gradually with the sensory-motor maturation of children. Stöckel et al. [24] investigated anticipatory motor planning and the development of cognitive representations of body postures in children. Interestingly, the sensitivity toward comfortable end-states was related to the mental representation of certain grasp postures.

To learn about building blocks of motor performance, anticipation in human memory, and underlying brain structures, the Neurocognition and Action Biomechanics Research Group (NCA) at Bielefeld University (Germany) investigates biological motion in natural and artificial (e.g., Virtual Reality) environments. The main focus of this research is the neurocognitive architecture of human motor action and its adaptability under various conditions. For this purpose, we use state-of-the-art research methods to investigate the cognitive-perceptual organization and kinematic parameters of anticipation in human motor actions.

On the one hand, understanding the neurocognitive architecture of actions based on empirical research is an important step for applied fields such as mental coaching of athletes in high-performance sports or rehabilitation. On the other hand, it is a fundamental aspect of the growing field of cognitive robotics, particularly in relation to its central goal of elevating the still rigid action repertoire of robots to a level that allows robots to select and adjust their actions flexibly according to the varying demands of real-world scenarios [18, 20].

3 New Approaches for Anticipatory Systems. Building Bridges Between Biological and Technical Systems

To facilitate smooth interactions with humans, a robot or virtual avatar should be able to establish and maintain a shared focus of attention with its human partner or instructor. Furthermore, it should be able to react to commands delivered in a "natural" way, such as speech, gestures and demonstration. To address research questions arising from these requirements, thirty research groups from five faculties have jointly established the Excellence Center Cognitive Interaction Technology (CITEC) at Bielefeld University, which offers the infrastructure that allows pertinent research topics to be approached from an interdisciplinary perspective. Among the key issues being addressed concern how anticipatory skills and related structured representations can arise during skill acquisition, and how the underlying processes can be replicated on robotic platforms. Working towards this common goal, we translate our findings from studies of human movements and related representation into theoretical models that can guide the implementation of corresponding features on cognitive robot architectures.

The development of appropriate action representation in memory plays a central role in the control of actions and interactions between humans and technical systems by enabling agents to select and combine effective sources of information. Regardless of whether a surgeon has to select the appropriate instrument for an operation, a mechanic has to find a suitable tool for repairing an engine, or a basketball player has to remember which member of the team to pass the ball to, agents use their mental representations to identify functionally relevant sensory inputs.

The results of our experimental studies support the hypothesis that voluntary movements are planned, executed, and stored in memory as representations of their anticipated perceptual effects. We investigate shared mental action representations in order to design intelligent technical systems with improved anticipation and interaction capabilities, particularly in the area of medicine (rehabilitation, every-day support of the elderly), and sports.

3.1 Assistive Glasses

In a particular research project, we are supporting anticipation by "Seeing the World through Assistive Glasses." This project, called Adaptive and Mobile Action Assistance in Daily Living Activities (ADAMAAS) focuses on the development of a mobile adaptive assistance system in the form of intelligent glasses that provide unobtrusive, anticipatory, and intuitive support in everyday situations. The system will identify problems during ongoing action processes, warn of errors, and provide context-related assistance in textual, pictorial, or avatar-based formats superimposed on a transparent virtual display. The technical platform is provided by the eye-tracking specialists SensoMotoric Instruments (SMI, www.smivision.com). This project integrates mental representation analysis, eye-tracking, physiological measures (pulse, heart rate), computer vision (object and action recognition), and augmented reality with modern diagnostics and corrective intervention techniques. The major perspective distinguishing ADAMAAS from stationary diagnostic systems and conventional head-mounted displays will be its ability to react to errors in real-time, provide individualized feedback for action support, and learn from expert models as well as the individual user's behavior. We are further planning to use this device in the context of telemedicine. In such a context, an expert in the USA could interact with a surgeon in Germany to assist during a clinical operation. Therefore the glasses and an expert could support the perception, anticipation, and the motor performance of a surgeon within a clinical setting.

3.2 The Role of Technical Systems

Human performance, rooted in biological evolution, has matured to a point where it can profit from technical systems. The lines of research presented here not only help us to understand the cognitive background of human performance, they also provide a basis for building artificial cognitive systems that can interact with humans in an intuitive way and even acquire new skills by learning from the user. In this context, it is clearly advantageous for a real or virtual coach to know how mental representation structures are formed, stabilized and adapted in daily actions. This knowledge enables a coach or technical system (such as intelligent glasses) to address individual users or trainees concerning their current level of learning and performance, and to shape instructions to optimize learning processes and maximize performance.

Taking into consideration all the different problems of relevance to anticipation in medicine—various treatments or a system for diagnosis ranging from specialization to holistic understanding—and exploring further development of technical devices (assistive technologies) as another independent (to a certain degree) dimension, we would place technical devices such as ADAMAAS in a coordinate system to complete different traditional healing ceremonies. Traditional healing ceremonies are more related to a holistic understanding of human beings and healing than to the conceptual definition of high level assistive systems. With this remark as an introduction, we shall now investigate the extent to which centuries-old traditional healing ceremonies (maintained and practiced in our time) evince aspects of anticipation as a central element of human action, life, and culture; and the extent to which we have lost some of the beneficial aspects of traditional healing methods through cultural, religious, and educational development, as a result of which medicine has become more and more specialized, ignoring or abandoning the holism of life.

4   Traditional Healing Ceremonies

At first glance, there seems to be many interesting phenomena regarding antici-pation in the context of traditional healing ceremonies. Phenomena such as altered states of consciousness (ASC) have been a topic of serious scientific research for the past few decades [9, 10, 25]. Shamans (healers) seem to be regularly able to anticipate future states of individuals, social groups, or physical events while experiencing such ASC. Researchers such as Winkelman addressed the point that during a healing ceremony and process, shamanism (traditional healing) uses cer-tain techniques to access different modes of consciousness and to integrate deeper (older, in terms of evolution) brain levels for perception and cognitive processing. It would be possible to link such phenomena to the concept of *predictive anticipatory activity* (PAA), as described by Mossbridge et al. [26] and within additional frameworks by Radin (see Radin[1]) and others [27, 28]. Here we shall explain the functional meaning and place of anticipation for participants and healers within traditional healing ceremonies.

Our presentation is based on literature and on participant observations of dif-ferent healing ceremonies over the last few years (e.g., Native American sweat lodges, Santería, Palo Monte). We limit ourselves to a small number of traditional healing methods since we are not familiar with the many varieties of traditional healing ceremonies. Neither can we discuss such ceremonies in general, because many of these practices are specific to the particular native (American) tribe, or group (e.g., family), or house (e.g., *cabildo*). Nevertheless, our aim here is to identify some basic features and regularities of such ceremonies.

Ceremonies are a central part of traditional native healing. They integrate dif-ferent techniques for communication, goal setting, and meditation, for example. Usually they start with an introduction, during which a particular (physiological, sensory) contact to nature and a (mental) contact to the ancestors of the participants (by recalling and invoking their names) are carried out. In general, ceremonies

---

[1]Radin, D.: Intuition in Medicine: Orthodox and Unorthodox In: Nadin, M.: (ed.) Anticipation and Medicine, pp. 258–269. Springer, Cham (2016).

create a framework for interaction with the healer and with different members of the family or community. They open up a space for (re-)connecting different levels of the participants' identity, social-motor functioning, and consciousness. They are designed to create a space for identity transformation and for work on participant anticipation. Religious elements are not necessarily involved (depending on the sensitivity and motivation of participants) but are usually used to create a common space of symbols and communication beyond the physical reality between the healer and the participant(s).

Concerning the healing process itself, we would like to claim two main features of traditional ceremonies: responsibility and a bio-psychosocial perspective.

Responsibility: the healer takes the responsibility for the healing process in cooperation with the patient. Because the patient is mainly responsible for his health and healing behavior, the healer is assisting the patient. The healing process is created as a goal setting process.

Bio-psychosocial perspective with a historical dimension: From the healer's perspective, health is based on the ability of self-regulation within a social, natural, and supernatural environment. The healing process transcends the recent physiological and psychological status of an individual patient. It tries to link the past, recent goals, and the potential future of a participant and to offer a developmental perspective. Furthermore, it takes into account the historical dimension of the individual in the form of ancestors, family, and the social network. Basic (neuro-) biological dimensions are activated with the help of particular transformation techniques, such as sweating, dancing or rhythm of music.

Ceremonies are intended to reconnect participants to individual, natural, supernatural sources of healing, and life. They are techniques for improving (work on, actualize, realize) personal and social anticipation.

4.1 Native American Ceremonies: Sweat Lodges

Sweat lodges have been utilized for thousands of years and in different forms by many North American indigenous tribes (e.g., Inuits, Navajo, Lakota, Sioux, Muskogee), as well as by tribes that have disappeared as such (Aztecs, Olmecs) in the Americas. Old European tribes (e.g., Teutons, Celts) have also used them. (The European tradition is partially alive in the Nordic—Finland, Estonia, Sweden—sauna.) Sweat lodges (stone-keepers lodges) for healing ceremonies are mostly ritual places where participants meet, talk over recent situations, problems, and plans (identity). A medicine council is created and a healing space opens up in which (identity) transformation in the context of nature, especially fire, stones, herbs, and steam becomes possible. The lodge itself is constructed of natural materials (e.g. stones, wood, plants). There are several structures and processes used in different tribes, but they all have the following in common: a lodge with a hole or a particular place in the center for heated stones; a fireplace to heat stones and hallowed (sacred) areas to prepare and carry out the ceremony.

If we examine the sequential stages of the sweat lodge ceremony—in this case, typical of the Muskogee, we can differentiate the stages of the whole: introduction, medical council, the ceremonial fire, the preparation of tobacco prayer ties, the time spent in the sweat lodge itself, and a closing communal meal with seeds, nuts, fruits and vegetables. During the period in which participants remain in the sweat lodge, four different directions (east, south, west, and north) are respected in a precise functional manner. These directions have particular meanings, such as birth (east), growth (south), completeness (connection to the ancestors, west), and readiness for continued life and future tasks (north). In the lodge, participants honor these developmental steps with praying and singing. Herbs (like sweetgrass) are also used in the context of the ritual. Participants speak of related topics while sitting on the ground and facing red stones in the darkness of the lodge. Four ceremonial elements are presented in detail below.

4.1.1 Respect for Time and Nature

Much attention is given to Mother Earth during the introduction, but also during other parts of the ceremony. Participants sit on the bare earth in order to connect with Mother Earth. The stones are claimed and greeted as grandfathers as the fire keeper moves them from the fire to the middle of the lodge. Stones are the bones of Mother Earth and hold the information of times long ago. When they are heated, they start to "speak" (especially when they come in contact with water): "The ancestors are speaking." In the understanding of the indigenous Muskogee (the "stone people"), seven generations of ancestors are motivated to contact living human beings. In view of this seven future generations must be considered in advance.

4.1.2 Creating a Healing Space

Participants and healer meet in what is called the medicine council. By bringing relevant problems and needs into the medicine council, a conceptual grounding is created that connects all those present. Preparation for prayer is of particular interest to anticipation. Participants create tied bundles by placing an amount of tobacco into different colored cloths. Each participant explains his (only men attend sweat lodges ceremonies) reasons and aims for entering the sweat lodge. Participants must be clear about why they would spend hours in the heated lodge. They must also reflect upon how their aims fit in the network of social and natural connections (web of life) and try to formulate them as precisely as possible. Justifications based on family, nature, work, social network, health, future, developmental aspects, etc. will have to generate enough energy and motivation for staying in the sweat lodge for hours on end. This process of goal setting in a healing context is an important feature of the ceremony.

4.1.3 Transformation

The sensible phase for transformation is the actual stay in the sweat lodge. Participants have to find and create a place for their own healing. In the heat of the lodge they actualize their aims (holding the prayer ties in their hand), participate in singing and praying, and may formulate specific requests related to their future development. This is a permanent and focused (mostly non-verbal) communication with all members in the lodge, in particular the healer. Because the lodge is related to the four directions (east, south, west, north), it is therefore opened and closed four times for bringing in new stones and fresh water. The heat in the lodge increases. In order to bear the heat and pain, participants may die a symbolic death, and then are born anew out of the womb of Mother Earth (the lodge), and then leave the lodge. The process of dying and being born anew is a symbol of the transformation taking place inside the participant. This transformation includes changes on various levels: the biological (heat, sweat and hours of sitting); the psychological (inner conversation, actualizing aims); and a social (among the members of the "family" in the lodge and in respect regard to the mentally present social structures of the participant). This provides many opportunities for clarifying personal matters and developing new perspectives on one's own problems. Gaining health means to come into contact and harmony with oneself, others, nature, and the universe.

4.1.4 Spirit Guides and Animal Spirits

A particular option for change is offered in the third round of the sweat lodge ceremony, focusing on the direction of west. Here the ancestors could send one or more spirit guides to help participants to fulfill their tasks. In the understanding of most native tribes, spirit guides manifest themsleves in the form of an animal (e.g., eagle, bear, wolf, fox, butterfly, etc.). It is important to understand the strength and behavior (the way of moving) of the animal and the extent to which these features could help the participant to develop a new perspective.

In summary: the healing process taking place in the sweat lodge ceremony has been addressed as bio-psychosocial development, where the past and future states, together with goal setting concerning recent situations in life, play a central role. Participants work in cooperation with other participants and the healer on a redefinition of personal anticipations at different (health, family, career etc.) dimensions.

4.2 Afro-American and Afro-Cuban Ceremonies: Santería, Palo

Many recent African (Yoruba), Afro-Cuban (Santería, Palo), and Afro-American (Condomblé, Voodoo, Hoodoo) healing ceremonies had already been partially developed in the former Yoruba Kingdom in West Africa (today's Nigeria, Benin, Ghana, etc.). Between 1650 and 1860 about 15 million black Africans were transported, under barbarous conditions, from Africa to the Americas to work as slaves. The Africans brought their ceremonies with them and were practiced by the slaves who survived in the new world. Through the generations, the slaves adopted their ancient ceremonies to their new circumstances: discontinuity of family and social ties, religious prohibitions imposed by Christian religions. For instance, in Santería (in the Spanish-speaking lands), they combined their Yoruba religious traditions with elements of Catholicism in order to convince plantation owners that they accepted the Catholic faith. Afro-Cubans maintained their native ceremonies and religions alive by syncretizing their spirit guides with Catholic saints.

In order to focus on Santeria and related aspects in Palo ceremonies, we have to clarify that these ceremonies are mainly healing ceremonies, based on different techniques of predictions, incorporating the use of herbs and other natural ingredients for healing. The center for these ceremonies is the house of the priest and the related "spiritual" family. Ceremonies always start by honoring the ancestors. A central role in the ceremony are spirit guides, so-called *Osha* (Orisha). Although these spirit guides are depicted as angels, featuring a particular human-like character and look, they are related to natural phenomena (river, ocean, wind, fire, etc.). Oshas are believed to take care of the development of related human beings and to communicate with them. The main ceremonies (e.g., initiation rites) are secret. The following aspects are comparable to the native American sweat lodge-ceremonies:

4.2.1 Nature

Nature and respect for nature play an important role in all ceremonies. Many ceremonies are performed for reconnection with nature, using plants, herbs, soil, and tobacco. (Tobacco use is common among Native American tribes; it is the plant endowed with the strongest holy meaning.)

4.2.2 Healing Space

A main part of the ceremony is the meeting between the participant and the healer to discuss the participant's current situation and problems. Relevant needs are identified, but also considered in the social framework (family, friends), and situated in a context that comprises past (ancestors, personal development) and future. In that context, different types of divination might also be resorted to. In contrast to other forms of divination, Santería divination systems make use of a clearly defined corpus of calculations based on the use of shells (*diloggun*), coconut parts (*obi*), or different natural elements familiar to the high priests (*Ifa*).[2] The purpose of these divination systems is to help participants develop and align themselves with future possibilities and to connect

---

[2]The *Ifa* divination system was entered in UNESCO's list of *Masterpieces of the Oral and Intangible Heritage of Humanity* in 2008.

them as much as possible with natural and spiritual (supernatural) dimensions. In the context of such sessions, participants are able to learn how to bring their lives into harmony with themselves in respect to recent circumstances ("web of life": family, nature, work, etc.).

4.2.3 Transformation

Transformation is facilitated by actions such as drumming, meditating, and dancing, as well as by meeting natural entities like the ocean, a river, or a tree. Particular offerings for the spirit guides in the context of nature are supplied. Within such ceremonies, shamans (healers) go through ASC as they represent particular spirit guides. This experience is clearly accompanied by so-called non-epileptic attacks. For the healer and the participants, it is a good sign when such ACSs occur. In their understanding, the spirit guides (Oshas) are taking part in the ceremony and supporting the healing process. Indeed, the character of the person (voice, movement characteristics, etc.) changes drastically, giving the impression that the spirit guides are dancing, drinking, and singing with the participants. More important is the fact that the Oshas "speak" with the participants about the current situation, ancestors, and future options in their lives. The detailed knowledge expressed is impressive and sometimes seems to surpass the knowledge that the healer has about the participants.

4.2.4 Spirit Guides

As already mentioned, spirit guides are mainly Oshas (or Orishas), energy—entities (angels) with a kinship to nature (thunder and lightning, a river) and with a human-like appearance and character.

   Anticipation aspects can be observed as central in the described Afro-American and Afro-Cuban ceremonies at different levels. The past is respected and activated (ancestors), the current situation and the goals aims are discussed with the healer. Furthermore, with the help of (to some degree) objective divination systems, participants and healers redefine the current and the future situations within a social and spiritual context. This is dramatically supported by the healer's ASC, which lead to an extensive work on the participants' future perspectives.

5 Conclusions

The traditional healing ceremonies described herein work with humans as anticipatory systems, grounded in their past (ancestors, history) and address the "open-ended ever-changing space of possibilities" [3]. Working on the whole human being (holistic level) seems to be a central part of the healing process.

Traditional ceremonies also address the healing process from a bio-psychosocial perspective. The transformation process involves different but connected (bio-psychosocial) levels of life and therefore of action organization. It seems that within this rich healing space, the healer opens a zone of proximal development [4, 7] for the participant.

Whereas highly technological modern medicine focuses mainly on methods for optimizing action organization in a specified context (e.g., assistive technologies provided for a surgeon during operation), traditional ways take the multi-dimensionality of life into account (referring to past and future aspects, as well as to social, psychological, biological, spiritual, and natural dimensions). It is highly likely that traditional healers would not conceive of modern and traditional healing as alternatives, but would see them as complementary.

Independent of recent and interesting approaches that see Shamanism in the context of Neuro-Ethology and Evolutionary Psychology [9, 10], in addition to ACS, some phenomena manifest themselves (e.g., clear and surprising predictions) that cannot be explained within the frame of our scientific understanding of the world. This will not pose any problem for traditional healing—because, independent of modern medicine's problems in understanding anticipation in its full meaning, traditional healers understand that it is essential for their ceremonies.

Acknowledgments We wish to express gratitude to the German Excellence Initiative of the Deutsche Forschungs Gemeinschaft (German National Science Foundation) for its support of the work of CITEC; and to the German Federal Ministry for Education and Research for its support of the ADAMAAS research. We acknowledge especially the help and input of Tom Blue Wolf (Georgia/USA), spiritual leader of the Muskogee nation, charter member of the World Council of Elders, and the Indigenous Healers Association; and Alexander Borges Fernandez (Matanzas, Cuba), head of Santeria-Institution (Cardenas); and so many other healers for providing insight into their practices and an understanding of the world, life, and anticipation.

References

1. Bernstein, N.A.: The Co-ordination and Regulation of Movements. Pergamon Press Ltd., Oxford (1967)
2. Nadin, M.: Prolegomena: what speaks in favor of an inquiry into anticipatory processes? In: Klir, G. (ed.) Anticipatory Systems. International Book Series on Systems Science and Systems Engineering, pp. xv–lx, 2nd edn. Springer, London (2012)
3. Nadin, M.: Concerning the knowledge domain of anticipation—awareness of early contributions in the context of defining the field. Int. J. Gen. Syst. **44**(6), 621–630 (2015)
4. Land, W., Volchenkov, D., Bläsing, B., Schack, T.: From action representation to action execution: exploring the links between cognitive and biomechanical levels of motor control. Front. Comput. Neurosci. **7**(127), 1–14 (2013)
5. Marken, R.S.: The hierarchical behaviour of perception. In: Marken, R.S. (ed.) More Mind Readings: Methods and Models in the Study of Purpose, pp. 84–116. New View Publications, Chapel Hill (2002)
6. Schack, T.: Relation of knowledge and performance in action. J. Knowl. Manag. **4**(8), 38–53 (2004)

7. Vygotsky, L.S.: Mind in Society—The Development of Higher Psychological Processes. Harvard University Press, Cambridge (1978)
8. Schack, T., Schütz, C., Krause, A., Seegelke, C.: Representation and anticipation in motor action. In: Nadin, M. (ed.) Anticipation Across Disciplines, pp. 203–215. Springer, New York (2016)
9. Winkelman, M.: Shamanism as neurotheology and evolutionary psychology. Am. Behav. Sci. **45**(12), 1873–1885 (2002)
10. Winkelman, M.: Shamanism—A Biopsychosocial Paradigm of Consciousness and Healing, 2nd edn. Praeger, Oxford (2010)
11. Nadin, M., Kurismaa, A.: From Russia with love/Russian experimental and empirical contributions informed by an anticipatory perspective. Int. J. Gen. Syst. **44**(6), 615–620 (2015) (Taylor and Francis, London)
12. Schack, T.: Vygotsky und Bernstein: Psychologie und Bewegungskoordination. In: Hirtz, P., Nüske, F. (eds.) Bewegungskoordination und sportliche Leistung integrativ betrachtet, pp. 99–104. Czwalina, Hamburg (1997)
13. Hoffmann, J.: Die Welt der Begriffe. Verlag der Wissenschaften, Berlin (1986)
14. Hoffmann, J.: Anticipatory behavioral control. In: Butz, M.V., Sigaud, O., Gérard, P. (eds.) Anticipatory Behavior in Adaptive Learning Systems, pp. 44–65. Springer, Berlin (2003)
15. Rosch, E.: Principles of categorization. In: Rosch, E., Lloyd, B. (eds.) Cognition and Categorization, pp. 27–48. Lawrence Erlbaum, Hillsdale (1978)
16. Prinz, W.: A common coding approach to perception and action. In: Neumann, O., Prinz, W. (eds.) Relationships Between Perception and Action, pp. 167–201. Springer, Berlin (1990)
17. Jeannerod, M.: The Cognitive Neuroscience of Action. Wiley-Blackwell, Oxford (1997)
18. Schack, T., Ritter, H.: The cognitive nature of action—functional links between cognitive psychology, movement science, and robotics. In: Raab, M., Johnson, J.G., Heekeren, H.R. (eds.) Progress in Brain Research, pp. 231–250. Elsevier, Amsterdam (2009)
19. Schack, T.: The cognitive architecture of complex movement. Int. J. Sport Exerc. Psychol. **2** (4), 403–438 (2004)
20. Schack, T., Ritter, H.: Representation and learning in motor action—bridges between experimental research and cognitive robotics. New Ideas Psychol **31**(3), 258–269 (2013)
21. Rosenbaum, D.A., Marchak, F., Barnes, H.J., Vaughan, J., Slotta, J.D., Jorgensen, M.J.: Constraints for action selection: overhand versus underhand grips. In: Jeannerod, M. (ed.) Attention and Performance XIII: Motor Representation and Control, pp. 211–265. Erlbaum, Hillsdale (1990)
22. Rosenbaum, D.A., Meulenbroek, R.G.J., Vaughan, J., Jansen, C.: Posture-based motion planning: applications to grasping. Psychol. Rev. **10**, 709–734 (2001)
23. Weigelt, M., Schack, T.: The development of end-state comfort planning in preschool children. Exp. Psychol. **57**, 476–482 (2010)
24. Stöckel, T., Hughes, C., Schack, T.: Representation of grasp postures and anticipatory motor planning in children. Psychol. Res. **76**(6), 768–776 (2012)
25. Tart, C.: Putting the pieces together: A conceptual framework for understanding discrete states of consciousness. In: Zinberg, N. (ed.) Alternate States of Consciousness. Free Press, New York (1977)
26. Mossbridge, J.A., Tressoldi, P., Utts, J., Ives, J.A., Radin, D., Jonas, W.B.: Predicting the unpredictable: critical analysis and practical implications of predictive anticipatory activity. Front. Hum. Neurosci. **8**, 1–10 (2014)
27. Maier, M., Büchner, V.: (2016). Time and consciousness. In: Nadin, M. (ed.) Anticipation Across Disciplines, pp. 93–104. Springer, Cham (2016)
28. Tressoldi, P.: Anticipation of random future events. In: Nadin, M. (ed.) Anticipation Across Disciplines, Cognitive Systems Monographs, pp. 11–17. Springer, Cham (2016)

At the Meeting Point Between Anticipation and Chiropraxis

Jean Paul Pianta

Abstract Most people might think that chiropractic has mainly to do with ailments affecting the muscular-skeletal system. Today, chiropractic looks beyond the immediate "dis-ease" to see how this might affect physical, mental, and spiritual development. Not "How do you feel?" but "How do you want to feel?" Anticipation-informed chiropractic focuses on maintaining the human being's anticipatory expression. This is comprised of motoric performance: sense of balance, spatial and temporal navigation, adaptation to tasks, etc., and the associated cognitive activity. The chiropractic way of thinking is compared to classical, conventional, localized therapeutic approaches promoted by the mechanistic medical model, and to the reductionist views implicit in specialized medicine.

Keywords Anticipatory expression · Holism · Posture · Motoric · Mechanistic medicine

None other than Dr. Ralph Gay, Director of the Spine Biomechanics Research Group at the Mayo Clinic, stated: The evidence supporting spinal manipulation for back pain…is very good. […] Chiropractic treatment… has just as much evidence supporting it as any medical treatment of back pain [1]. Dr. Heidi Haarvik, Director of Research at the New Zealand College of Chiropractic, has a PhD in neurophysiology, a significant qualification for a researcher trying to build the base of scientific evidence for chiropractic care. She investigated adjustments of dysfunctional spinal segments (vertebral subluxations), as well as somatosensory processing, sensorimotor integration, and motor cortical output. From among many practitioners in healthcare, these two professionals received more public attention

J.P. Pianta (✉)
Pianta Chiropractic Center, 30163 Hannover, Germany
e-mail: jppianta@aol.com

J.P. Pianta
German Chiropractic Academy, Herforder Strasse 2, Bad Oeynhausen 32545, Germany

© Springer International Publishing Switzerland 2017
M. Nadin (ed.), *Anticipation and Medicine*, DOI 10.1007/978-3-319-45142-8_21

because their arguments are aligned with those that certify medical procedures. As we shall see, such competent views do not settle the conflict between practitioners of medicine in the classical sense and practitioners in chiropractic healing. Nevertheless, we are experiencing more of a coming together, a convergence, than intolerance. In Germany, a great number of internists have adopted chiropractic methods in the last five years. A chiropractor's education, in turn, integrates more advanced knowledge of anatomy, physiology, cognitive science, as well as molecular biology and genetics.

Against this background, I would like to proceed with an example, the details of which belong to empirical evidence. Nobody would create an accident in order to see how it echoes in the life of a young person, and further to their offspring. Such "experiments" have no place in medicine. But we all pursue horizontal studies with many patients and keep evidence (the medical record) of desired or undesired outcomes.

1 A Not Unusual Example

A 13-year-old girl is rollerblading in a parking lot near the apartment where she lives with her family. One day, after falling, she bursts into tears from pain. She goes home and tells her mother what happened. The mother comforts her: "Don't worry, the pain will go away in a few days."

A few years later, the girl is a pretty teenager, 17 cm taller, but she does not have very good posture. Pain in her lower back often causes her to complain. She also has recurrent headaches and tries to find a connection between her state and weather patterns, food, fatigue. Every now and then, her mother gives her a painkiller.

At 23 years of age, the young woman becomes pregnant and she notices more pain in her low back, noticeably worse during the last three months of pregnancy. Her mother and other women explain to her that this is not unusual. She is not the first woman to expect a baby who complains of such symptoms; all will be wonderful once the baby is born. The birth is not easy. After the woman spent a longer than usual time in labor, the obstetrician decides to use forceps to pull the baby out. This leads to trauma for the baby and trauma for the mother. She continues her complaints of recurrent lower back pain, now associated with sciatica in her right leg. And as frequently occurs after a birth assisted with forceps, the baby presents a narrow cranium at the temporal level and some red spots around his head.

A few weeks later, the redness is gone and the baby's face assumes a normal expression. But the baby is restless, cries a lot, and wakes up several times during the night. He appears very sensitive to noise and tends to overreact when the light is turned on. A couple of years later, the mother still suffers from pain in her spine, headaches, and sciatica. She is always tired and has to consult a physician. He begins his treatment by taking X-rays that reveal a slightly tilted and rotated pelvis and slight scoliosis. He prescribes a pain-killer and exercises. The young mother has to keep taking pain-killers on a regular basis, which negatively affects her digestive system. Carrying the baby does not help her back. The young boy is hyperactive, very nervous. Some time after starting school, his teacher calls the mother to explain that her son presents all the symptoms of ADHD (Attention Deficit Hyperactivity Disorder) [2]. Ritalin could be a good solution.[1]

The timeline of this case already suggests events of predictable consequences of the fall: developmental peculiarities, problematic pregnancy, difficulty in giving birth, raising a problematic baby. This is where anticipation also comes into the picture. A certain decision—using painkillers or birth with forceps—leads to a development that could have been avoided. The case described, and similar mishaps, are not rare events. One can possibly predict the negative mechanical and structural consequences of a trauma, but it leaves out what we cannot fully assess: the amount of psychological stress, the disturbed neurochemistry that plays such an important role on communication between structures and organs, the weight of the emotional burden, the hormonal changes, the possible neurotransmitters variations, the acidity, the impact on the digestive tract, the impact on breathing, the affected sensory perception of reality, the fear of experiencing another trauma, a new consciousness, and finally the social and economical impact. If we could precisely study all effects, can we be convinced that all of these outcomes would remain permanent, or last only a week, months, or three years later?

The short summary (based on a real case) of how a certain event affects an individual points to the need to see every symptom within the larger framework of the entire body—and even beyond, in relation to others (family, friends, employers, doctors, etc.).

If we could perfectly study the case of one person, collecting detailed evaluations, could we then with certitude draw the exact same conclusions for another young girl or young boy who has also suffered a bad fall when rollerblading? The answer is a definite "No!"—because we are not studying the objective material damage to a car after an accident, for example, where you find the faulty part and replace it. Yet we would attempt to evaluate damages to a *unique* human being, permanently changing, adapting and improvising in relation to internal and external factors. We tend to understand, in simple terms, that life reproduces life; but it may be more accurate to describe life as a permanent creative and original process.

Is there a relationship between the muscular-skeletal system, the nervous system, and other systems in the body? This begs the question pertinent to this study: Can chiropractic have a positive effect on organic functions, physiology, biochemistry, health and well-being—on the whole human being? Or does it pertain only to muscles and bones and tendons? I will try to answer such questions.

---

[1]Ritalin sales have been skyrocketing for the last decade. 83 % increase in the drug use between 2006 and 2010. The Methylphenidate belongs to the amphetamine family.

2   The Importance of the Vertebral Column

The spine has been called the "Tree of Life," and has been studied for thousands of years. From the time beginning in utero until the end of life, the spinal column protects the spinal cord and goes through constant change. From the C primary curve fetal position, a secondary cervical curve, known as *lordosis* (concave posteriorly), starts to develop at around six months, when the baby starts to sit and the cervical muscles are strong enough to carry the weight of the head. Later, another secondary lumbar lordotic curve will develop when the baby starts standing up and walking. A teenager's spine will adapt during a period of rapid growth; the curves compensate one another in the best possible way to adapt to gravity. In older ages, the spine and all directly related structures such as muscles, tendons, ligaments, fascia, lose elasticity, mobility and the vertebral column might tend to stiffen and to go back to a postural flexion that can lead to a C posture known as kyphosis (concave anteriorly).[2]

The shingles that doctors hang out to identify themselves to possible patients tell the story: the body's organs are the names of each medical specialty: neurology, orthopedics, cardiology, pneumology (pulmonology), endocrinology, podiatry, hematology, etc., etc. Since the time Descartes set forth his view of the human body, cutting, reducing, and separating became engrained in our way of thinking about the organism. It might be useful to keep in mind that the word "individual" literally means "that cannot be divided". However, in more recent times, science is finding out that life is ruled by interdependences and interrelationships. Looking specifically at a cell of wood under the microscope is interesting, but it cannot replace the amount of information our senses and nervous system pick up when we walk in a forest. Despite this, physicians continue to act in the Cartesian spirit.

Without a doubt, specialized and technological medicine has made huge progress saving and helping lives. However, as far as chronic or non-transmissible conditions are concerned, there might be a lot more that needs to be studied and understood. Often blinded by having to treat the complaints of patients who ask for help when symptoms or pain is most obvious, medical providers rely on a "fix the part" service, reductionist in its approach. The patient as well as the provider is guilty in such a case because healthcare should be a matter of anticipation, of considering possible outcomes of a condition before it reaches the stage where the symptoms are treated, not the whole person.

Two-thirds of all medical consultations have to do with pain. This quite understandable priority relies on deducing the immediate symptoms as efficiently as possible. This attitude prevails in our medical system but it seems to leave aside or underestimate two factors that play a key role in health and well-being: gravity and

---

[2]"Posture affects and moderates every physiological function, from breathing to nervous system. Despite the considerable evidence that posture affects physiology and function, the significant influence of posture on health is not addressed by most physicians." American Journal of Pain Management.

evolution. Our body simultaneously fights and adapts to gravity. Here anticipation is at work in the entire motoric expression [3]. Our posture reflects the way all structures of our body manage to deal with gravity.

> Better than 90 % of the energy output of the brain is used in relating the physical body in its gravitational field. The more mechanically distorted a person is the less energy available for thinking, metabolism and healing [4].

The proprioceptors under the feet send information enabling the brain to choose proper motor answers. Stress information can only generate stress adaptive response. All muscles of the spine form an aponeurosis[3] attached to the cranium at the level of the occipital line. The relationship between the sacrum and the occiput, connected by the spinal column, the spinal cord, meninges (Dura Matter), muscles, ligaments and tendons is rather intimate. The slightest movement of one structure affects the entity [4]; a weakness in one area can weaken the whole.

The fight against gravity is a full-time job. Cartilage, bone structure, elasticity of tendons, ligaments, fascia and muscles will tend to progressively lose their elasticity that will objectively translate into a diminished mobility. Studying the posture and the walk of a person allows understanding the way various structures work in harmony with one another.

3 The Propeller Mechanism—An Anticipatory Function

When walking, the right leg moves together with the left arm and the left leg will move together with the right arm. With the right leg forward, the lower part of the body could be seen as the blade of a propeller turned to the left, at the same time the upper trunk will tend to turn to the right with the left arm forward. The center of the propeller will be located in the lower lumbar spine. It is responsible for 75 % of back problems. In terms of relationship or interdependencies, we can see that the left foot, left knee, and left hip move together with the right hand, right elbow and right shoulder. Interestingly, these leg and arm structures share great anatomical similarities. *They cultivate a close functional physiological relationship for a lifetime.* I am grateful that Eberhard Loosch made Nikolai Bernstein's *Atlas des Ganges und Laufes des Menschen* available on the internet. This will further help us learn about how movement is related to the motoric system.

Considering the upper trunk and head when the right arm moves forward for example, the shoulder girdle will tend to turn to the left, and at the same time the head will rotate to the right because of the tension of the right trapezius and the tension of the left sterno-cleido-mastoid muscles. This mechanism constitutes a second propeller: the lower blade being the shoulder girdle twisted to the left; the upper blade being the cervical region and the head. The center of the upper

---

[3]Simply put, aponeurosis is a a sheetlike tendinous expansion, mainly serving to connect a muscle with the parts it moves (cf. http://medical-dictionary.thefreedictionary.com/aponeurosis.

propeller is at the junction between cervical and upper dorsal spine. It is responsible for 25 % of back problems.

The body moves forward through the air according to the same physical principles as the propeller of a small airplane. There are two propellers: a lower—the pelvic girdle-shoulder girdle—and an upper—the shoulder girdle-cervical and head. These two centers of the propellers should work exactly one above the other. If the centers of the two propellers are not properly aligned, a person will be walking with a sway, head tilting more on one side than the other, or one longer step on one side, or one arm balancing more than the other.

We notice the intimate relation of all joints working together, making the same movements millions of times, sharing information, communicating with one another and staying permanently connected. Studying posture and gait provides valuable information for assessing the quality of the relationships among the structures. Any stress at any specific area can affect the whole, and studying the whole will allow to understand any specific stress in one area. We anticipate the firm or unstable surface on which we step and proceed with our movement based on this anticipation.

Appreciating the interrelation of all structures working together allows anticipation of the fact that a minor stress in one joint of the right foot, for example, could create over some time a chain of negative reactions in the entire body associated to various symptoms and pathologies. Anticipation pertains to the understanding of the relationships among all elements involved in motoric expression.

Considering the human body as the sum of separated parts cannot allow an appreciation for the weight of interrelationships and interdependences.

3.1 The Role of Gravity

One more word about gravity, a physical law that applies to the living and the non-living: 18 % of our food intake is spent on gravity. Bad posture can only increase this percentage, creating extra tension on the muscle, spasms, hypertonicity, added pressure on certain joints, as well as more oxygen and sugar consumption, more lactic acid production (possibly leading joint inflammation), and accelerating the ageing process [5].

3.2 The Role of Change

The second factor often underestimated in crisis intervention, acute, emergency medicine, is the outcome of change. To focus on the current situation to better patient quality of life is normal. Anticipating possible secondary effects of our interventions that could affect a person's future is no less important. From conception until death, the human body does not know stability. Life is not a single event; it is an ongoing dynamic process. Permanent changes are key concepts for all living organisms. Anticipation-guided medicine has to consider how a current, timely individual intervention can evolve, i.e., change, over the years. There are no two identical individuals, but we can anticipate that all individuals are affected by time, with a tendency for all structures of the human body to lose tonicity, elasticity, and mobility.

Taking into account several factors, such as a patient's history, genetics, hereditary background, way of life, activity, and environment can be interesting in terms of accumulating data. However, the past cannot determine with certainty the future of one particular person. Each is affected by his own perception of information, his ideas, his conscience, the weight of gravity, and the effects of change that keep the door open to new possibilities.

3.3 Avoid the Error of Simplifications

From conception, each individual has a unique history and a unique destiny. Accumulating data about machines leads to artificial intelligence. In the health system, imposing the same medicine on all individuals presenting similar symptoms or analogous conditions—i.e., considering all individuals as the same—is an error of simplification. Diversity, adaptability, complexity nourish life.

Medicine was considered the little sister of physics. This kind of medicine tends to think in a mechanistic, linear, reproducible way. The paradigms still taught in medical schools are based on rules elaborated by Descartes and Newton during the 17th century. In 1657, Christiaan Huygens and Salomon Coster developed the first pendulum clock. Watch- and puppet-makers were quite fashionable at the time. A healthy body is still understood as a clock or a well-made puppet, although today we know better.

Specialization seems to be resulting into more and more specialization. The family doctor of the past, aware of the specific bio-psycho-social components of each of his patients, aware of the global picture, had to give up his role to specialists. An early and proper diagnosis on the teenage girl would have spotted the multitude of possible consequences for the pelvis from the rollerblading accident. The stress on the pelvic girdle affects the lumbar spine, the dorsal spine and the cervical spine. The twisted rotated pelvis cannot allow the full elasticity necessary for an easier delivery. Having to use forceps is a huge stress on the cranium, on the atlas-occiput relationship and on the baby's central nervous system [5, 6].

The sacrum-coccyx structure, the 24 vertebrae and the occiput work very closely together. It is possible to anticipate that whiplash and stress to the cervical spine, due to a car accident, for instance, could produce lower back pain as a secondary effect weeks or months later; shock to the coccyx-sacrum could later produce headaches or episodes of vertigo. A traumatic stress factor perceived by the vertebral column and the spinal cord plays a role on the function of the hypothalamic-pituitary-adrenal (HPA) axis, leading to reduced reactivity and enhanced negative feedback sensitivity [7].

4 Chiropractic: The Art of Healing Informed by Science

Chiropractic, since its birth in 1895 in the USA, has opened the door to much controversy outside and within the profession. Let us notice that over centuries, in various cultures, manipulation of the body was part of healing. With chiropractic, strict mechanical, structural explanation was accepted. Many colleagues in medicine around the world integrate the therapeutic mechanistic aspects. For many others, new horizons found in chiropractic are opening. They are directly related to new scientific discoveries and new imaging technologies, all of which allow for better understanding of human brain functioning. Chiropractic science has still much to do in order to better appreciate the impacts of the adjustment on one individual. But every day, new scientific studies allow looking at chiropractic with an open perspective that goes much beyond the strictly physics-based mechanical-structural explanation of the living. Life is ruled by dynamic interdependences and interrelationships that could not be fully appreciated during the centuries when the prevalent science strategy was to separate the whole into parts.

The concepts that Daniel David Palmer proposed are well known; but this does not mean that they are all accepted. There is no need to revisit the history of chiropractic. Many authors have dealt with Palmer's ideas. Let us revisit only a few basic principles.

(1) Rather than separating each part of the individual, Palmer introduced a holistic vision of the human being.
(2) Palmer broke away from the dualistic distinction of *pain-no pain*, *healthy-sick*, *good treatment-bad treatment*, and opened the way to more nuanced perceptions of a patient's state.
(3) Instead of giving the front role to diseases and symptoms, Palmer attempted to place the individual and his resistance as the very top of priorities to be considered. Palmer stated that studying pathologies has little effect on producing health.

Palmer suspected a relation between the subluxation concept and overall health and well-being. He considered subluxation to be a stress factor that negatively affected the spinal column and spinal cord. In other words, an undetected stress area forces various structures of the body to function with increased efforts. It must be born in mind that this has an effect on the spinal cord as it carries information to the brain (afferent) and from the brain (efferent) to everywhere in the human body. It is conceivable that if an area of the spine cannot work properly and efficiently, it might expand to other areas of the spine and to structures depending on the integrity of the central and peripheral nervous system.

The negative effects related to stress were first studied by Selye [8], who in 1935 defined the GAS, General Adaptation Syndrome, having three phases:

(1) The alarm reaction, where the body tries to mobilize his defense mechanism.
(2) The resistance stage, leading to the possibility to adapt.
(3) The exhaustion stage, when the stress factors are strong enough or last long enough to annihilate the individual's capacity to adapt.

According to Palmer's ideas, the subluxation mechanism leading to possible dysfunction seems to work in a similar pattern as the General Adaptation Syndrome that Selye described.

In 2005, the World Health Organization defined the chiropractic subluxation as "a lesion or dysfunction in a joint or motion segment in which alignment, movement integrity and/or physiological function are altered, although contact between joint surfaces remains intact." To a chiropractor, a vertebral subluxation is a signal corresponding to a specific area of the spine that cannot work at its optimal potential, thus negatively affecting various neighboring structures, which in turn have to compensate or to produce extra work, affecting the nervous system, physiological functions, health, and well-being. The Chiropractor analyses the spine to find the subluxation (sometimes called "fixation" or "spinal dysfunction" or "disturbed segmental joint"). He will then "adjust"[4] or attempt to correct the subluxation. That is, the chiropractor will manually attempt to release tension in a precise way at a specific point.

> The process of spinal adjustment is like rebooting a computer. The signals that the adjustments send to the brain, via the nervous system, reset muscles behavioral patterns. By stimulating the nervous system we can improve the function of the whole body [9].

Since 2008 there has been a great number of papers looking at spinal manipulation therapy and its effect on brain function. This research has demonstrated sympathetic relaxation and corresponding metabolic changes in the brain, as well as reduced muscle tightness and decreased pain intensity following a chiropractic adjustment. The spinal adjustment of dysfunctional joints then modifies transmission in neuronal circuitries, not only at a spinal level, as indicated by previous research, but also at a cortical level, possibly reaching deeper brain structures such as the basal ganglia. People with chronic back pain might display decreased grey matter and fewer brain cells in many parts of the brain, such as the prefrontal cortex, thalamus, brain stem, and somato-sensory cortex [10].

---

[4]The word "adjustment" belongs to the chiropractic vocabulary. When the word manipulation refers more to an idea of imposed force, implying a certain wish to control associated to a pejorative meaning the adjustment concept implies a gentle, very precise move, well tolerated, agreed.

4.1 Intelligence and Energy

Palmer also considered the role played by intelligence and energy. Given the goal of this paper—to see how chiropractic care considers patient health concerns—we will focus on the definition given by Chiampo [11]: "Intelligence is the general capacity of an organism to meet novel situations or stimuli (information) by improvising a unique adaptive response, in particular to resolve a survival problem." This is important for practitioners of chiropractic because it suggests that healing is the outcome of the shared effort of patient and chiropractor.

Innate intelligence is quite busy digesting food, controlling how salty or sweet our blood is, reproducing, renewing millions of cells per second, keeping the body temperature within constant limits in the mountains at −20 °C or swimming in water at 25 °C. Feeling a pulse on a wrist could lead one to conclude that there is some intelligence at work in the body. When there is no noticeable pulse, it could mean that innate intelligence has deserted the body.

If we consider that an intelligent life creates new intelligent life every single second in response to various and numerous amounts of information, internal and external stimuli, we can understand that absolute reproducible certainty cannot belong to the image of the living organism. We must admit that life is a dynamic process, permanently improvising its evolution. If second after second we could re-create the exact same answers to information stimuli, we could hope to stay forever young—and possibly live forever. The quality of the individual answer varies according to factors such as the quality-intensity of the stimulus, differently perceived according to the individual, age, elasticity, nervous system, resilience, structural, organic integrity, and energy. The concept of intelligence is important in the sense that it allows breaking free from mechanistic perspective, opening the door to anticipation. Flexibility, mobility, freedom, and the capacity to adapt all play a key role to define survival prerequisites allowing optimal physiological vital functions. Anticipation and preparedness (for what the body performs) go hand in hand.

4.2 The Turning Point

Considering the concept of anticipation, Epicure used the word *prolepsis*, which means certain knowledge of an object in our mind that we need to have in order to stimulate understanding, generating a desire to discuss and to study. This appetite to understand, bound to fundamental human curiosity we might say, is not strictly based on acquired science, but belongs to some "prescience" mechanism; pre-science literally translating to "knowing before." It is expressed in the Latin *antecapere* that describes anticipation.

The three main factors governing life have to do with: information-communication-connection. All structures exchange information permanently, communicate with one another, and remain connected for a lifetime. Misinformation, bad transmission of information, bad communication, or any threat to the connection quality could only translate into a waste of energy and into various pathological conditions.

The chiropractor makes an analysis, a postural and a mobility assessment, and will look for what he considers to be a specific stress area in the spine. The chiropractic adjustment triggers a variety of processes in the body. This adjustment-information will spread through the entire body. We can expect that the nervous system will notice the new and introduced adjustment-information-stimulus, and will use it to its best ability in order to propose an adapted survival answer.

Neither medicine nor chiropractic can claim that we know why some remedies work and others do not. Dr. Ralph Gay (mentioned at the beginning of this paper) was asked: "How does chiropractic care work?" His answer: "Why does any form of treatment work?" Gay also took note of the increasing number of medical doctors who refer patients to a chiropractor (and even go for treatment themselves). Spinal fusion is by many orders of magnitude riskier than an adjustment.

5   A Difficult Science

Chiropractic is a difficult course of treatment to fully estimate. But there is progress, as already mentioned. Chiropractors recognize the complexity and the multitude of interdependences expressed by the various functions of the human being. We are aware that the adjustment-information cannot be overlooked by a highly perceptive, sensitive and intelligent central nervous system in charge of calculating every second for a lifetime: a unique, individual, punctual, and best possible survival answer to any information-stimulus. Time is irreversible; the same adjustment cannot be exactly repeated, just as one cannot twice swim in the same river. The chiropractor and the patient change continuously.

Reasoning from a purely restricted and structural mechanical standpoint, one might consider a mechanical correction could affect the structure in a desired way. But can we still reduce the human body to a clock or a puppet? Can we separate the structure from the nervous system, the hormonal system, the metabolism, possible biological variations, or a psychological effect? Linear methods of research do not allow for the various fields of possible responses.

Edward Lorenz, the father of the Chaos Theory, showed that even minute changes in the input data can create enormous changes in the output. His findings translated into the famous "Butterfly Effect."

An adjustment of the spine might be such a minute change in the input. Modern mathematics states that complex patterns can arise from simple causes; simple mathematical equations can produce a wide variety of complex patterns that resemble those seen in living nature.

Falling on the pelvis can affect a whole life, possibly the life of relatives and even the life of a child to come. Subnormal mobility of the vertebral column due to trauma can create negative effects, objectively measurable or not, now or in the future. The chiropractic adjustment constitutes a way to anticipate the worsening of a given fragility at a given time. One can assume that an adjustment, although directed at the structure, has the ability to stimulate various and complex adaptive answers selected by the central nervous system.

For decades, chiropractors adjusting patients have reported life-changing experiences. But it would be more beneficial to all of us to better understand what happens with one adjustment. More studies are needed if we want to remain committed to treatments that preserve life.

6   Concluding Considerations

The greatest danger comes from those who are convinced they have the right and definite answers. The "common sense" view maintains that more spending on healthcare systems, more healthcare, and ever more sophisticated drugs and treatments will lead to better health. It is time to challenge such so-called "common sense." Does more healthcare lead to better health? Instead of reacting to symptoms, diseases, and pain (reaction will remain necessary), why not mine the innate intelligence of life in order to promote an anticipatory approach to health and medical care?

> It should be recognized that the most fundamental question in medicine is why disease occurs rather than how it operates after it has occurred; that is to say, conceptually, the origins of disease should take precedence over the nature of the disease process [12].

The anticipatory systems perspective facilitates the consideration of several possibilities, transcending the falsely reassuring system that has been in place for centuries. "We might be living a revolution at the beginning of the 21st century, a second Cartesian revolution" [13].

References

1. Deardorff, J.: A medical doctor's view of chiropractic care. Chicago Tribune, August 4 (2009) http://featuresblogs.chicagotribune.com/features_julieshealthclub/2009/08/a-medical-doctors-take-on-chiropractic-care.html
2. Obstetric problems. Difficult labor causing lack of oxygen to the brain, babies with very low weight have an increased risk of developing ADHD. http://www.patient.info/health/attention-deficit-hyperactivity-disorder-leaflet
3. Bernstein, N.A.: Atlas des Ganges und Laufes des Menschen. Deutsche Gesellschaft für die Geschichte der Sportwissenschaft, Dortmund (1929)
4. Sperry, R.: Quoted in, Posture: Are You Energy Efficient? http://atlaschiro.com/posture-are-you-energy-efficient/

5. Flanagan, M.F.: The Role of the Craniocervical Junction in Craniospinal Hydrodynamics and Neurodegenerative Conditions. Neurology Research International, vol. 2015, Article ID 794829 (2015). doi:10.1155/2015/794829
6. Sterner,Y., Toolanen, G., Gerdle, B., Hildindsson, C.: The incidence of whiplash trauma and the effects of different factors on recovery. J. Spinal Disord. Tech. **16**, 195–199 (2003)
7. Gaab, J., Baumann, S., Budnoik, A., Hottinger, N., Ehlert, U.: Reduced reactivity and enhanced negative feedback sensitivity of the hypothalamus-pituitary-adrenal axis in chronic whiplash-associated disorder. Pain **119**(1–3), 219–224 (2005). http://www.ncbi.nlm.nih.gov/pubmed/16298068
8. Selye, H.: Stress and the general adaptation syndrome. Br. Med. J., June 17 (1950). http://www.ncbi.nlm.nih.gov/pmc/articles/PMC2038162/pdf/brmedj03603-0003.pdf
9. Haavik, H.: The Reality Check: A Quest to Understand Chiropractic from the Inside Out. Haavik Research, New Zealand (2014)
10. Missing Link Found Between Brain, Immune System; Major Disease Implications. https://www.sciencedaily.com/releases/2015/06/150601122445.htm
11. Chiampo, L. (ed.): Dizionario Enciclopedico di Medicina. Zanechelli/Mc Graw-Hill, bologna/New York (1988)
12. Palmer, B.J.: The Bigness of the Fellow Within, vol. 22. Palmer College, Davenport, IA (1949)
13. Nadin, M.: Anticipation—The Ends Is Where We Start From. Lars Muller Verlag, Basel (2003)

Anticipation in Motion-Based Games for Health

Rainer Malaka, Marc Herrlich and Jan Smeddinck

Abstract Digital motion-based games with affordable tracking methods for full body tracking allow users to play computer games controlled with body movements. They can be used as so called *exergames*, combining exercises and games. Exergames have three major benefits: (1) they can raise motivation for doing exercises through immersive interactive game play; (2) they can give users feedback regarding physiologically beneficial movements and aggregate performance over time; (3) they can be adapted to the individual user. This is a great chance, but also a great challenge in the development of exergames. When employing exergames as games for health, e.g. in physiotherapy, many patients (players) display individual predispositions and abilities with temporal variations. Anticipating and adapting to physical ability and individual training goals on various timescales require subtle mechanisms that capture differences for individual users. We discuss and analyze adaptivity requirements and implementation approaches for anticipatory techniques in this context.

Keywords Adaptivity · Anticipation · Exergames · Therapy · Human-computer interaction · Games for health · Rehabilitation

1 Introduction

Motion-based games for health (MGH) are a kind of serious games aiming for involving users in gameplay that often use full-body input in order to implement interactive physical exercises. In recent years, such games have gotten growing

R. Malaka (✉) · M. Herrlich · J. Smeddinck
TZI, Digital Media Lab, University of Bremen, Bremen, Germany
e-mail: malaka@tzi.de

M. Herrlich
e-mail: mh@tzi.de

J. Smeddinck
e-mail: smeddinck@tzi.de

© Springer International Publishing Switzerland 2017
M. Nadin (ed.), *Anticipation and Medicine*, DOI 10.1007/978-3-319-45142-8_22

351

scientific and commercial attention. These games can either address just general fitness of healthy players or specific needs of people who need individual training or treatment [1–5].

One reason for the success of MGH is tied to the fact that computer games are one of the most successful genres of digital media. They address the fundamental human desire to engage in playful activities. Huizinga's concept of the *homo ludens* [6] identifies game play not only as something innate in mankind, but also as a principle that leads to cultural evolution. If we enter the magic circle of a game, we obviously follow a quite natural instinct [7]. The fun of gameplay can be turned into a motivational driver for involving people in activities with a serious purpose. Games for health and fitness (exergames) are based on this principle.

In combination with new game console technology and advanced tracking methods, games for health and fitness have turned into a mass-market phenomenon. In combination with the automated feedback and the evaluation of the player's objective activity, custom programs and personalized sessions can be designed to act as a virtual personal trainer. With this technology, not only young and healthy gamers can be addressed. Patients undergoing rehabilitation, older adults, and people with special needs can benefit from custom exercises with individual monitoring and feedback, especially in situations in which a therapist is not available and/or due to prohibitive cost [8].

A central challenge for MGH is adaptability, that is, to adapt exergames to the specific abilities and needs of individual users from heterogeneous target groups [9–11]. Recent research and development aims for automated adaptation (adaptivity) [10]. An adaptive digital system can take numerous contextual factors into account, such as physiological, cognitive, and situational parameters [12, 13]. Adaptation for MGH is required with respect to several different layers and timescales. First, there are principal, more or less constant, differences between individual users and specific target groups that have to be addressed, e.g., the body height and other properties that do not change, or change very slowly (such as biological age). Second, the system has to support medium-term changes, i.e., over several days, weeks, or months, which are often aligned to typical recovery cycles, but may also result from acute disease or acute episodes of chronic maladies. Third, the system has to anticipate very dynamic short-term changes and fluctuations, i.e., daily changes in user performance (feeling tired, forgot to take medicine, etc.). As an additional complication, these timescales apply to multiple layers of heterogeneity (which introduces variance) that can be classified as follows:

- Heterogeneous general application areas, such as MGH for stroke patients, MGH for children with cerebral palsy, MGH for phantom pain, etc.
- Heterogeneity within those target groups. Commercial mass market and pure entertainment games can afford to target a comparatively thin slice of a normal distribution of their potential users which will cover a great proportion of the

users but leaves out the users deviating from this norm. Games for health cannot afford to leave somewhat deviating users behind and the variance with regard to individual capabilities and needs can be quite extreme.

- Heterogeneity within any individual user/patient (e.g. with regard to varying capabilities and requirements at different times, as illustrated above).

2 State of the Art

Exergames for physiotherapy, rehabilitation and prevention (PRP) have been a prominent research topic for a decade now, and the effort in scientific research is growing steadily. Systems focus on information; explanation and teaching or active support of physiotherapy, rehabilitation, and prevention (PRP) [14]. General exergames without PRP focus [15] are already available on the game market. A number of studies could prove the effectiveness of exergames for a number of target groups (e.g. stroke patients [16, 17], Parkinson's disease patients [1] or older adults who face rather general age-related challenges [18]).

Exergames involve users in dual-task scenarios [19, 20]: they have to both act cognitively in the game (learning the rules that the game-mechanics entail, developing and enacting strategies, tactics, etc.) and to practice physical exercises. In classical PRP, such conditions are appreciated and considered to be especially effective. However, the mappings and appearance have to be designed with respect to the target group and the needs of the user [21]. For older adults, for instance, age-related constraints have to be taken into account [22]. Next to an individual adaptation to skills, the psycho-physiological condition of the user is key to reaching a certain game experience for all users [23, 24]. Automatic adaptation using modern sensor technology can be used to achieve dual flow, i.e. matching both skills in the game and exertion to the difficulty level of the game and the exercises [25].

In contrast to conventional instructions in PRP based on exercise instruction sheets, exergames have a number of advantages, in particular support for motivation, feedback, and analysis and can lead to a more efficient performance of the exercises [11, 26]. Some motivational dimensions and therapeutic effects have even been shown to be significantly in favor of MGH in comparison to instructions by a live physiotherapist [27]. Another benefit of MGH is that, unlike a real therapist, the instructions are always available, and they may also serve to gather a consistent objective view of one's therapeutic progression. Contrary to a widespread misconception, playing computer games is not an activity only for young people, but also for older adults [28], who show growing interest in computer games that can contribute to their physical and cognitive well-being [10, 29].

The challenge in PRP is to keep a constantly high level of effective treatment quality. This is difficult due to several problems, in particular when patients practice their exercises without supervision. The self-controlled adherence to the correct execution and right number of repetitions is a dominant problem [26, 30]. Even life-threatening diagnoses do not ensure a high quota of adherence to the exercise plan, as was shown with cardio patients [31]. Feedback, encouragement, social interaction and emotional engagement in particular can make a positive difference [32]. Hence, exergames can be ideal vehicles, as they can include such factors. The wealth of sensor devices for the game market, such as vision- or motion-based controllers, made it possible to use devices originally designed for entertainment purposes also for PRP with positive results [33].

In addition to sensing and proper game control mappings, user- and context-dependent adaptation is necessary. Even though early user-adapted systems were received quite critically [34], today many software systems successfully employ subtle adaptation mechanisms. Particularly in games, it is an aspect of balancing, which aims for leveling out differences of individual players [35] and allows for similar player experience, even though the skills may vary [36]. Such concepts are also discussed for exergames [37, 38], where physiological data can be used for online adaptivity and customization [39, 40].

User models for continuous adaptation for individual players have been proposed [41] and used, e.g., for games for depression prevention [42]. More generic ways of adaptation for exergames have been discussed in recent research [2, 42, 43], but the present approaches are still largely reactive in their adaptivity. The following section provides an example of a contemporary approach to adaptive exergames (or MGH).

3   Towards Adaptive Ad Personalized Interactive MGH

How should an ideal MGH adapt to a user? This is a complex questions and we want to illustrate this claim with the example of a rather simple exergame for Parkinson's disease (PD) patients [1]. In the game *StarMoney*, the patient stands in front of a monitor and is tracked with a vision-based sensor. The task is to make wide movements with the arm in order to follow a trace of stars that is dynamically generated. Each star that is hit can be caught and in turn can be collected as a point for the user's score (Fig. 1).

There are basically three game-related parameters that can be adjusted:

1. The scaling of the mapping of the user's hand to the position of the hand on the screen. This could, e.g., compensate for various arm and body sizes of the users.
2. The size of the virtual hand. This would relate to the required accuracy for hitting the stars.
3. The only dynamic parameter would be the timing, i.e., speed of the trail of stars.

StarMoney is a simple game with only very limited game mechanics. Other games might have a multitude of adaptable parameters. If we consider a certain user, we could measure the arm length or, even better, the area that can be reached,

Fig. 1 Exergame for PD patients: *StarMoney*

response time, and accuracy and set the parameters accordingly. However, this will not suffice if we consider dynamic aspects. These play a role on various time scales [10]:

- During a game session: in many exercise sessions it takes some warm-up or activation times until patients reach their level of physical ability. In particular PD patients often suffer from a remarkable stiffness, which is reduced after some activation exercises
- Positive effect due to training: If the PRP is successful, the exergames can lead to increased performance from session to session
- Negative effect due to aging or progression of a disease: In particular for PD, but also for aging, we have to expect a progressive decline of physical ability on a long term time scale
- Individual performance variances: due to many factors, the day-to-day performance can vary a lot. This can be due to infections, other treatments, medications, etc.

These temporal aspects show that dynamic variations are a challenge to the automatic adaptation of exergames—in particular because the users themselves adapt and anticipate the dynamics of the system. Such co-adaptation of both user and system can lead to mutually unpredictable effects that can be positive or negative [44, 45]. Human adaptation to exergames has been observed in a number of commercial sports games: for example, in Wii Tennis, where after a while some users no longer execute full arm movements, but rather just move the hand with the controller, resulting in the same in-game effect. Thus, if users learn—either consciously or unconsciously—that the system adapts to low performance by lowering the physical demand of an exercise, they might trigger a spiral of succeeding easier exercises.

This negative dual-adaptive process however has not been observed in our studies with adaptive exergames. In contrast to sports games that focus on a game and not on a sustainable training effect of the physical movements, exergames with clear physiological goals seem to be less prone to this effect. Patients typically state that they are aware of the "serious" aspect of their interaction with the system. They can therefore be more willing to accept sub-optimal scores [10] and may be hesitant to exploit adaptive systems. However, since it is likely that not all users are like this, it seems advisable to implement maximal relative offsets per adjustment cycle [27] and to boost/decrease extreme positive or negative developments with the motivational tool set of game design. It also seems likely that therapists will continue to play at least a role as sporadic advisors/supervisors in adaptive systems for MGH in order to prevent undesired positive or negative spirals.

4 Components of Adaptive Exergames

Considering a simple overarching model and building on the work of Adams et al. [46], in order to realize an adaptive exergame, we need to consider the following elements:

(a) Sensors
(b) Models for estimating the user's psycho-physical state
(c) Adaption strategies

4.1 Sensor Technologies for Adaptive Exergames

In order to build an adaptive exergame, the system must reliably detect the state of the user and assess the user's performance and fitness. Many exergames are vision based and thus the visual input is the key for tracking the user and the exercises. Alternative tracking is often done with motion sensors (e.g., with Wii controllers or mobile phones). Less common are sensor mats or devices like the Wii balance board. Even less common are complex medical devices, such as gait rehabilitation devices or rehabilitation robotics. Additional sensors can also be used to directly get data related to the user's physical condition. Such sensors can be heart rate or blood pressure monitors, multiple degree-of-freedom motion-trackers, among others. In principle, many combinations of sensors are possible and as a result of sensor data fusion, the user's state can be measured more and more comprehensively.

4.2 Psycho-Physical User Models

User modeling has been a research topic for the last 20 years. Usually, user modeling aims for classifying users in groups according to user data such as demographic attributes (e.g., age, gender) or behavioral patterns (e.g., activities with an interactive system). Such methods use, for instance, classification techniques such as cluster algorithms, nearest neighbor classifiers, or neural networks. For exergames, such user models can be a starting point. However, they have to be extended with respect to physiological and psychological aspects, both reflecting static as well as dynamic aspects. Examples for factors related to static aspects are:

- Physical abilities and limitations (e.g., arm flexibility, vision, …)
- Diagnosed diseases (e.g., PD, diabetes, …)
- Risks (e.g., high blood pressure)
- Treatment/Therapy (e.g., exercises against back pain)
- Training goals
- Game-related factors (e.g., high score)

More dynamic aspects are:

- Accuracy of the exercises (on the basis of the tracking data)
- Physiological data (e.g., heart rate)
- Exertion (relative to personal condition)
- Performed repetitions
- Game-related factors (e.g., score)

Based on these factors, user profiles can be compared with those from other users (or prototypical users), with historical data and with pathological models, in order to assess training progress. In many models, the user model consists of rather simple data sets. More subtle models would not just collect flat data, but include deep models. Thus therapy or treatment could be an elaborate model of an actual physiotherapeutic treatment for the individual user.

4.3 Adaptation Strategy

The adaptation of motion-based games for health needs to be discussed along a number of basic dimensions, such as:

- Automated versus manual adaptation
- Adaptation with regard to exercise and game
- Temporal aspects of adaptation
- Granularity of adaptation, complexity, etc.

With manual adaptation, users have full control on all settings. Moreover, other persons, such as physicians or therapists, can use adaptation interfaces in order to customize exergames for their patients. But even though adapting parameters by hand allows for control of all settings, it has two main disadvantages: (1) it can be cumbersome if too many parameters have to be adjusted; and (2) dynamic changes cannot be followed easily, thus requiring frequent updates [27]. The first issue may lead to over reliance on the default parameter settings, and thus the power of adaptability is wasted. The second issue can lead to precise settings at the beginning, but might entail a continuous drift towards maladjustment or mismatch of the parameters. Therefore, automatic adaptation methods are necessary. They can, in turn, be combined with manual methods. Often this combination is used such that in early phases manual methods are used, which are replaced over time by automated techniques based on observations. In particular, direct manual adaptation can often be interpreted as corrections and can be used as reward/punishment cues for machine learning methods that improve the automated system.

The adaptation can influence both the physical exercises and the game-play. For automated adaptivity, it is necessary to define goals towards which parameters and settings are adjusted. Such goals can accordingly relate to both the game-play and the exercises. In many games, the designers likely want to achieve a good balancing such that the difficulty of the game matches the player's skills. This relates to the *flow theory* [47], which assumes that players reach a flow state that correlates with a good player experience when skills and difficulty are well balanced. Similarly, for exercises, exertion level and exercise intensity have to be balanced in order to avoid overexertion or ineffective training. The requirement of balancing both aspects has been formulated as the dual-flow theory [25]. Obviously both aspects are independent when contrasting any two persons. A player can be physically very fit but novice to the game, or vice versa.

Modeling the complex dynamic fluctuations of user performance is a very difficult task. Advanced models will take medical knowledge into account that would for instance describe the flexibility and fitness of users depending on age and diseases during individual sessions (e.g., before and after warm-up) and between sessions (e.g., with respect to training goals). But even simple models that merely aim to keep the user roughly in the double-flow state can be helpful [10].

Lastly, the goal of motion-based games for health is usually not just to motivate the players to occasionally perform exercises, but to support lasting behavior change. According to the well-established *Fogg Behavior Model* [48], MGH can support both the factor of *motivation*, and the factor of *ability*, when players are frequently motivated to exercise on an occasional basis; this in effect contributes towards moving individuals beyond their "action line," which in turn facilitates lasting behavior change.

5   Building Adaptive Exergames

The ultimate goal of adaptive games for health is to integrate all aspects and components discussed above. This has not been achieved for several reasons. In particular, there are not yet models that cover all temporal, physiological, and psychological aspects. The good news, however, is that not everything has to be done automatically, and even simple models can serve as proof-of-concept. In an adaptive version of the *StarMoney* game for PD patients (Fig. 1), we considered adaptation of timing, accuracy, and the distance that can be reached by the patient's

Fig. 2 Adaption possibilities for the *StarMoney* game: timing, accuracy, range of reach

arm (Fig. 2). Using a simple threshold heuristics for the motion (amplitude) of the arms, we wanted to see in particular, what the effect of an adaptive game would be over a longer period of time [10].

With a small number of participants, but a study lasting over three weeks, we could show that the system works and was well received. Adaptation did not confuse the users. The perceived difficulty correlates with performance. The patients expected a high level of challenge and were not very confident of their own success. The study showed that the amplitude (range of motion) actually developed positively over successive game rounds, and we observed an objective increase for all participants.

In a more recent development, results from a medium-term study of exergames for physiotherapy and rehabilitation with 30 participants in the situated context of a physiotherapy practice indicated that automated adaptive versions of exergames—even when building on simple rubber-banding heuristics—can be roughly on a par with manually adaptable versions regarding user performance and experience. At the same time, therapists appreciate the adaptive versions due to the lessened need for manual effort [27].

6 Towards Anticipatory Exergames

The approaches summarized in the state-of-the-art sections and introduced in the examples from our own practical implementations have in common that these contemporary adaptive exergames are still largely performance-based and reactive. This means that the system reacts after deviations from a more optimal course of developments are detected. In many cases this means that a system even made an adjustment that worsened the situation, only in order to then readjust. Bringing true anticipation into practical adaptive MGH (e.g., by relying on more predictive user and/or group models, likely in combination with known models for average therapy/rehabilitation progressions and context models), might help to eliminate this systematic over- or under-shooting and produce settings progressions that meet the patients' needs much more precisely. From the perspective of a general model, this means moving from a system based on a *performance evaluation* and an *adjustment mechanism* that takes only the most recent performance evaluation into

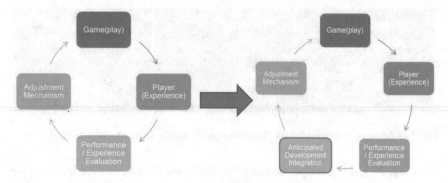

Fig. 3 Moving from reactive adaptivity to anticipatory customization

account to an adjustment mechanism that also takes an anticipated reaction to upcoming changes and anticipated development into account. This follows the main aim of preventing overstraining or under challenging the players (cf. Fig. 3).

Anticipation in relation to games for health has been discussed by Nadin et al. in the context of the project Seneludens [49] and Amazing Grace.[1] According to the focus of that work, anticipation is expressed in action. It is an indicator of the ability to adapt to changes, an ability which decreases with aging. Motion-based games can arguably play a beneficial role in retaining or improving the anticipatory profiles of individual players. The view presented in this paper highlights the potential an anticipatory component in the adaptivity of a motion-based game system to the needs and capabilities of a user, thereby providing an additional perspective to the role of anticipation in the context of games for health.

7   Conclusions and Discussion

Exergames are a very promising approach for increasing the motivation and adherence of people who need to perform exercises on a regular basis, and to increase the likelihood of achieving effective and safe exercising. The users can be healthy individuals with general fitness goals or patients who need to do exercises as part of some rehabilitation or physiotherapy program. Exergames generally have the potential to provide three positive aspects: motivation, feedback, and customization. Considering the human being's natural playfulness, motivation can be increased and thus may lead to a more sustainable training program and long-lasting adherence to the scheduled exercises. Tracking with modern sensor technology allows for precise guidance on how to perform certain exercises correctly. Moreover, the system can use the tracking for feedback. This can be live feedback

---

[1]Nadin, M., Personal Communication. See also: http://www.youtube.com/watch?v=RaUJ_ Hpfwm4 and http://www.anteinstitute.org/index.php?page=seneludens.

during a training session, reflective feedback after a session, and development feedback on the changes in performance over a longer period of time.

Adaptive exergames are an important step forward in personalizing exergames to the needs of users. In principle, many temporal, contextual, physiological and other personal parameters can be used to either manually or automatically adapt the game and/or the exercise program. So far we have seen some first adaptive exergames. More subtle models will allow for more complex adaption strategies. In the long run, we expect exergames to become reliable personal trainers that may anticipate many factors in order to guide users through a personalized and optimized training program, avoiding the "need of prior mistakes" that is inherent to many contemporary adaptation strategies.

Acknowledgments We would like to thank all members of the projects *WuppDi!*, *Spiel Dich fit und gesund!*, and *Adaptify*, as well as all participants of associated research studies, and our contributing researchers and authors. This work was generously supported with grants from the Federal Ministry of Education and Research of Germany (Bundesministerium für Bildung und Forschung: BMBF), the Klaus Tschira Foundation (KTS), the Wirtschaftsförderung Bremen, (Department of Economic Development—Bremen), and the EU Fund for Regional Development (EFRE).

References

1. Assad, O., Hermann, R., Lilla, D., Mellies, B., Meyer, R., Shevach, L., Siegel, S., Springer, M., Tiemkeo, S., Voges, J., Wieferich, J., Herrlich, M., Krause, M., Malaka, R.: Motion-based games for Parkinson's disease patients. In: Entertainment Computing–ICEC 2011, pp. 47–58. Springer, Heidelberg (2011)
2. Göbel, S., Hardy, S., Wendel, V., Mehm, F., Steinmetz, R.: Serious games for health: personalized exergames. In: Proceedings of the International Conference on Multimedia, pp. 1663–1666. ACM, New York (2010)
3. Ijsselsteijn, W., Nap, H.H., de Kort, Y., Poels, K.: Digital game design for elderly users. In: Proceedings of the 2007 Conference on Future Play, pp. 17–22 (2007)
4. Malaka, R.: How computer games can improve your health and fitness. In: Games for Training, Education, Health and Sports, pp. 1–7. Springer, Heidelberg (2014)
5. Walther-Franks, B., Wenig, D., Smeddinck, J., Malaka, R.: Exercise my game: turning off-the-shelf games into exergames. In: Anacleto, J.C., et al. (eds.) Entertainment Computing —ICEC 2013, pp. 126–131. Springer, Heidelberg (2013)
6. Huizinga, J.: Homo ludens: proeve eener bepaling van het spel-element der cultuur. Amsterdam University Press (2008)
7. Crawford, C.: The Art of Computer Game Design. (1984)
8. Schuler, T., Drehlmann, S., Kane, F., von Piekartz, H.: Abstract virtual environment for motor rehabilitation of stroke patients with upper limb dysfunction. A pilot study. In: 2013 International Conference on Virtual Rehabilitation (ICVR), pp. 184–185. IEEE (2013)
9. Aarhus, R., Grönvall, E., Larsen, S.B., Wollsen, S.: Turning training into play: embodied gaming, seniors, physical training and motivation. Gerontechnology **10**, 2 (2011)
10. Smeddinck, J., Siegel, S., Herrlich, M.: Adaptive difficulty in exergames for Parkinson's disease patients. In: Proceedings of Graphics Interface 2013. Regina, Canada (2013)
11. Smeddinck, J.D., Gerling, K.M., Malaka, R.: Anpassbare Computerspiele für Senioren. Inform. Spektrum. Special Issue on Entertainment Computing (2014)

12. Porzel, R., Zorn, H.-P., Loos, B., Malaka, R.: Towards a separation of pragmatic knowledge and contextual information. In: Proceedings of ECAI-06 Workshop on Contexts and Ontologies (2006)

13. Porzel, R., Gurevych, I., Malaka, R.: In context: integrating domain-and situation-specific knowledge. In: SmartKom: Foundations of Multimodal Dialogue Systems, pp. 269–284. Springer, Heidelberg (2006)

14. Gekker, A.: Health games. In: Ma, M., et al. (eds.) Serious Games Development and Applications, pp. 13–30. Springer, Heidelberg (2012)

15. Yim, J., Graham, T.C.N.: Using games to increase exercise motivation. In: Proceedings of the 2007 Conference on Future Play, pp. 166–173. ACM, New York (2007)

16. Alankus, G., Proffitt, R., Kelleher, C., Engsberg, J.: Stroke therapy through motion-based games : a case study. Therapy., 219–226 (2010)

17. Burke, J.W., McNeill, M.D.J., Charles, D.K., Morrow, P.J., Crosbie, J.H., McDonough, S.M.: Optimising engagement for stroke rehabilitation using serious games. Vis. Comput. 25(12), 1085–1099 (2009)

18. Gerling, K.M., Livingston, I.J., Nacke, L.E., Mandryk, R.L.: Full-body motion-based game interaction for older adults. In: Proceedings of the 30th International Conference on Human factors in Computing Systems, CHI'12 (2012)

19. de Bruin, E.D., Reith, A., Dorflinger, M.: Feasibility of strength-balance training extended with computer game dancing in older people; Does it affect dual task costs of walking? J. Nov. Physiother, 1 (2011)

20. Swanenburg, J., de Bruin, E.D., Favero, K., Uebelhart, D., Mulder, T.: The reliability of postural balance measures in single and dual tasking in elderly fallers and non-fallers. BMC Musculoskeletal Disorders 9(1) (2008)

21. Geurts, L., Vanden Abeele, V., Husson, J., Windey, F., Van Overveldt, M., Annema, J.-H., Desmet, S.: Digital games for physical therapy: fulfilling the need for calibration and adaptation. In: Proceedings of the Fifth International Conference on Tangible, Embedded, and Embodied Interaction, pp. 117–124 (2011)

22. Gerling, K.M., Schulte, F.P., Smeddinck, J., Masuch, M.: Game design for older adults: effects of age-related changes on structural elements of digital games. In: Herrlich, M., et al. (eds.) Entertainment Computing—ICEC 2012, pp. 235–242. Springer, Heidelberg (2012)

23. Drachen, A., Nacke, L.E., Yannakakis, G., Pedersen, A.L.: Correlation between heart rate, electrodermal activity and player experience in first-person shooter games. In: Proceedings of the 5th ACM SIGGRAPH Symposium on Video Games, pp. 49–54 (2010)

24. Mandryk, R.L., Inkpen, K.M., Calvert, T.W.: Using psychophysiological techniques to measure user experience with entertainment technologies. Behav. Inf. Technol. 25(2), 141–158 (2006)

25. Sinclair, J., Hingston, P., Masek, M.: Exergame development using the dual flow model. In: Proceedings of the Sixth Australasian Conference on Interactive Entertainment, pp. 11:1–11:7. ACM, New York (2009)

26. Uzor, S., Baillie, L.: Exploring & designing tools to enhance falls rehabilitation in the home. In: Proceedings of the SIGCHI Conference on Human Factors in Computing Systems, pp. 1233–1242. ACM, New York (2013)

27. Smeddinck, J.D., Herrlich, M., Malaka, R.: Exergames for physiotherapy and rehabilitation: a medium-term situated study of motivational aspects and impact on functional reach. In: Proceedings of the 33rd Annual ACM Conference on Human Factors in Computing Systems, pp. 4143–4146. ACM, New York (2015)

28. Ijsselsteijn, W., Nap, H.H., de Kort, Y., Poels, K.: Digital game design for elderly users. In: Proceedings of the 2007 Conference on Future Play, pp. 17–22 (2007)

29. Nap, H.H., de Kort, Y.A.W., IJsselsteijn, W.A.: Senior gamers: preferences, motivations and needs. Gerontechnology 8, 247–262 (2009)

30. Stenström, C., Arge, B., Sundbom, A.: Home exercise and compliance in inflammatory rheumatic diseases–a prospective clinical trial. J. Rheumatol. 24(3), 470–476 (1997)

31. Ice, R.: Long-term compliance. Phys. Ther. 65(12), 1832–1839 (1985)

32. Sluijs, E.M., Kok, G.J., van der Zee, J.: Correlates of exercise compliance. Phys. Ther. **73**(11), 771–782 (1993)
33. Gerling, K.M., Schild, J., Masuch, M.: Exergame design for elderly users: the case study of SilverBalance. In: Proceedings of the 7th International Conference on Advances in Computer Entertainment Technology, pp. 66–69 (2010)
34. Horvitz, E., Breese, J., Heckerman, D., Hovel, D., Rommelse, K.: The Lumiere project: Bayesian user modeling for inferring the goals and needs of software users. In: Proceedings of the Fourteenth Conference on Uncertainty in Artificial Intelligence, pp. 256–265 (1998)
35. Mueller, F., Vetere, F., Gibbs, M., Edge, D., Agamanolis, S., Sheridan, J., Heer, J.: Balancing exertion experiences. In: Proceedings of the 2012 ACM Annual Conference on Human Factors in Computing Systems, pp. 1853–1862 (2012)
36. Andrade, G., Ramalho, G., Gomes, A.S., Corruble, V.: Dynamic game balancing: an evaluation of user satisfaction. In: AIIDE'06, pp. 3–8 (2006)
37. Gerling, K.M., Miller, M., Mandryk, R.L., Birk, M.V., Smeddinck, J.D.: Effects of balancing for physical abilities on player performance, experience and self-esteem in exergames. In: Proceedings of the 2014 CHI Conference on Human Factors in Computing Systems, CHI'14 (2014)
38. Rego, P., Moreira, P.M., Reis, L.P.: Serious games for rehabilitation: a survey and a classification towards a taxonomy. In: 5th Iberian Conference on Information Systems and Technologies (CISTI), pp. 1–6 (2010)
39. Lindley, C.A., Sennersten, C.C.: Game play schemas: from player analysis to adaptive game mechanics. In: Proceedings of the 2006 International Conference on Game Research and Development, pp. 47–53 (2006)
40. Liu, C., Agrawal, P., Sarkar, N., Chen, S.: Dynamic difficulty adjustment in computer games through real-time anxiety-based affective feedback. Int. J. Hum. Comput. Interact. **25**(6), 506–529 (2009)
41. Missura, O., Gärtner, T.: Player modeling for intelligent difficulty adjustment. In: Gama, J., Costa, V.S., Jorge, A.M., Brazdil, P.B. (eds.) Discovery Science, pp. 197–211. Springer, Heidelberg (2009)
42. Janssen, C.P., Van Rijn, H., Van Liempd, G., Van der Pompe, G.: User modeling for training recommendation in a depression prevention game. In: Proceedings of the First NSVKI Student Conference, pp. 29–35 (2007)
43. Hardy, S., Göbel, S., Gutjahr, M., Wiemeyer, J., Steinmetz, R.: Adaptation model for indoor exergames. Int. J. Comput. Sci. Sport. **11**, 1 (2012)
44. Fischer, G.: User modeling in human–computer interaction. User Model. User-Adapt. Interact. **11**(1), 65–86 (2001)
45. Mackay, W.E.: Responding to cognitive overload: co-adaptation between users and technology. Intellectica **30**(1), 177–193 (2000)
46. Adams, E.: Fundamentals of Game Design. New Riders (2010)
47. Csikszentmihalyi, M.: Flow: The Psychology of Optimal Experience. Harper & Row, New York (1990)
48. Fogg, B.: A behavior model for persuasive design. In: Proceedings of the 4th International Conference on Persuasive Technology, pp. 40:1–40:7. ACM, New York (2009)
49. Nadin, M.: Play's the thing: a wager on healthy aging. In: Cannon-Bowers, J., Bowers, C. (eds.) Serious Game Design and Development, pp. 150–177. IGI Global, Hershey, NY (2010)

Erratum to: Anticipatory Mobile Digital Health: Towards Personalized Proactive Therapies and Prevention Strategies

Veljko Pejovic, Abhinav Mehrotra and Mirco Musolesi

Erratum to:
Chapter "Anticipation Mobile Digital Health: Towards Personalized Proactive Therapies and Prevention Strategies" M. Nadin (ed.), *Anticipation and Medicine,*
DOI 10.1007/978-3-319-45142-8_15

The original version of the chapter was inadvertently published without the updated corrections in the chapter "Anticipatory Mobile Digital Health: Towards Personalized Proactive Therapies and Prevention Strategies". The erratum chapter has been updated with the changes.

The updated original online version this chapter can be found at http://dx.doi.org/10.1007/978-3-319-45142-8_15